Simple Theories

Volume 503

Simple Theories

by

Frank O. Wagner

Institut Girard Desargues,
Université Claude Bernard (Lyon-1),
Villeurbanne, France

KLUWER ACADEMIC PUBLISHERS
DORDRECHT / BOSTON / LONDON

A C.I.P. Catalogue record for this book is available from the Library of Congress.

ISBN 978-90-481-5417-3

Published by Kluwer Academic Publishers,
P.O. Box 17, 3300 AA Dordrecht, The Netherlands.

Sold and distributed in North, Central and South America
by Kluwer Academic Publishers,
101 Philip Drive, Norwell, MA 02061, U.S.A.

In all other countries, sold and distributed
by Kluwer Academic Publishers,
P.O. Box 322, 3300 AH Dordrecht, The Netherlands.

Printed on acid-free paper

Für meine Eltern

Contents

Preface

The class of simple theories extends that of stable theories, and contains some important structures, such as the random graph, pseudo-finite fields, and fields with a generic automorphism. Following Kim's proof that Shelah's notion of "forking independence" is symmetric in a simple theory, this area of model theory has been a field of intense study, placing stability theory into a wider framework, and thus serving to explain a phenomenon previously noticed in connection with the model theory of pseudo-finite fields, namely the appearance of stability-theoretic behaviour in an unstable context.

The generalization has required some important new tools, in particular the model-theoretic treatment of hyperimaginaries (classes modulo type-definable equivalence relations). While many of the results known for stable theories have been generalized to simple structures, some fundamental results from stability theory are as yet only conjectural for the simple case (e.g. the group configuration theorem, or the binding group theorem).

In this book, I shall present the (or rather: my) present knowledge of simplicity theory, viewing it not as an appendix to stability theory, but as a general theory of which the stable structures form a particular case. It is intended both as an introduction for the graduate student, and as a reference for research in this area.

Acknowledgements

Parts of this book were written while I held a Heisenberg-Stipendium of the Deutsche Forschungsgemeinschaft (Wa 899/2-1) at the University of Oxford. I should like to thank the DFG for its financial support, the Mathematical Institute for its hospitality, and St. Catherine's College for looking after me so well. Other parts were written during various sojourns in Tokyo; I am grateful to the Kanto Model Theory group and Waseda University for the friendly welcome extended to me. Finally, the manuscript was finished at the Université Claude Bernard at Lyon, where again I was received warmly. Further thanks are due to the Fields Institute for Research in the Mathematical Sciences at Toronto and to the Mathematical Sciences Research Institute at Berkeley, both of which I had the opportunity to visit for extended periods, and where many of the results in this book originated.

I am indebted to Angus Macintyre for starting it all off, to Anand Pillay for sharing his extensive knowledge and insight, and to Ambar Chowdhury, Bradd Hart, and Byunghan Kim for many discussions on simplicity theory. Thanks also to Steven Buechler, Enrique Casanovas, Zoé Chatzidakis, David Evans, Ehud Hrushovski, Masanori Itai, Hirotaka Kikyo, Daniel Lascar, Chris Laskowski, Dugald McPherson, Keishi Okamoto, Bruno Poizat, Akito Tsuboi, Martin Ziegler, and Boris Zil'ber, all of whom contributed in some form or other.

F.O.W.

Chapter 1

PRELIMINARIES

1.1 INTRODUCTION

Mathematics often proceeds from the specific to the general, and the development of simplicity theory is no exception to the rule. It began with Michael Morley's study of uncountably categorical theories, where he defined ω-stability, and for some time remained in the categorical context. This changed when Saharon Shelah embarked on an ambitious programme of classifying the models of a complete first-order theory, using his newly invented notion of "forking" and Rowbottom's "stability". Unstable theories have the maximal number of models and are thus considered unclassifiable; nevertheless, he tried in [149, 152] to extend the framework, defining a well-behaved class of unstable first-order theories which he called "simple unstable". However, symmetry of forking for those theories eluded him, and at the time those papers did not receive the attention they deserved.

Meanwhile, Lascar and Poizat in Paris rewrote Shelah's results, starting from the notions of heir and coheir. In their approach, definability of types in a stable theory plays a major rôle; it is used to derive all the other basic properties of forking. As types in a simple theory are in general not definable, this approach, albeit very elegant, tied the forking theory to the theory of multiplicity. It took 20 years until Kim re-examined [149] and managed to prove symmetry of forking for simple theories; from this, he and Pillay quickly developed the basic theory. The logic year 1996/97 at the Fields Institute in Toronto and the logic semestre 1998 at MSRI Berkeley saw an acceleration of the pace, with Buechler, Evans, Hart, Lascar, Scanlon and myself joining the quest to find suitable generalizations of known stable phenomena to the simple context.

But even before Kim's result, stable behaviour had been observed in unstable structures, and particular cases of the general theory had been shown. Most notably, Hrushovski treated pseudo-finite fields (sitting inside an algebraically closed field), or more generally "pseudo-finite" non-elementary substructures of a strongly minimal set. This was applied to groups definable in pseudo-finite fields and led to the analysis of existentially closed fields with automorphism by Chatzidakis and Hrushovski (and Peterzil for characteristic $\neq 0$), shown also independently by Macintyre, and used in Hrushovski's proof of the Manin-Mumford conjecture. In another direction, Cherlin and Hrushovski studied "smoothly approximable structures", a generalization of the ω-categorical ω-stable structures analyzed by Cherlin, Harrington and Lachlan. Finally, methods of constructing new simple theories were explored by Chatzidakis, Evans, Hrushovski, Kikyo, Pillay, Pourmahdian and Tsuboi.

Simple theories not only form a unified setting containing many particular (unstable) structures, whose properties had previously been described somewhat *ad hoc*, using local stability-theoretic or other methods (like S_1-rank), but can be considered the right class in which to develop pure forking theory, as separate from phenomena which derive from definability of types or stationarity. In this sense, they should also help us understand *stable* theories better. For instance, while an ω-categorical strongly minimal set has a locally modular geometry, Hrushovski has constructed an ω-categorical structure of SU-rank 1 whose geometry is not locally modular (Example 6.2.27). So local modularity is not implied by ω-categoricity and the exchange property for algebraic closure, but also needs multiplicity 1.

In the remainder of this chapter, we shall quickly review the model-theoretic prerequisites and provide the main examples of simple theories. In Chapter 2, after an introduction into the customary set-up of working inside a universal domain in Section 2.1, we shall earnestly start in Section 2.2 by defining *dividing* and *forking*. The next three sections define simplicity and develop the basic properties of forking in simple theories, including the Independence Theorem; it is also shown that simplicity of a theory is equivalent to symmetry or to transitivity of non-forking or non-dividing. Section 2.6 is somewhat separate: here we show that forking independence in a simple theory can be abstractly characterized as a relation satisfying some elementary properties; this will be applied in Section 6.3 to prove simplicity of the structures obtained there. Section 2.7 expands on the notion of Lascar strong type encountered in Section 2.5 and serves as a precursor to Chapter 3. After a characterization of simplicity in terms of counting partial types reminiscent of stability theory in Section 2.8, the last section introduces model-theoretic stability from a simplicity-theoretic point of view.

Chapter 3 develops the model theory of hyperimaginary elements, i.e. classes modulo type-definable equivalence relations. It turns out that although one cannot simply expand the structure by adding new sorts for these objects and treat them as ordinary elements, as one does for classes modulo definable equivalence relations, one can nevertheless in many respects just treat them as if they were real elements. The main difference is that they are intrinsically infinite objects; e.g. if a hyperimaginary element a lies in the definable closure of an infinite set A, there need not be some $a_0 \in \mathrm{dcl}(a)$ and a finite subset $A_0 \subseteq A$ with $a_0 \in \mathrm{dcl}(A_0)$ (as would be the case if a were a tuple, even of infinite length). Using hyperimaginaries, we can then transfer many tools and techniques from stability theory, in particular the notion of a *canonical base* (Section 3.3) and *internality* or *analysability* of a type in some family of types (Section 3.4). The last two sections of Chapter 3 treat the question under which circumstances we can eliminate hyperimaginary elements; while Section 3.6 is a down-to-earth formula based approach, Section 3.7 expands on Section 2.7 and takes a more Galois-theoretic line. It is worth pointing out at this point that all known examples of simple theories *do* eliminate hyperimaginaries, but that nevertheless so far they are a necessary feature of the general theory.

Groups (and fields) not only furnish important examples and applications for the general theory, they also play an important rôle in the general analysis of arbitrary simple theories, via interpretation and definability theorems. We therefore study groups in a simple theory in Chapter 4, starting first with type-definable groups in Section 4.1, before progressing to groups whose domains consist of hyperimaginaries (Section 4.3). Chain conditions, in particular the chain condition on centralizers, play a major part in the analysis of stable groups; although they fail in the general simple case, we shall recover them up to finite, or bounded, index in Sections 4.2 and 4.4, where we also define the connected and other components. Stabilizers follow in Section 4.5 and conclude the exposition of the basic tools. Section 4.6 presents the internality and analysability machinery of Section 3.4 in the context of groups, while Section 4.8 deals with local modularity introduced in Section 3.5. Finally, we generalize the Weil-Hrushovski theorem on the reconstruction of a group from generically given data in Section 4.7. In this context, we should point out that two important group existence theorems have so far resisted proof in the simple context: the *group configuration theorem*, which constructs a group from geometric data, and the *binding group theorem*, which asserts the hyperdefinability of a certain Galois group (see e.g. [126] or [161]).

In Chapter 5 we define supersimplicity, and the two main ranks associated with it: Lascar rank SU and Shelah rank D. However, it is no longer true that Shelah rank necessarily witnesses forking, as it does in a stable theory. Moreover, there is no obvious analogue for Morley rank, and no suitable definition for *ω-simplicity*, or *total simplicity*, has been suggested to date. Section 5.2 applies the results of 3.4 and lays the foundations for the fundamental result of Section 5.3, elimination of hyperimaginaries for supersimple theories. No direct proof is known for this fact: one first has to establish the full theory for hyperimaginaries in a supersimple theory, only to eliminate them afterwards! In the last three sections, we turn towards groups in supersimple theories. We first generalize the analysis for superstable groups to the simple context in Section 5.4, including the appropriate version of Zil'ber's Indecomposability Theorem 5.4.5. In Section 5.5 we show that a type-definable group in a supersimple theory is the intersection of definable groups; this may be considered a form of elimination of hyperimaginaries for coset spaces. Note, however, that the most general question is still open: is a hyperimaginary group the projective limit of type-definable groups? Since elimination of hyperimaginaries only works on a complete type, this does not follow simply from Section 5.3. Finally, we prove in Section 5.6 that a supersimple division ring is commutative. Alas, a full classification (as in the superstable case: algebraically closed fields) is only conjectural.

Chapter 6 presents various additional aspects of the theory. Section 6.1 deals with small theories, i.e. theories with few pure types. They have particularly nice properties in the stable case, and this is again true in simple theories: a small simple theory eliminates hyperimaginaries, a small one-based simple theory, or more generally a small simple theory with finite coding, has finite weights. We thus obtain Lachlan's Conjecture for simple theories with finite coding: they are either ω-categorical, or have infinitely many non-isomorphic countable models.

Section 6.2 treats ω-categorical structures (which of course are small). We explain in Subsection 6.2.1 the modifications to Hrushovski's amalgamation construction needed to obtain simple, rather than stable, ω-categorical structures, thus obtaining exotic geometries of rank 1. One should point out that the construction is not restricted to the ω-categorical case: one can obtain similar results under the hypothesis that the closure relation of the geometry is locally finite. However, it is not clear what happens in the general case, when the closure relation is not locally finite, and whether one can modify the construction so that it no longer eliminates hyperimaginaries (and thereby produce the first example of such a theory). The next two sections try to generalize the known properties of ω-categorical superstable theories, which are known to be locally modular of finite rank. While the situation is satisfactory in the case of groups (Subsection 6.2.2), which are finite-by-Abelian-by-finite of finite rank, the only other case we can treat is that of CM-trivial theories (a property which holds for the amalgamation constructions from Subsection 6.2.1), and even here the proof is not totally satisfactory: in analogy to the superstable case, one should only have to assume CM-triviality for types of finite rank, rather than for a whole set of infinite rank, as we do in the proof of Corollary 6.2.49.

Section 6.3 deals with ways of obtaining new examples of simple theories. In particular we show that the union of any two theories in disjoint languages $\mathfrak{L}_1$ and $\mathfrak{L}_2$ (up to equality) which eliminate infinite quantifiers has a model-companion, and study the question as to when this model-companion is simple.

The matter is more complicated in Subsection 6.3.2, where we seek to add a generic automorphism to a simple theory (and hence there must be compatibility between the original structure and the new bijection). Again we only obtain partial results; in particular, no necessary and sufficient criterion is known as to when a simple structure allows a simple expansion by an automorphism.

Finally, the last section treats low theories (which include all known *natural* examples of simple theories). We prove that in a low theory, Lascar strong type is the same as strong type. However, the question whether a low theory allows full elimination of hyperimaginaries, or even of bounded hyperimaginaries, is still open.

All chapters are followed by a short section *Bibliographical remarks*, which collects references and attributions.

1.2 NOTATION AND MODEL-THEORETIC PREREQUISITES

We shall assume some knowledge of basic model theory (in particular familiarity with the compactness theorem), but no stability theory — although of course that will be helpful, as we shall frequently compare the simple to the stable case. However, we shall develop simplicity from scratch, and in fact will define stable theories as a subclass of the simple ones.

Following set-theoretic convention, we shall denote the set of natural numbers by ω; higher ordinal numbers will usually be denoted by α, β, etc. (but sometimes also by i). We identify cardinal numbers with the smallest ordinal number of their cardinality and denote them by λ, κ, etc.; κ^+ will stand for the successor cardinal of κ (whereas $\alpha + 1$ is the *ordinal* successor).

Formulas will be denoted by Greek letters $\varphi, \psi, \vartheta, \ldots$, sometimes we exhibit the variables or parameters, as in $\varphi(\bar{x}, \bar{a})$. A formula $\varphi(\bar{x}, \bar{y})$ is a formula with free variables $\bar{x}\bar{y}$; the distinction indicates that the variables $\bar{x}$ and the variables $\bar{y}$ will be used differently. If $\varphi(\bar{x})$ is a formula and $\mathfrak{M}$ a model, $\varphi^{\mathfrak{M}}$ will denote the set $\{\bar{m} \in \mathfrak{M} : \mathfrak{M} \models \varphi(\bar{m})\}$.

DEFINITION 1.2.1 Let $\mathfrak{M}$ be an $\mathfrak{L}$-structure. A subset X of $\mathfrak{M}^k$ is

1. *definable* if there are a tuple $\bar{c} \subset \mathfrak{M}$ of parameters and an $\mathfrak{L}(\bar{c})$-formula $\varphi(x_1, \ldots, x_k, \bar{c})$ such that $X = \varphi^{\mathfrak{M}}$, and

2. *type-definable*, if X is the intersection of an arbitrary family of definable subsets.

If all the parameters used for some (type-)definable set X are contained in a subset A of M, we say that X is (type-)definable *over* A, or A-(type-)definable. We shall often identify a formula with the set it defines (where the model is given implicitly).

If $\mathfrak{M}$ is a model of some theory T we write $\mathfrak{M} \models T$; conversely the *theory of* $\mathfrak{M}$, denoted $\mathrm{Th}(\mathfrak{M})$, is the collection of all $\mathfrak{L}(\mathfrak{M})$-sentences true in $\mathfrak{M}$. The *diagram* $\mathrm{diag}(\mathfrak{M})$ is the collection of all quantifier-free $\mathfrak{L}(\mathfrak{M})$-sentences true in $\mathfrak{M}$.

If $A \subseteq \mathfrak{M}$ and $\bar{a} \in \mathfrak{M}$, then $\mathrm{tp}_{\mathfrak{M}}(\bar{a}/A)$ is the type of $\bar{a}$ over A in $\mathfrak{M}$; we shall write $S_n(A)$ for the collection of all types in n variables over A, and $S(A)$ for the set of all types over A. If the ambient model is clear, but we want to emphasize the language $\mathfrak{L}$ used, we write $\mathrm{tp}_{\mathfrak{L}}(\bar{a}/A)$. Types are usually denoted by $p, p', p_0, q, r, \ldots$.

DEFINITION 1.2.2 Let $\mathfrak{M}$ be a model of a theory T, and $A \subseteq \mathfrak{M}$. A type $p(\bar{x}) \in S(A)$ is *isolated* by a formula $\varphi(\bar{x}) \in p$ if $\varphi(\bar{x})$ proves $p(\bar{x})$, i.e. any realization of φ is also a realization of p. A subset B of $\mathfrak{M}$ is *atomic* over A if the type of every tuple $\bar{b} \in B$ over A is isolated; we just say *atomic* instead of atomic over $\emptyset$.

REMARK 1.2.3 It is clear that every algebraic formula (i.e. with only finitely many realizations) contains an isolated type. If $\mathrm{tp}(\bar{m})$ is isolated by $\varphi(\bar{x})$ and $\mathrm{tp}(\bar{n}/\bar{m})$ is isolated by $\psi(\bar{y}, \bar{m})$, then $\varphi(\bar{x}) \wedge \psi(\bar{y}, \bar{x})$ isolates $\mathrm{tp}(\bar{m}, \bar{n})$. Conversely, if $\varphi(\bar{x}, \bar{y})$ isolates $\mathrm{tp}(\bar{m}, \bar{n})$, then $\varphi(\bar{x}, \bar{n})$ isolates $\mathrm{tp}(\bar{m}/\bar{n})$ and $\exists \bar{x}\, \varphi(\bar{x}, \bar{y})$ isolates $\mathrm{tp}(\bar{n})$.

Clearly, an isolated type over A is realized in any model containing A. Conversely, one has:

REMARK 1.2.4 OMITTING TYPES THEOREM A countable theory T has a model which does not realize a type $p \in S(\emptyset)$ if and only if p is not isolated.

DEFINITION 1.2.5 Let κ be some cardinal. A theory T is *κ-categorical* if all models of cardinality κ are isomorphic.

There is a good characterization of ω-categoricity, which we shall prove in Chapter 6 (Corollary 6.1.6):

COROLLARY 1.2.6 RYLL-NARDZEWSKI THEOREM *A countable complete theory T is ω-categorical if and only if $S_n(\emptyset)$ is finite for all $n < \omega$.*

Opposed to atomic models are saturated structures:

DEFINITION 1.2.7 A model $\mathfrak{M}$ is κ-saturated for some cardinal κ if every type over a subset of cardinality less than κ is realized in $\mathfrak{M}$. A model $\mathfrak{M}$ is *saturated* if it is $|\mathfrak{M}|$-saturated.

By a union-of-chains argument, it is easy to see that every model has a κ-saturated elementary extension, for every cardinal κ; moreover, any partial isomorphism between two subsets of cardinality less than κ of two saturated models of cardinality at least κ can be extended to an isomorphism of the models. However, the existence of saturated models in general depends on set-theoretic assumptions. We shall therefore do with less:

DEFINITION 1.2.8 A model $\mathfrak{M}$ of some theory T is *κ-homogeneous* if any partial automorphism (elementary map) between two subsets of cardinality less than κ can be extended to an automorphism of $\mathfrak{M}$.

In particular, if $\text{Aut}_A(\mathfrak{M})$ denotes the group of automorphisms of $\mathfrak{M}$ fixing A pointwise, then in a κ-homogeneous model $\mathfrak{M}$ any two realizations of a type p (in less than κ variables) over a set A of cardinality less than κ are conjugate under $\text{Aut}_A(\mathfrak{M})$. We shall see in the next chapter that every complete theory has a κ-saturated and κ-homogeneous model, for all cardinals κ.

The following combinatorial principle is often very useful:

THEOREM 1.2.9 RAMSEY'S THEOREM *Suppose X is an infinite set and the set of unordered n-tuples of X is painted in k different colours. Then there is an infinite monochromatic subset Y, i.e. a subset $Y \subseteq X$ whose n-tuples all have the same colour.*

PROOF: We shall use induction on n. If $n = 1$, then since X is coloured in finitely many colours, there must be an infinite monochromatic subset. So suppose the assertion holds for all colourings of n-tuples, and consider a colouring of the $(n+1)$-tuples of X.

Fix $x_0 \in X =: X_0$. This induces a colouring of the n-tuples of $X_0 - \{x_0\}$: an n-tuple $\bar{x}$ is ascribed the colour of $\{x_0, \bar{x}\}$, and by the inductive hypothesis there is an infinite monochromatic subset X_1. Pick any $x_1 \in X_1$, consider the colouring of n-tuples of $X_1 - \{x_1\}$ induced by x_1, and find an infinite monochromatic subset X_2; and repeat this ω many times to obtain a sequence $X_0 \supset X_1 \supset X_2 \supset \cdots$ of infinite subsets of X and elements $x_i \in X_i - X_{i+1}$ such that all subsets $\{x_i, \bar{x}\}$ with $\bar{x} \in X_{i+1}$ have the same colour c_i. As there are only finitely many colours, one colour c must have been used infinitely often, say whenever $i \in I$ for some infinite index set $I \subseteq \omega$.

We claim that $Y := \{x_i : i \in I\}$ is our monochromatic subset, of colour c. Indeed, if $\{x_{i_0}, \ldots, x_{i_n}\}$ is an $(n+1)$-tuple from Y with $i_0 < i_1 < \cdots < i_n$, then $x_{i_j} \in X_{i_0}$ for $1 \leq j \leq n$, and the tuple has colour c. ∎

REMARK 1.2.10 We should note that the compactness theorem now implies that for every triple (k, n, m) of natural numbers there is some natural number $R(k, n, m)$ such that whenever n-tuples of a set of size at least $R(k, n, m)$ are painted in k colours, then there is a monochromatic subset of size m. There are, however, constructive proofs of this finite version of Ramsey's Theorem which actually give bounds on the function $R(k, n, m)$, albeit not very good ones.

DEFINITION 1.2.11 Let A be a set of parameters, and $p \in S(A)$. A sequence $(a_i : i \in I)$ is a sequence *of type* p if $a_i \models p$ for all $i \in I$.
If the index set I is ordered, the sequence $(a_i : i \in I)$ is n-*indiscernible* over A if for all $i_1 < i_2 < \cdots < i_n$ the type $\text{tp}(a_{i_1} \ldots a_{i_n}/A)$ does not depend on the choice of indices; it is *indiscernible* over A if it is n-indiscernible over A for all $n < \omega$.

PROPOSITION 1.2.12 *Suppose* $\mathfrak{M}$ *is* $(|A|^+ + \kappa)$*-saturated. Then there is an indiscernible sequence of length* κ *over* A.

PROOF: Consider the type Σ in variables $(x_i : i < \kappa)$ saying that all the x_i are distinct, together with

$$\{\varphi(\bar{x}) \leftrightarrow \varphi(\bar{y}) : \varphi \in \mathfrak{L}; \bar{x}, \bar{y} \subset (x_i : i < \kappa) \text{ with increasing indices}\}.$$

CLAIM. This partial type is consistent.
PROOF OF CLAIM: Consider any finite subset Σ_0. This finite subset mentions only k distinct formulas for some $k < \omega$, which we may suppose all have n free variables (adding dummy variables if necessary). So we can colour the n-tuples of $\mathfrak{M}$ in 2^k different colours according to which of the k formulas they satisfy. By Ramsey's Theorem there is an infinite monochromatic subset of $\mathfrak{M}$, which must satisfy Σ_0. So Σ is consistent by compactness. ∎
By saturation we can realize Σ by some sequence in $\mathfrak{M}$. ∎
There actually is an ordinal version of Ramsey's Theorem which states that from a given sequence we can choose an n-indiscernible subsequence, provided the original sequence is long enough. However, its proof is rather more involved and will be omitted.

THEOREM 1.2.13 ERDÖS-RADO THEOREM *Suppose* κ *is a cardinal,* X *is an infinite set of cardinality* $(2^\kappa)^+$*, and the set of unordered* n*-tuples of* X *is painted in* κ *different colours. Then there is an infinite monochromatic subset* $Y \subseteq X$ *of size* κ^+.

Recall that a substructure $\mathfrak{M}$ of a structure $\mathfrak{N}$ is *elementary*, denoted $\mathfrak{M} \prec \mathfrak{N}$, if every sentence with parameters in $\mathfrak{M}$ true in $\mathfrak{M}$ is also true in $\mathfrak{N}$. The following criterion is useful in establishing the elementary nature of an inclusion:

THEOREM 1.2.14 TARSKI TEST *A substructure* $\mathfrak{M}$ *of a structure* $\mathfrak{N}$ *is elementary if and only if for every sentence of the form* $\exists x\, \varphi(x)$ *with parameters from* $\mathfrak{M}$ *and true in* $\mathfrak{N}$ *there is some* $m \in \mathfrak{M}$ *such that* $\mathfrak{N} \models \varphi(m)$.

PROOF: If $\mathfrak{M} \prec \mathfrak{N}$, then $\mathfrak{N} \models \exists x\, \varphi(x)$ implies $\mathfrak{M} \models \exists x\, \varphi(x)$, so there is $m \in \mathfrak{M}$ with $\mathfrak{M} \models \varphi(m)$, whence $\mathfrak{N} \models \varphi(m)$.

For the converse, assume that the criterion holds, and let φ be an $\mathfrak{L}(\mathfrak{M})$-sentence true in $\mathfrak{M}$. We may assume that φ is of the form $\forall x\, \psi(x)$ (the case of the existential quantifier being trivial); by induction on the number of quantifiers, we may assume that for all $m \in \mathfrak{M}$ we have $\mathfrak{N} \models \neg\psi(m)$ if and only if $\mathfrak{M} \models \neg\psi(m)$. But this holds for no $m \in \mathfrak{M}$; by assumption $\mathfrak{N} \models \psi(n)$ for all $n \in \mathfrak{N}$, whence $\mathfrak{N} \models \forall x\, \psi(x)$, and $\mathfrak{N} \models \varphi$. ∎

A theory T is *model-complete* if whenever $\mathfrak{M} \subseteq \mathfrak{N}$ are models of T, then $\mathfrak{M} \prec \mathfrak{N}$.

THEOREM 1.2.15 ROBINSON TEST *Let T be a theory. The following are equivalent:*

1. *T is model-complete.*

2. *Every formula is equivalent modulo T to an existential formula.*

3. *Every formula is equivalent modulo T to a universal formula.*

4. *Whenever $\mathfrak{M} \subseteq \mathfrak{N}$ are models of T and $\varphi(\bar{x})$ is a quantifier-free $\mathfrak{L}(\mathfrak{M})$-formula with $\mathfrak{N} \models \exists \bar{x}\, \varphi(\bar{x})$, then there is $\bar{m} \in \mathfrak{M}$ such that $\mathfrak{M} \models \varphi(\bar{m})$.*

PROOF: 1. ⇒ 4. and 2. ⇒ 1. are trivial, as is 2. ⇔ 3.; we shall show 4. ⇒ 3.. So let $\exists \bar{y}\, \varphi(\bar{x}, \bar{y})$ be an existential formula, with φ quantifier-free, and consider the collection $\Psi(\bar{x})$ of all universal formulas $\psi(\bar{x})$ implied by $\exists \bar{y}\, \varphi(\bar{x}, \bar{y})$ modulo T. Let $\bar{a}$ be a new tuple of constant symbols, and $(\mathfrak{M}, \bar{a})$ any model of $T \cup \Psi(\bar{a})$.

CLAIM. $T \cup \mathrm{diag}(\mathfrak{M}) \cup \{\exists \bar{y}\, \varphi(\bar{a}, \bar{y})\}$ is consistent.

PROOF OF CLAIM: Otherwise by compactness there is a quantifier-free $\mathfrak{L}(\mathfrak{M})$-sentence $\vartheta(\bar{m}, \bar{a})$ true in $\mathfrak{M}$, such that $T \cup \{\exists \bar{y}\, \varphi(\bar{a}, \bar{y})\}$ proves $\forall \bar{z}\, \neg\vartheta(\bar{z}, \bar{a})$. But then $\forall \bar{z}\, \neg\vartheta(\bar{z}, \bar{a}) \in \Psi(\bar{a})$, contradicting $\mathfrak{M} \models \Psi(\bar{a})$. ∎

Hence there is $\mathfrak{N} \models T$ containing $\mathfrak{M}$ and satisfying $\exists \bar{y}\, \varphi(\bar{a}, \bar{y})$. By assumption we find $\bar{m} \in \mathfrak{M}$ with $\mathfrak{M} \models \varphi(\bar{a}, \bar{m})$; as $\mathfrak{M}$ was arbitrary, $T \cup \Psi(\bar{a}) \vdash \exists \bar{y}\, \varphi(\bar{a}, \bar{y})$. By compactness, there is a finite subset of $\Psi(\bar{x})$ equivalent to $\exists \bar{y}\, \varphi(\bar{x}, \bar{y})$.

It now follows by induction on the quantifier complexity that every formula is equivalent to a universal and to an existential formula. ∎

In Chapter 6 we shall also make use of the notion of a model-companion.

DEFINITION 1.2.16 Let T be any theory. A *model-companion* for T is a model-complete theory T^* such that any model of T embeds into a model of T^*, and any model of T^* embeds into a model of T.

LEMMA 1.2.17 *If T has a model-companion, it is unique.*

PROOF: If T_1 and T_2 are two model-companions for T, we get a chain of structures $\mathfrak{M}_0 \subseteq \mathfrak{M}_1 \subseteq \cdots$ such that $\mathfrak{M}_i \models T$ for odd i, and $\mathfrak{M}_{4n+2i} \models T_i$, for $i = 1, 2$ and $n < \omega$. By model-completeness of T_1 and T_2, the chains $(\mathfrak{M}_{4n} : n < \omega)$ and $(\mathfrak{M}_{4n+2} : n < \omega)$ are elementary, and both $\mathfrak{M}_2$ and $\mathfrak{M}_4$ are elementary substructures of $\bigcup_{n<\omega} \mathfrak{M}_n$. Therefore $T_1 = T_2$. ■

1.3 EXAMPLES

Proving simplicity for a theory usually follows one of two strategies: either one proves a quantifier-elimination result in a suitable language, which yields a good description of the definable sets and allows a direct proof, say of finiteness of the local ranks $D(., \varphi, k)$, or of symmetry of forking. Or one defines a notion of independence, shows that it satisfies the Independence Theorem, and invokes Theorem 2.6.1.

We shall now present the most important examples of simple theories, and refer the reader to the later chapters for the relevant definitions.

EXAMPLE 1.3.1 ■ The theory of a set without any structure is strongly minimal.

- The theory of ω or $\mathbb{Z}$ with the successor function is strongly minimal.

EXAMPLE 1.3.2 Let X be a set, and f a bijection from X to $X \times X$ without any cycles. Then $\mathrm{Th}(X, f)$ is stable, not superstable.

EXAMPLE 1.3.3 Let $\mathfrak{E}$ be a family of pairwise commuting equivalence relations on a set X. Then $\mathrm{Th}(X, \mathfrak{E})$ is stable.

EXAMPLE 1.3.4 An *Abelian structure* is an Abelian group A with some predicates for subgroups of powers of A. Any module over a ring R (in the ring language $\mathfrak{L}_R = \{0, +, \lambda_r : r \in R\}$, where λ_r is scalar multiplication by r) is an Abelian structure, as λ_r may be identified with the subgroup $\{(a, \lambda a) : a \in A\}$ of A^2. In an Abelian structure, every definable set is equal to a Boolean combination of cosets of $\mathrm{acl}(\emptyset)$-definable subgroups. It is then easy to see that Abelian structures are stable one-based, whence simple.

EXAMPLE 1.3.5 Let V be an infinite vector space over a finite field, and f a non-degenerate bilinear form on V. Then (V, f) is unstable, supersimple of SU-rank 1 and ω-categorical.

EXAMPLE 1.3.6 Examples of simple fields:

1. An *algebraically closed field* K (of given characteristic) has full elimination of quantifiers in the natural language of fields, and is strongly minimal. Conversely, every superstable field is algebraically closed.

2. A field K is *separably closed* if it has no proper separable field extension. In characteristic zero a separably closed field is algebraically closed, but in characteristic $p \neq 0$ we get a chain of subfields $K \geq K^p \geq K^{p^2} \geq \cdots$. The transcendence degree tr.deg.(K/K^p) is called the *Eršov invariant* of K. Separably closed fields have quantifier elimination in a suitably enriched language, and are stable.

3. A field K is *pseudo-algebraically closed* if every variety over K which is irreducible over the algebraic closure of K has a K-rational point. A pseudo-algebraically closed field with only finitely many extensions of a given degree is simple, unstable (unless it is separably closed) and not supersimple (unless it is perfect).

4. A field is *pseudo-finite* if it satisfies the theory of all finite fields. This happens if and only if it is perfect, pseudo-algebraically closed, and has exactly one extension of every degree. A pseudo-finite field is unstable supersimple of SU-rank 1; more generally, Hrushovski has shown that a perfect pseudo-algebraically closed field with only finitely many extensions of any given degree is supersimple of SU-rank 1.

5. A *differential field* is a field K with an additional function δ, which is interpreted as a *derivation* on K and satisfies $\delta(x+y) = \delta(x) + \delta(y)$ and $\delta(xy) = \delta(x)y + x\delta(y)$. It is *differentially closed* if every system consisting of a differential equation in x of degree n and a differential inequality of degree less than n, both with coefficients in K, has a solution in K. The theory DCF_0 of a differentially closed field of characteristic zero has quantifier elimination and is ω-stable of SU-rank ω. Moreover, both K and the subfield $C = \{k \in K : \delta(k) = 0\}$ of constants are algebraically closed, and C carries no structure other than the field structure.

   Hrushovski has constructed superstable differential fields which are not differentially closed.

6. A *difference field* K is a field K with an additional function σ, which is interpreted as an automorphism of K. It is *existentially closed* if every quantifier-free formula (with parameters in K) which has a solution in a difference field containing K has a solution in K itself. The theory ACFA_0 of existentially closed difference fields of characteristic 0 is unstable supersimple, of SU-rank ω; the field is algebraically closed and the subfield of fixed points of σ is pseudo-finite.

7. The theory DCFA_0 of a differentially closed field of characteristic 0 with a generic automorphism is supersimple (Hrushovski).

However, the following classes are not elementary (and hence do not yield simple fields):

- Existentially closed fields with two commuting automorphisms.
- Existentially closed fields with an automorphism whose fixed field is algebraically closed.
- Pseudo-finite fields with an automorphism.

Another method to obtain simple structures is a direct construction from finite (or finitely generated) substructures, via an amalgamation process initially due to Ehrenfeucht, Fraïssé, and Jonsson. It was adapted by Hrushovski to give stable relational structures, and again to yield simple unstable structures (see Section 6.2.1). Particular examples of this are the *random graph*, the *random bipartite graph*, or the *random n-ary hypergraph*, all of which have SU-rank 1. Note, however, that the random triangle-free graph does not satisfy the Independence Theorem and cannot be simple.

Baudisch has adapted the (stable case of the) construction to obtain a totally categorical non-(Abelian-by-finite) group which does not interpret a field; in he has extracted a stable, non-superstable subgroup out of it. Hrushovski has also managed to adapt his amalgamation method to amalgamate two strongly minimal theories (with the definable multiplicity property).

In certain theories it is also possible to establish stability by counting types. Baudisch proved that the free nilpotent group (on infinitely many generators) of exponent p^n and class $c < p$ is ω-stable; Chapuis has constructed centreless soluble ω-stable groups of derived length n for every $n < \omega$. I know of no example, though, where the results of Section 2.8 have been used to establish simplicity by counting partial types.

Mekler associates to every theory a nilpotent group of class 2 and prime exponent which shares many of the model-theoretic properties of the original structure. In particular, simplicity and stability are preserved, although finiteness of rank is not. If we allow extra structure on the group, such an interpretation is actually quite easy:

EXAMPLE 1.3.7 Let $\mathfrak{M}$ be any structure in a language $\mathfrak{L}$, and consider $\mathfrak{M}$ to be a basis for an Abelian group A of exponent p. Let P be a new unary predicate, put $\mathfrak{L}_A := \mathfrak{L} \cup \{P, 0, +\}$, and turn A into an $\mathfrak{L}_A$-structure by interpreting P as $\mathfrak{M}$, with the natural interpretation of $\mathfrak{L}$ on $P^A = \mathfrak{M}$. Then for any elementary superstructure $\mathcal{A}$ the subset $P^{\mathcal{A}}$ is an elementary superstructure of $\mathfrak{M}$ and $P^{\mathcal{A}}$ is a linearly independent subset. Hence any automorphism of $P^{\mathcal{A}}$ can be extended to an automorphism of $\mathcal{A}$. There are two kinds of elements: those algebraic over $P^{\mathcal{A}}$, which are determined by finitely many elements in $P^{\mathcal{A}}$, and a unique *generic* type which is transcendental over $P^{\mathcal{A}}$. It follows that $\mathfrak{M}$ is simple (stable) if and only if A is. However, $SU(A) = SU(\mathfrak{M}) \cdot \omega$ (which may of course be ∞).

Similarly, we may take $\mathfrak{M}$ to be a transcendence basis for an algebraically closed field K and obtain analogous results.

Finally, any structure interpretable in a simple structure is itself simple.

DEFINITION 1.3.8 An $\mathfrak{L}$-structure $\mathfrak{M}$ is *interpretable* in an $\mathfrak{L}'$-structure $\mathfrak{N}$ if there is an $\mathfrak{L}'$-definable subset M' of $\mathfrak{N}^{eq}$, for every constant of $\mathfrak{L}$ an element of M', for every j-ary function of $\mathfrak{L}$ an $\mathfrak{L}'$-definable function $(M')^j \to M'$, and for every j-ary relation of $\mathfrak{L}$ an $\mathfrak{L}'$-definable subset of $(M')^j$ such that all this induces on M' a structure isomorphic to $\mathfrak{M}$.

If T has a model which is interpretable in a model of T', we say that T is interpretable in T'.

In particular, structures interpretable in the fields of Example 1.3.6 are simple. In fact, for groups interpretable in these fields quantifier elimination yields even better results:

1. A group interpretable in an algebraically closed field is an algebraic group.
2. A group interpretable in a differentially closed field is a differentially algebraic group.
3. A group interpretable in a separably closed field is an algebraic group over that field.

1.4 BIBLIOGRAPHICAL REMARKS

The classic introduction to model theory is the book by Chang and Keisler [22]; Hodges' *Model Theory* [45] is entertaining and offers a sometimes unusual perspective. Poizat's white book *Cours de Théorie des Modèles* [125] in addition covers some stability theory; more advanced material can be found in the recent book by Pillay, *Geometrical Stability Theory* [106]. The big encyclopaedia on stability is of course Shelah's *Classification Theory* [147]; Baldwin [3] aims at the same comprehensiveness. More accessible are Buechler [16], Lascar [85] and Pillay [103].

Morley proved his famous theorem in [100]; Rowbottom defined λ-stability in [134]. Baldwin, Lachlan and Macintyre study ω- and uncountably categorical structures in [2, 6, 12, 73, 74, 75, 76, 92, 93]; the breakthrough in the description of ω-categorical ω-stable structures comes in Zil'ber's work [167, 169, 168, 170, 171, 172], partially reproved on the Western side of the iron curtain by Cherlin, Harrington and Lachlan [29]. Shelah developed his programme in [140, 141, 142, 143, 144, 146, 147, 148, 150, 151]; the Paris approach to forking appeared in [78, 79, 80, 81, 83, 84, 85, 91, 116, 117, 119, 118, 120, 122, 123, 124, 125]. Simple theories were studied by Shelah in [149, 152]; Kim and Pillay took off in [64, 65, 67, 66, 69, 68, 70, 71, 108, 111] and were joined by Buechler, Evans, Hart, Lascar and Scanlon in [17, 19, 36, 40, 41, 89, 90, 114, 160, 163].

For an introduction to the model theory of fields one should consult Marker, Messmer and Pillay [95]. Elimination of quantifiers for algebraically closed fields was proved by Tarski [155], algebraic closedness of superstable fields by Macintyre [93] and Cherlin [28]. Eršov proves elimination of quantifiers for separably closed fields in [34]; Wood shows stability in [166]. Differential fields were studied by Seidenberg [137], Shelah [145], Hrushovski, Pillay, Sokolovič [115, 59] and Itai [55]; pseudo-finite fields byAx [1] and Hrushovski [49], bounded pseudo-algebraically closed fields by Chatzidakis and Pillay [27]; existentially closed difference fields by Chatzidakis, Hrushovski and Peterzil [24, 25, 26]. For groups interpretable in theories of fields, one should look at Weil [165], van den Dries [157] Poizat [126], Hrushovski and Pillay [57, 58, 107], and Messmer [97]. The recent interaction between the model theory of fields and algebraic geometry is explained in Bouscaren [15], following Hrushovski's proofs of the Mordell-Lang [52] and Manin-Mumford [54] conjectures.

The original amalgamation method can be found in Henson [42], Hrushovski's adaptations in [50, 51], and Baudisch's groups in [10, 9]. Generic simple constructions were studied by Chatzidakis, Evans, Hrushovski, Kikyo, Pillay Pourmahdian and Tsuboi in [27, 35, 53, 62, 63, 110, 130, 156]; Hrushovski [53] and Pillay[109] also leave the first-order set-up and consider forking in the category of existentially closed structures. Smoothly approximable structures (such as Example 1.3.5) are introduced by Cherlin and Hrushovski in [30]

The Omitting Types Theorem was shown by Grzegorczyk, Mostowsky and Ryll-Nardzewski [39], the Ryll-Nardzewsky Theorem independently by Engeler [32], Ryll-Nardzewsky [135] and Svenonius [153]. Ramsey's Theorem comes from [132], and the Erdös-Rado Theorem 1.2.13 from [33]. Szmielev [154], Monk [98] and Baur [11] prove pp-elimination for abelian groups and modules; a good introduction to the model theory of modules is Prest [131]. Various other groups are studied by Mekler [96], Baudisch [8] and Chapuis [23].

Chapter 2

SIMPLICITY

2.1 THE MONSTER MODEL AND IMAGINARIES

Throughout this book, we shall work inside a big, κ-saturated κ-homogeneous model $\mathfrak{C}$ of a complete first-order theory T in a language $\mathfrak{L}$; we call $\mathfrak{C}$ the *monster model*. Every model we consider will have cardinality less than κ and be an elementary substructure of $\mathfrak{C}$; every set of parameters will have cardinality less than κ and be a subset of $\mathfrak{C}$, and every tuple will have length less than κ and come from $\mathfrak{C}$. Since every model of T of cardinality less than κ can be elementarily embedded into $\mathfrak{C}$, this is no restriction, but it simplifies the arguments and helps the intuition, as everything takes place inside a fixed "universal domain".

LEMMA 2.1.1 *Every complete first-order theory T has a κ-saturated κ-homogeneous model, for arbitrarily large κ.*

PROOF: Clearly we may assume that $\kappa \geq |\mathfrak{L}|$.
CLAIM. Let $\mathfrak{M} \models T$ with $|\mathfrak{L}| \leq |\mathfrak{M}| = \lambda$. Then there is an λ^+-saturated $\mathfrak{N} \succ \mathfrak{M}$ with $|\mathfrak{N}| \leq 2^\lambda$.
PROOF OF CLAIM: If $\mathfrak{M}' \models T$ with $|\mathfrak{M}'| \leq 2^\lambda$, then $\mathfrak{M}$ has at most $(2^\lambda)^\lambda = 2^\lambda$ subsets of cardinality λ, and over every such subset there are at most 2^λ types. Hence there is an $\mathfrak{M}'' \succ \mathfrak{M}$ realizing all types over all subsets of $\mathfrak{M}'$ of size λ. Starting from $\mathfrak{M} = \mathfrak{M}_0$, we construct an elementary chain $(\mathfrak{M}_i : i < \lambda^+)$ such that $|\mathfrak{M}_i| \leq 2^\lambda$ for all $i < \lambda^+$, all types over subsets of size λ in $\mathfrak{M}_i$ are realized in $\mathfrak{M}_{i+1}$, and $\mathfrak{M}_\alpha = \bigcup_{i<\alpha} \mathfrak{M}_i$ for limit ordinals $\alpha < \lambda^+$. Put $\mathfrak{N} = \bigcup_{i<\lambda^+} \mathfrak{M}_i$, an elementary superstructure of $\mathfrak{M}$. Then $\mathfrak{N}$ is λ^+-saturated by regularity of λ^+, and $|\mathfrak{N}| \leq \lambda^+ 2^\lambda = 2^\lambda$. ■
Now let $(\mathfrak{M}_i : i < \kappa^+)$ be an elementary chain of models of T with $|\mathfrak{M}_0| \geq \kappa$, such that $\mathfrak{M}_\alpha = \bigcup_{i<\alpha} \mathfrak{M}_i$ for limit ordinals $\alpha < \kappa^+$, and $\mathfrak{M}_{i+1}$ is $|\mathfrak{M}_i|^+$-

saturated for all $i < \kappa^+$. Put $\mathfrak{N} = \bigcup_{i<\kappa^+} \mathfrak{M}_i$, so $\mathfrak{N} \succ \mathfrak{M}_0 \models T$. Then $\mathfrak{N}$ is κ^+-saturated, by regularity of κ.

We claim that $\mathfrak{N}$ is κ-homogeneous. So let A and B be two subsets of $\mathfrak{N}$ of size κ, with $\mathrm{tp}_{\mathfrak{N}}(A) = \mathrm{tp}_{\mathfrak{N}}(B)$. By regularity of κ^+ there is some $i < \kappa^+$ such that $A \cup B \subseteq \mathfrak{M}_i$.

CLAIM. If $A \cup B \subseteq \mathfrak{M}_i$ with $\mathrm{tp}_{\mathfrak{N}}(A) = \mathrm{tp}_{\mathfrak{N}}(B)$, then there is an automorphism σ of $\mathfrak{M}_{i+\omega}$ with $\sigma(A) = B$,

PROOF OF CLAIM: By $|\mathfrak{M}_i|^+$-saturation of $\mathfrak{M}_{i+1}$ there is $\mathfrak{M}'_i \prec \mathfrak{M}_{i+1}$ with $\mathrm{tp}(\mathfrak{M}_i, A) = \mathrm{tp}(\mathfrak{M}'_i, B)$. In particular $B \subseteq \mathfrak{M}'_i$. Similarly, there is $\mathfrak{M}'_{i+1} \prec \mathfrak{M}_{i+2}$ with $\mathrm{tp}(\mathfrak{M}'_{i+1}, \mathfrak{M}_i) = \mathrm{tp}(\mathfrak{M}_{i+1}, \mathfrak{M}'_i)$, and $\mathfrak{M}'_{i+2} \prec \mathfrak{M}_{i+3}$ with $\mathrm{tp}(\mathfrak{M}_{i+2}, \mathfrak{M}_{i+1}) = \mathrm{tp}(\mathfrak{M}'_{i+2}, \mathfrak{M}_{i+1})$. Iterating this construction yields models $\mathfrak{M}'_j$ for $i \leq j < i + \omega$ with $\mathfrak{M}'_j \prec \mathfrak{M}_{j+1} \prec \mathfrak{M}'_{j+2}$ for all j, and a sequence of isomorphisms $\sigma_j : \mathfrak{M}_j \to \mathfrak{M}'_j$. Then $\sigma = \bigcup_{n<\omega} \sigma_{i+2n}$ is an isomorphism from $\bigcup_{n<\omega} \mathfrak{M}_{i+2n} = \mathfrak{M}_{i+\omega}$ to $\bigcup_{n<\omega} \mathfrak{M}_{i+2n} = \mathfrak{M}_{i+\omega}$ with $\sigma(A) = B$. ■

If τ is an automorphism of $\mathfrak{M}_j$ for some $j < \kappa^+$, then $\mathfrak{M}_{i+\omega} \prec \mathfrak{N}$ implies $\mathrm{tp}_{\mathfrak{N}}(\mathfrak{M}_{i+\omega}) = \mathrm{tp}_{\mathfrak{N}}(\tau(\mathfrak{M}_{i+\omega}))$, so τ can be extended to $\mathfrak{M}_{j+\omega}$ by the claim above. By the claim there is an automorphism σ of $\mathfrak{M}_{i+\omega}$ with $\sigma(A) = B$; we can then extend this automorphism to $\mathfrak{M}_{i+\omega 2}$, then to $\mathfrak{M}_{i+\omega 3}$, etc. Taking unions at limits, we eventually build an automorphism of $\mathfrak{N}$ mapping A to B. ■

From now on, fix κ and a monster model $\mathfrak{C}$. We shall denote subsets of $\mathfrak{C}$ (of cardinality less than κ) by $A, B, C, \ldots$, elements of $\mathfrak{C}$ by $a, b, c, \ldots$, and tuples from $\mathfrak{C}$ by $\bar{a}, \bar{b}, \bar{c}, \ldots$. Then in particular for any set A and any tuple $\bar{a}$ the type $\mathrm{tp}(\bar{a}/A)$ will mean the type inside $\mathfrak{C}$, and for a sentence φ with parameters in $\mathfrak{C}$ we shall write $\models \varphi$ instead of $\mathfrak{C} \models \varphi$. By an automorphism we shall mean an automorphism of $\mathfrak{C}$; and A-automorphism will be an automorphism which stabilizes A pointwise.

We shall now introduce Shelah's *eq*-construction, which allows us to deal with equivalence classes (modulo a definable equivalence relation) just as if they were real elements (or tuples) in the structure.

DEFINITION 2.1.2 Let $\mathfrak{M}$ be a model of a theory T in a first-order language $\mathfrak{L}$. For every $\emptyset$-definable equivalence relation $E(\bar{x}, \bar{y})$ we add a new unary predicate $P_E(x)$ to our language, together with a new function π_E. This will form the language $\mathfrak{L}^{eq}$; the model $\mathfrak{M}^{eq}$ is the disjoint union of domains $\mathfrak{M}_E$ consisting of all the equivalence classes of tuples in $\mathfrak{M}$ modulo E, for all $\emptyset$-definable E. We shall identify the original model $\mathfrak{M}$ with the set of classes modulo equality, and call it the *home sort*; the original constants, functions and relations will live there. All other elements will be *imaginary sorts.* Finally, we shall interpret P_E as $\mathfrak{M}_E$, and π_E as the projection from a tuple in the home sort to its equivalence class modulo E. The theory T^{eq} is then the $\mathfrak{L}^{eq}$-theory of $\mathfrak{M}^{eq}$.

It can be checked that T^{eq} does not depend on the choice of our original model $\mathfrak{M}$. In general, a model of T^{eq} will have elements satisfying no predicate P_E; these elements form the *superfluous sort* and carry no structure at all. In particular, the structure obtained by simply omitting superfluous elements will be an elementary substructure. Particular sorts of T^{eq} will be

- *permutation sorts*, i.e. classes of M^j modulo E, where $(m_1, \ldots, n_j)$ and $(n_1, \ldots, n_j)$ are equivalent modulo E if and only if one tuple is a permutation of the other, and
- *tuple sorts*, i.e. classes modulo E, where $(m_1, \ldots, m_j)E(n_1, \ldots, n_j)$ holds if and only if the tuples are the same.

It is clear that any model of T can be uniquely expanded to a model of T^{eq} without superfluous elements; conversely the restriction to the home sort of a model of T^{eq} is a model of T. By considering pre-images, it is easy to see that if $\mathfrak{C}$ is a monster model for T, then $\mathfrak{C}^{eq}$ is a monster model for T^{eq} (where we disregard the superfluous sort, with respect to which $\mathfrak{C}^{eq}$ is of course not saturated).

REMARK 2.1.3 Another way of looking at this is to introduce many-sorted logic (whence the "sorts"). Instead of adding predicates, for every $\emptyset$-definable equivalence relation E we add a new sort S_E, and functions $\pi_E : S^n_= \to S_E$, where n is the length of the tuples in the domain of E. There will be no superfluous sort (and $\mathfrak{C}^{eq}$ is truly κ-saturated), but variables will carry a sort, and functions and relations will have to specify the sorts they live on.

REMARK 2.1.4 If $E_{\bar{m}}(\bar{x}, \bar{y})$ is an $\bar{m}$-definable equivalence relation, we may consider the $\emptyset$-definable equivalence relation $(\bar{x}, \bar{s})E(\bar{y}, \bar{t})$ iff $\bar{s} = \bar{t}$, and either $E_{\bar{s}}$ is an equivalence relation and $E_{\bar{s}}(\bar{x}, \bar{y})$, or else $E_{\bar{s}}$ is not an equivalence relation and $\bar{x} = \bar{y}$. Then the quotient structure $\mathfrak{M}^n/E_{\bar{m}}$ is isomorphic to $\{e \in \mathfrak{M}^{eq} : \exists \bar{x} \in \mathfrak{M}^n \; e = \pi_E(\bar{x}, \bar{m})\}$, where n is the length of $\bar{x}$. Thus we can also talk in $\mathfrak{M}^{eq}$ about quotients modulo parameter-definable equivalence relations.

For the rest of this section, we shall work in $\mathfrak{C}^{eq}$. As finite tuples in the home sort are imaginary elements in a tuple sort, we shall omit the bar when denoting tuples.

DEFINITION 2.1.5 Let A be a set of parameters.

- An element a is *definable* over A if it is fixed under all A-automorphisms.
- An element a is *algebraic* over A if it has only finitely many images under A-automorphisms.

- A set X is *A-invariant* if it is stabilized setwise under all A-automorphisms.
- Two sets X and Y are *A-conjugate* if there is an A-automorphism mapping X to Y.

The *algebraic closure* of A, denoted by $\mathrm{acl}(A)$, is the set of all elements algebraic over A; the *definable closure* $\mathrm{dcl}(A)$ of A is the set of all elements definable over A.
If we want to emphasize that we take the algebraic or definable closure in $\mathfrak{C}^{eq}$, we denote this by $\mathrm{acl}^{eq}(.)$ and $\mathrm{dcl}^{eq}(.)$.

LEMMA 2.1.6 *Let A and a be as above. Then a is definable over A if and only if there is an $\mathfrak{L}(A)$-formula $\varphi(x)$ whose sole realization is a; it is algebraic over A if and only if there is an $\mathfrak{L}(A)$-formula true of a with finitely many realizations.*

PROOF: $\Leftarrow$ is obvious in both cases, as $\varphi^{\mathfrak{C}}$ is invariant under $\mathrm{Aut}_A(\mathfrak{C})$. For the other direction, suppose $a \in \mathrm{dcl}(A)$ and consider $p(x) = \mathrm{tp}(a/A)$. By κ-saturation of $\mathfrak{C}$, it must be inconsistent to say $p(x) \cup p(x') \cup \{x \neq x'\}$. But this means that $\pi(x) \cup \pi(x') \cup \{x \neq x'\}$ is inconsistent for some finite part π of p, and $\varphi(x) = \bigwedge \pi(x)$ will do. The case $a \in \mathrm{acl}(A)$ is similar. ■
Hence a is algebraic over A if and only if the stabilizer S_a of a in $\mathrm{Aut}_A(\mathfrak{C})$ has finite index; and definable over A if and only if S_a equals $\mathrm{Aut}_A(\mathfrak{C})$. It is now easy to see that acl and dcl are idempotent.

Given a formula $\varphi(\bar{x}, \bar{m})$, we consider the equivalence relation $E_\varphi(\bar{y}, \bar{z})$ given by $\forall \bar{x}\, [\varphi(\bar{x}, \bar{y}) \leftrightarrow \varphi(\bar{x}, \bar{z})]$, and we define the *canonical parameter* $\mathrm{Cb}(\varphi(\bar{x}, \bar{m}))$ of $\varphi(\bar{x}, \bar{m})$ to be the definable closure of the class of $\bar{m}$ modulo E_φ. It is easy to see that an automorphism of $\mathfrak{C}$ stabilizes the set X of realizations of $\varphi(\bar{x}, \bar{m})$ (in $\mathfrak{C}$) setwise if and only if it fixes $\mathrm{Cb}(\varphi(\bar{x}, \bar{m}))$ pointwise. So if $\psi(\bar{x}, \bar{n})$ is another formula defining X, then $\mathrm{Cb}(\varphi(\bar{x}, \bar{m})) = \mathrm{Cb}(\psi(\bar{x}, \bar{n}))$, and we can just talk about $\mathrm{Cb}(X)$ without specifying a formula.

2.2 DIVIDING AND FORKING

DEFINITION 2.2.1 Let $k < \omega$. A formula $\varphi(\bar{x}, \bar{a})$ *k-divides* over A if there is a sequence $(\bar{a}_i : i < \omega)$ of type $\mathrm{tp}(\bar{a}/A)$ such that $\{\varphi(\bar{x}, \bar{a}_i) : i < \omega\}$ is k-inconsistent, i.e. any finite subset of size k is inconsistent.
A partial type $\pi(\bar{x})$ *k-divides* over A if there is a formula $\varphi(\bar{x})$ implied by $\pi(\bar{x})$ which k-divides over A. A formula or a partial type *divide* over A if they k-divide for some $k < \omega$.
A partial type $\pi(\bar{x})$ *forks* over A if there are $n < \omega$ and formulas $\varphi_0(\bar{x}), \ldots, \varphi_n(\bar{x})$ such that $\pi(\bar{x})$ implies $\bigvee_{i \leq n} \varphi_i(\bar{x})$, and each φ_i divides over A.
If $A \subseteq B$ and $p(\bar{x}) \in S_n(B)$ does not fork over A, we call p a *non-forking extension* of $p\lceil A$.

More or less immediately from the definition, we see:

LEMMA 2.2.2 *1. Dividing implies forking.*

2. *If π and π' are two partial types which fork over A, so does $\pi \vee \pi'$.*

3. *If $p \vdash q$ and q divides (forks) over A, so does p.*

4. *φ k-divides over A if and only if it k-divides over all finite $\bar{a} \in A$.*

5. *In the definition of dividing, we may require the sequence $(a_i : i < \omega)$ to be indiscernible over A.*

6. *A partial type $\pi(\bar{x})$ k-divides (forks) over A if and only if there is a finite conjunction $\varphi(\bar{x})$ of formulas in π which k-divides (forks) over A.*

7. *No $p \in S_n(A)$ divides over A.*

8. *Let $A \subseteq B \subseteq C$. If $\mathrm{tp}(\bar{a}/C)$ does not divide (fork) over A, then it does not divide (fork) over B, and $\mathrm{tp}(\bar{a}/B)$ does not divide (fork) over A.*

9. *$\mathrm{tp}(\bar{a}/A, \bar{a})$ divides (forks) over A if and only if $\bar{a} \notin \mathrm{acl}(A)$.*

PROOF: For 5. use Ramsey's Theorem and compactness. For 6. consider a formula $\varphi(\bar{x}, \bar{a})$ implied by π which k-divides over A, as witnessed by a sequence $(\bar{a}_i : i < \omega)$. Then there is a finite part $\pi_0(\bar{x}, \bar{b})$ such that $\pi_0(\bar{x}, \bar{b}) \vdash \varphi(\bar{x}, \bar{a})$; since $\mathrm{tp}(\bar{a}_i/A) = \mathrm{tp}(\bar{a}/A)$, there is a sequence $(\bar{b}_i : i < \omega)$ with $\mathrm{tp}(\bar{b}_i/A) = \mathrm{tp}(\bar{b}/A)$ and $\pi_0(\bar{x}, \bar{b}_i) \vdash \varphi(\bar{x}, \bar{a}_i)$. Then this sequence witnesses that $\bigwedge \pi_0$ k-divides over A; the case of forking is obvious. Clearly 6. implies 7. For 9. use that any infinite indiscernible sequence over A must be indiscernible over $\mathrm{acl}(A)$. The rest is immediate from the definition. ■

The importance of forking (as opposed to dividing) comes from:

LEMMA 2.2.3 *Let $A \subseteq B$, and $\pi(\bar{x})$ be a partial type over B which does not fork over A. Then there is a complete type $p \in S(B)$ extending π which does not fork over A.*

PROOF: We show that for any B-formula $\varphi(\bar{x}, \bar{b})$ either $\pi \cup \{\varphi\}$ or $\pi \cup \{\neg\varphi\}$ does not fork over A. So suppose otherwise. Then there are formulas $\varphi_1(\bar{x}), \ldots, \varphi_m(\bar{x})$ and $\psi_1(\bar{x}), \ldots, \psi_n(\bar{x})$, all of them dividing over A, such that $\pi \cup \{\varphi\} \vdash \bigvee_i \varphi_i$ and $\pi \cup \{\neg\varphi\} \vdash \bigvee_j \psi_j$. Hence $\pi \vdash \bigvee_i \varphi_i \vee \bigvee_j \psi_j$, so π forks over A, a contradiction.

As non-forking over A is a local property by Lemma 2.2.2.6, it is closed under unions of chains. The existence of a non-forking completion of π now follows from Zorn's Lemma. ■

EXAMPLE 2.2.4 Let $\mathfrak{L}$ consist of a single ternary relation $R(x, y, z)$, and $\mathfrak{M}$ be the circle, where $R(x, y, z)$ holds if and only if y lies on the shorter arc between x and z including the endpoints (in particular, z is not opposite x). Let a, b, c be three equidistant points on the circle. Then $\mathfrak{M} \models \forall x\, R(a, x, b) \vee R(b, x, c) \vee R(c, x, a)$. Let $a, a_0, b_0, a_1, b_1, \ldots, b$ be a sequence of consecutive points on the circle. As $\mathrm{Aut}(\mathfrak{M})$ acts transitively on pairs of non-opposite points, we have $\mathrm{tp}(a, b) = \mathrm{tp}(a_i, b_i)$ for all $i < \omega$; since $\mathrm{Aut}(\mathfrak{M})$ also acts n-transitively on the short arc between a and b for all $n < \omega$, the sequence $(a_i b_i : i < \omega)$ is indiscernible over $\emptyset$ (it is even indiscernible over a, b). As the $R(a_i, x, b_i)$ are disjoint, the formula $R(a, x, b)$ 2-divides over $\emptyset$, as do $R(b, x, c)$ and $R(c, x, a)$. Therefore the true formula $x = x$ forks over $\emptyset$, as does every completion $p \in S_1(\emptyset)$. But neither $x = x$ nor any $p \in S_1(\emptyset)$ divide over $\emptyset$.

The following definition gives particular examples of non-forking extensions.

DEFINITION 2.2.5 Let $A \subseteq B$ and π be a partial type over B. We say that π is *finitely satisfiable* in A if every finite conjunction of formulas in π is satisfied by some tuple in A.
If $\mathfrak{M}$ is a model of T, $p \in S(\mathfrak{M})$ and $B \supseteq \mathfrak{M}$, then an extension $q \in S(B)$ of p is a *coheir* of p if it is finitely satisfiable in $\mathfrak{M}$.

LEMMA 2.2.6 *Let $A \subseteq B$ and π be a partial type over B which is finitely satisfiable in A. Then π has a completion $p \in S(B)$ which is finitely satisfiable in A. If $\bar{b}, \bar{b}' \in B$ with $\mathrm{tp}(\bar{b}/A) = \mathrm{tp}(\bar{b}'/A)$ and $\bar{c} \models p$, then $\mathrm{tp}(\bar{c}\bar{b}/A) = \mathrm{tp}(\bar{c}\bar{b}'/A)$. Furthermore, p does not fork over A.*

PROOF: As finite satisfiability is closed under unions of chains, it is enough to show that for every B-formula φ either $\pi \cup \{\varphi\}$ or $\pi \cup \{\neg\varphi\}$ is finitely satisfiable in $\mathfrak{M}$. So suppose not. Then there are finite bits $\pi_0 \subseteq \pi$ and $\pi_1 \subseteq \pi$ such that $\pi_0 \cup \{\varphi\}$ and $\pi_1 \cup \{\neg\varphi\}$ are both not satisfied by any tuple in A. But then $\pi_0 \cup \pi_1$ is not satisfied by any tuple in A, a contradiction. So π can be completed to a type which is finitely satisfiable in A.

Now suppose $\bar{b}, \bar{b}' \in B$ with $\mathrm{tp}(\bar{b}/A) = \mathrm{tp}(\bar{b}'/A)$ and $\bar{c} \models p$. If there is a formula φ with $\models \varphi(\bar{c}, \bar{b}) \wedge \neg\varphi(\bar{c}, \bar{b}')$, then by finite satisfiability there is $\bar{a} \in A$ with $\models \varphi(\bar{a}, \bar{b}) \wedge \neg\varphi(\bar{a}, \bar{b}')$, contradicting $\mathrm{tp}(\bar{b}/A) = \mathrm{tp}(\bar{b}'/A)$.

Finally, let π be finitely satisfiable in A, and suppose $\pi(\bar{x}) \vdash \bigvee_{i<n} \varphi_i(\bar{x}, \bar{b}_i)$. By finite satisfiability, there must be some $\bar{a} \in A$ and some $i < n$ with $\models \varphi_i(\bar{a}, \bar{b}_i)$. But then for any A-indiscernible sequence I with $\bar{b}_i \in I$ we have $\models \varphi_i(\bar{a}, \bar{b}')$ for all $\bar{b}' \in I$, so $\varphi_i(\bar{x}, \bar{b}_i)$ cannot divide over A. ∎

If p is a type over a model $\mathfrak{M}$, then p is finitely satisfiable in $\mathfrak{M}$ by consistency of the type, and can be considered as a partial type over any superset $B \supseteq \mathfrak{M}$. Therefore types over models always have coheirs. However, not every type over B which does not divide (or fork) over $\mathfrak{M}$ is a coheir of its restriction to $\mathfrak{M}$.

PROPOSITION 2.2.7 *The following are equivalent:*

1. $\mathrm{tp}(\bar{a}/A\bar{b})$ *does not divide over* A.

2. *For any* A*-indiscernible sequence* I *with* $\bar{b} \in I$*, there is a tuple* $\bar{a}'$ *realizing* $\mathrm{tp}(\bar{a}/A\bar{b})$ *such that* I *is indiscernible over* $A\bar{a}'$.

3. *If* I *is an* A*-indiscernible sequence over* A *with* $\bar{b} \in I$*, then there is an* $A\bar{b}$*-automorphic image* J *of* I *which is indiscernible over* $A\bar{a}$.

PROOF: The equivalence of 2. and 3. follows by taking an $A\bar{b}$-automorphism mapping $\bar{a}$ to $\bar{a}'$ and J to I.

Suppose $\mathrm{tp}(\bar{a}/A\bar{b})$ does not divide over A, and let I be an A-indiscernible sequence with $\bar{b} \in I$. Write $\mathrm{tp}(\bar{a}/A\bar{b}) = p(\bar{x}, \bar{b})$.

CLAIM. $\bigcup_{\bar{b}' \in I} p(\bar{x}, \bar{b}')$ is consistent.

PROOF OF CLAIM: If not, then for some formula $\varphi(\bar{x}, \bar{b}) \in p(\bar{x}, \bar{b})$ the conjunction $\bigwedge_{\bar{b}' \in I} \varphi(\bar{x}, \bar{b}')$ is inconsistent, and hence k-inconsistent by compactness and indiscernibility of I, for some $k < \omega$. So $p(\bar{x}, \bar{b})$ divides over A, a contradiction. ∎

By Ramsey's Theorem and by indiscernibility of I over A, for every finite set Δ of formulas we can find an $\bar{a}_\Delta \models \bigcup_{\bar{b}' \in I} p(\bar{a}_\Delta, \bar{b}')$ such that I is Δ-indiscernible over $A\bar{a}_\Delta$. Note that automatically $\bar{a}_\Delta \models p(\bar{x}, \bar{b}) = \mathrm{tp}(\bar{a}/A\bar{b})$. By compactness, we find the required $\bar{a}'$.

The converse follows from Lemma 2.2.2.5. ∎

COROLLARY 2.2.8 *Let* $I = (\bar{a}_i : i < \omega)$ *be an indiscernible sequence over* A *such that* $\mathrm{tp}(\bar{a}_n/A, \bar{a}_i : i < n)$ *does not divide over* A *for all* $n < \omega$*. Let* $(\bar{a}_0^j, \ldots, \bar{a}_n^j : j < \omega)$ *be an* A*-indiscernible sequence in* $\mathrm{tp}(\bar{a}_0, \ldots, \bar{a}_n/A)$*. Then there is an* A*-automorphic image* J *of* I *such that*

1. $(\bar{a}_0^j, \ldots, \bar{a}_n^j)^\frown J$ *is indiscernible over* A *for all* $j < \omega$*, and*

2. $(\bar{a}_0^j, \ldots, \bar{a}_n^j : j < \omega)$ *is indiscernible over* AJ.

PROOF: We shall find inductively $\bar{b}_i$ for $i < \omega$ such that

1. $\mathrm{tp}(\bar{a}_0^j, \ldots, \bar{a}_n^j, \bar{b}_0, \ldots, \bar{b}_{m-1}) = \mathrm{tp}(\bar{a}_0, \ldots, \bar{a}_n, \bar{a}_{n+1}, \ldots, \bar{a}_{n+m})$, and

2. $(\bar{a}_0^j, \ldots, \bar{a}_n^j : j < \omega)$ is indiscernible over $A \cup \{\bar{b}_i : i < m\}$.

So assume we have already found those $\bar{b}_i$ for $i < m$. Put

$$p(\bar{x}, \bar{a}_i : i \leq n+m) = \mathrm{tp}(\bar{a}_{n+m+1}/A, \bar{a}_i : i \leq n+m);$$

this type does not divide over A. Since $(\bar{a}_0^j, \ldots, \bar{a}_n^j, \bar{b}_0, \ldots, \bar{b}_{m-1} : j < \omega)$ is an A-indiscernible sequence in $\mathrm{tp}(\bar{a}_i : i \leq n + m/A)$, Proposition 2.2.7 implies that there is some $\bar{b}_m$ realizing

$$\bigcup_{j<\omega} p(\bar{x}, \bar{a}_0^j, \ldots, \bar{a}_n^j, \bar{b}_0, \ldots, \bar{b}_{m-1}),$$

such that the sequence remains indiscernible over $A\bar{b}_m$. Then $\bar{b}_m$ will have the necessary properties.

It is clear that the sequence $J = (\bar{b}_i : i < \omega)$ is the required A-conjugate of I. ■

PROPOSITION 2.2.9 *Let $A \subseteq B$, and suppose* $\mathrm{tp}(\bar{a}_i/B\bar{a}_0, \ldots, \bar{a}_{i-1})$ *does not divide over* $A\bar{a}_0, \ldots, \bar{a}_{i-1}$ *for all $i \leq n$. Then* $\mathrm{tp}(\bar{a}_0 \ldots \bar{a}_n/B)$ *does not divide over A.*

PROOF: It is sufficient to show that $\mathrm{tp}(\bar{a}_0 \ldots \bar{a}_n/A\bar{b})$ does not divide over A for every $\bar{b} \in B$. Let I be an A-indiscernible sequence with $\bar{b} \in I$, and suppose that we have found $\bar{a}'_0, \ldots, \bar{a}'_{i-1} \models \mathrm{tp}(\bar{a}_0 \ldots \bar{a}_{i-1}/A)$ such that I is indiscernible over $A\bar{a}'_0 \ldots \bar{a}'_{i-1}$. If $\bar{a}'$ is such that $\mathrm{tp}(\bar{a}_0 \ldots \bar{a}_{i-1}\bar{a}_i/A\bar{b}) = \mathrm{tp}(\bar{a}'_0 \ldots \bar{a}'_{i-1}\bar{a}'/A\bar{b})$, then $\mathrm{tp}(\bar{a}'/A\bar{b}\bar{a}'_0 \ldots \bar{a}'_{i-1})$ does not divide over A by invariance of dividing under automorphisms. By Proposition 2.2.7 there is $\bar{a}'_i$ realizing $\mathrm{tp}(\bar{a}'/A\bar{b}\bar{a}'_0 \ldots \bar{a}'_{i-1})$ such that I is indiscernible over $A\bar{a}'_0 \ldots \bar{a}'_i$. Inductively, we thus find $\bar{a}'_0 \ldots \bar{a}'_n \models \mathrm{tp}(\bar{a} \ldots \bar{a}_n/A\bar{b})$ such that I is indiscernible over $A\bar{a}'_0 \ldots \bar{a}'_n$, and we finish by Proposition 2.2.7. ■

Note that in the above proof, the tuples $\bar{a}_i$ may be infinite.

2.3 SIMPLICITY

DEFINITION 2.3.1 A set A is *independent* of C over B, denoted $A \mathop{\smile\!\!\!\!\!\!|}_B C$, if $\mathrm{tp}(\bar{a}/BC)$ does not divide over B for any finite $\bar{a} \in A$.

A (complete first-order) theory is *simple* if independence is a symmetric notion.

We shall call a structure *simple* if its theory is.

REMARK 2.3.2 We shall see in Corollary 2.8.11 that simplicity is preserved under interpretations. In particular, T is simple if and only if T^{eq} is simple.

EXAMPLE 2.3.3 Let $\mathfrak{M}$ be the random graph. This is a graph on countably many vertices, such that for any finite disjoint sets A and B of vertices there is a vertex connected to all vertices in A and not connected to any vertex in B; it is easy to see by a back-and-forth argument that any two such graphs are isomorphic, so $\mathrm{Th}(\mathfrak{M})$ is complete and ω-categorical. Furthermore, $\mathrm{Th}(\mathfrak{M})$ has quantifier elimination, and the only forking formulas are of the form $x = a$ for some parameter a. Hence $\mathfrak{M}$ is simple.

DEFINITION 2.3.4 Let $\varphi(\bar{x},\bar{y})$ be a formula, and $k<\omega$. The rank $D(.,\varphi,k)$ is defined inductively on partial types as follows:

1. $D(\pi(\bar{x}),\varphi(\bar{x},\bar{y}),k)\geq 0$ if $\pi(\bar{x})$ is consistent.

2. $D(\pi(\bar{x}),\varphi(\bar{x},\bar{y}),k)\geq n+1$ if there is $\bar{b}$ such that $D(\pi(\bar{x})\wedge\varphi(\bar{x},\bar{b}),\varphi,k)\geq n$, and $\varphi(\bar{x},\bar{b})$ k-divides over the domain of π.

REMARK 2.3.5 Let π be a partial type over some parameters A. Then $D(\pi(\bar{x},A),\varphi,k)\geq n$ can be expressed by a partial type on A. Furthermore, if $D(\pi,\varphi,k)\leq n$, then there is a finite part π_0 of π with $D(\pi_0,\varphi,k)\leq n$.

We shall write $D(a/A,\varphi,k)$ for $D(\mathrm{tp}(a/A),\varphi,k)$.

LEMMA 2.3.6 *Let $\pi(\bar{x})$ be a partial type over A such that $D(\pi,\varphi,k)\geq n$. Then there is a completion p of π over A with $D(p,\varphi,k)\geq n$.*

PROOF: We use induction on n; the assertion being trivial for $n=0$. So suppose it holds, and $D(\pi,\varphi,k)\geq n+1$. Since $D(.,\varphi,k)\geq n+1$ is a local property and hence closed under unions of chains, by Zorn's Lemma it is sufficient to show that for any A-formula $\psi(\bar{x})$ either $D(\pi\cup\{\psi\},\varphi,k)\geq n+1$ or $D(\pi\cup\{\neg\psi\},\varphi,k)\geq n+1$.

By definition, there is an A-indiscernible sequence $(\bar{b}_i:i<\omega)$ such that $D(\pi\cup\{\varphi(\bar{x},\bar{b}_i)\},\varphi,k)\geq n$ for all $i<\omega$ and $\{\varphi(\bar{x},\bar{b}_i):i<\omega\}$ is k-inconsistent. By inductive hypothesis, there is a completion $q(\bar{x},\bar{b}_0)$ of $\pi(\bar{x})\cup\{\varphi(\bar{x},\bar{b}_0)\}$ with $D(q,\varphi,k)\geq n$. Let $\bar{a}_i$ realize $q(\bar{x},\bar{b}_i)$ for $i<\omega$. Then either infinitely many $\bar{a}_i$ realize ψ, or infinitely many realize $\neg\psi$. In the first case

$$D(\pi(\bar{x})\cup\{\psi(\bar{x}),\varphi(\bar{x},\bar{b}_i)\},\varphi,k)\geq D(q(\bar{x},\bar{b}_i),\varphi,k)\geq n$$

for infinitely many $i<\omega$, whence $D(\pi\cup\{\psi\},\varphi,k)\geq n+1$. In the second case, similarly $D(\pi\cup\{\neg\psi\},\varphi,k)\geq n+1$. ∎

LEMMA 2.3.7 *Let $D(\bar{x}=\bar{x},\varphi,k)\geq n$ for all $n<\omega$. Then for every linearly ordered index set I there are*

1. *an indiscernible sequence $(\bar{b}_i\bar{a}_i:i\in I)$ such that $\models\varphi(\bar{b}_i,\bar{a}_0)$ and $\varphi(\bar{x},\bar{a}_i)$ k-divides over $\{\bar{b}_j\bar{a}_j:j<i\}$, for all $i\in I$, and*

2. *a tuple $\bar{b}$ and a $\bar{b}$-indiscernible sequence $(\bar{a}_i:i\in I)$ such that $\models\varphi(\bar{b},\bar{a}_0)$ and $\varphi(\bar{x},\bar{a}_i)$ k-divides over $\{\bar{a}_j:j<i\}$ for all $i\in I$.*

PROOF: If $\pi(\bar{x})$ is a partial type over A and $D(\pi,\varphi,k)\geq n+1$, then by definition there are $\bar{b}\models\pi$ and $\bar{a}$ such that $\bar{b}\models\pi(\bar{x})\cup\{\varphi(\bar{x},\bar{a})\}$, the formula $\varphi(\bar{x},\bar{a})$ k-divides over A, and $D(\pi\cup\{\varphi(\bar{x},\bar{a}\},\varphi,k)\geq n$. Hence for every

$n < \omega$, since $D(\bar{x} = \bar{x}, \varphi, k) \geq n + 1$, we can inductively find $(\bar{b}_i^n \bar{a}_i^n : i \leq n)$ such that:

> If $A_i^n = \{\bar{b}_j^n \bar{a}_j^n : j < i\}$ and $\pi_i^n(\bar{x}) = \bigwedge_{j<i} \varphi(\bar{x}, \bar{a}_j^n)$, a partial type over A_i^n, then $\bar{b}_i^n \models \pi_i^n(\bar{x}) \cup \{\varphi(\bar{x}, \bar{a}_i^n)\}$, the formula $\varphi(\bar{x}, \bar{a}_i^n)$ k-divides over A_i^n, and $D(\pi_i^n \cup \{\varphi(\bar{x}, \bar{a}_i^n)\}, \varphi, k) \geq n - i$.

Thus for every $n < \omega$ there is a sequence $(\bar{b}_i^n \bar{a}_i^n : i < n)$ such that $\models \varphi(\bar{b}_i^n, \bar{a}_j^n)$ whenever $j \leq i < n$, and $\varphi(\bar{x}, \bar{a}_i^n)$ k-divides over $\{\bar{b}_j \bar{a}_j : j < i\}$ for all $i < n$. By compactness there is an infinite such sequence, which we may suppose to be indiscernible and of order type I by Ramsey's Theorem and compactness again. This proves 1.

Let $(\bar{b}_i \bar{a}_i : i \leq \omega)$ be a sequence as in 1., and put $\bar{b} = \bar{b}_\omega$. Then $(\bar{a}_i : i < \omega)$ is $\bar{b}$-indiscernible, $\models \varphi(\bar{b}, \bar{a}_0)$, and $\varphi(\bar{x}, \bar{a}_i)$ k-divides over $\{\bar{a}_j : j < i\}$ for all $i \leq \omega$. By compactness, we obtain 2. ■

PROPOSITION 2.3.8 *If dividing or forking is symmetric, then $D(\pi, \varphi, k) < \omega$ for all partial types $\pi(\bar{x})$, all formulas $\varphi(\bar{x}, \bar{y})$, and all $k < \omega$.*

PROOF: Suppose $D(\bar{x} = \bar{x}, \varphi, k) \not< \omega$ for some formula φ and some $k < \omega$. By Lemma 2.3.7 there is $\bar{b}$ and a $\bar{b}$-indiscernible sequence $(\bar{a}_i : i \leq \omega)$ such that $\models \varphi(\bar{b}, \bar{a}_\omega)$ and $\varphi(\bar{x}, \bar{a}_\omega)$ divides, and hence forks, over $\{a_i : i < \omega\}$. However, $\text{tp}(\bar{a}_\omega/\bar{b}, \bar{a}_i : i < \omega)$ is finitely satisfiable in $(\bar{a}_i : i < \omega)$ by $\bar{b}$-indiscernibility of $(\bar{a}_i : i \leq \omega)$, and therefore does not fork over $\{\bar{a}_i : i < \omega\}$. Hence neither dividing nor forking is symmetric. ■

PROPOSITION 2.3.9 *Let T be simple, $A \subseteq B$, and $p \in S(B)$. Then the following are equivalent:*

1. *p does not fork over A.*
2. *p does not divide over A.*
3. *$D(p, \varphi, k) = D(p\lceil A, \varphi, k)$ for all formulas φ and all $k < \omega$.*

PROOF: 1. ⇒ 2. is trivial.

3. ⇒ 1. Suppose p forks over A. Then there are $n < \omega$ and formulas $\varphi_i(\bar{x}, \bar{b}_i)$ which k_i-divide over A for $i < n$, with $p \vdash \bigvee_{i<n} \varphi_i(\bar{x}, \bar{b}_i)$. Let $\psi(\bar{x}, \bar{y}_i : i < n, \bar{z})$ be the formula $\bigvee_{i<n} \varphi_i(\bar{x}, \bar{y}_i) \wedge \bar{z} = \bar{y}_i$, and $k = \max\{k_i : i < n\}$. Clearly, we may replace k_i by k, and every $\varphi_i(\bar{x}, \bar{b}_i)$ by $\psi(\bar{x}, \bar{c}_i)$ (for suitable parameters $\bar{c}_i$ containing $\bar{b}_i$ at the ith block).

Choose a completion q of p over $B \cup \{\bar{c}_i : i < n\}$, with $D(q, \psi, k) = D(p, \psi, k)$. Then there is $i_0 < n$ such that $q \vdash \psi(\bar{x}, \bar{c}_{i_0})$, whence $D(p \cup \{\psi(\bar{x}, \bar{c}_{i_0})\}, \psi, k) = D(p, \psi, k)$. As $\psi(\bar{x}, \bar{c}_{i_0})$ k-divides over A, there is an A-indiscernible sequence $(\bar{c}_i' : i < \omega)$ with $\bar{c}_0' = \bar{c}_{i_0}$, such that $\{\psi(\bar{x}, \bar{c}_i') : i < \omega\}$

is k-inconsistent. Clearly this sequence witnesses $D(p \cup \{\psi(\bar{x}, \bar{c}'_0)\}, \psi, k) \geq D(p\lceil A, \psi, k) + 1$, whence $D(p, \psi, k) > D(p\lceil A, \psi, k)$.

2. $\Rightarrow$ 3. Suppose p does not divide over A. We shall show inductively on n that $D(p\lceil A, \varphi, k) \geq n$ implies $D(p, \varphi, k) \geq n$. This is clearly true for $n = 0$ by consistency of p, so assume it holds for n, and $D(p\lceil A, \varphi, k) \geq n+1$. Let $(\bar{b}_i : i < \omega)$ be an A-indiscernible sequence such that $D(p\lceil A \cup \{\varphi(x, \bar{b}_i)\}, \varphi, k) \geq n$ for all $i < \omega$, and $\{\varphi(x, \bar{b}_i) : i < \omega\}$ is k-inconsistent. By Lemma 2.3.6 there is a completion q of $p\lceil A \cup \{\varphi(\bar{x}, \bar{b}_0)\}$ with $D(q, \varphi, k) \geq n$; we may choose q such that there is a realization $\bar{a}$ of $p \cup q$. By 3. $\Rightarrow$ 1. trivially $\text{tp}(\bar{b}_0/A\bar{a})$ does not fork over $A\bar{a}$, and has a non-forking extension to $B\bar{a}$ by Lemma 2.2.3. Conjugating over $A\bar{a}$, we may assume that $\text{tp}(\bar{b}_0/B\bar{a})$ does not fork over $A\bar{a}$, whence $\bar{b}_0 \perp_{A\bar{a}} B$. As $\bar{a} \perp_A B$ by assumption, Lemma 2.2.9 yields $\bar{b}_0\bar{a} \perp_A B$, whence $\bar{a} \perp_{A\bar{b}_0} B$ by symmetry and Lemma 2.2.2.8. By inductive hypothesis,

$$D(\bar{a}/A\bar{b}_0, \varphi, k) = D(q, \varphi, k) \geq n$$

implies $D(\bar{a}/B\bar{b}_0, \varphi, k) \geq n$. Finally, since $B \perp_A \bar{b}_0$ by symmetry, there is a B-indiscernible $A\bar{b}_0$-conjugate $(\bar{b}'_i : i < \omega)$ of $(\bar{b}_i : i < \omega)$ by Proposition 2.2.7, witnessing $D(\bar{a}/B, \varphi, k) \geq D(\bar{a}/B\bar{b}_0, \varphi, k) + 1 \geq n + 1$. ∎

REMARK 2.3.10 Note that the implication 3. $\Rightarrow$ 1. needs simplicity only in the form $D(p, \varphi, k) < \omega$ for all formulas φ and all $k < \omega$. Furthermore, it also works for a partial type π instead of p.

DEFINITION 2.3.11 A formula $\varphi(x, y)$ has the *strict order property* if it defines a partial order with arbitrarily long chains.
A theory has the *strict order property* if some formula has it in some model of T.

LEMMA 2.3.12 *A simple theory does not have the strict order property.*

PROOF: Suppose $\varphi(x, y)$ is a formula with the strict order property. By compactness there is a model of T where φ orders a chain $(a_i : i \in \mathbb{Q})$. But then it is easy to see that $D(x = x, \varphi(y, x) \wedge \varphi(x, y'), 2)$ is infinite, so T is not simple by Proposition 2.3.8. ∎

We shall now collect the properties of non-forking (independence) in a simple theory.

THEOREM 2.3.13 PROPERTIES OF INDEPENDENCE *Suppose T is simple, and $A \subseteq B \subseteq C$. Then:*

1. EXISTENCE *If $p \in S(A)$, then p does not fork over A.*
2. EXTENSION *Every partial type over B which does not fork over A has a completion which does not fork over A.*

3. REFLEXIVITY $B \mathop{\smile\!\!\!\!\!\!\bot}_A B$ *if and only if* $B \subseteq \operatorname{acl}(A)$.

4. MONOTONICITY *If p and q are types with* $p \vdash q$ *and p does not fork over A, then q does not fork over A.*

5. FINITE CHARACTER $D \mathop{\smile\!\!\!\!\!\!\bot}_A B$ *if and only if* $\bar{d} \mathop{\smile\!\!\!\!\!\!\bot}_A B$ *for every finite* $\bar{d} \in D$.

6. SYMMETRY $D \mathop{\smile\!\!\!\!\!\!\bot}_A B$ *if and only if* $B \mathop{\smile\!\!\!\!\!\!\bot}_A D$.

7. TRANSITIVITY $D \mathop{\smile\!\!\!\!\!\!\bot}_A C$ *if and only if* $D \mathop{\smile\!\!\!\!\!\!\bot}_A B$ *and* $D \mathop{\smile\!\!\!\!\!\!\bot}_B C$.

8. LOCAL CHARACTER *For any* $p \in S(A)$ *there is* $A_0 \subseteq A$ *with* $|A_0| \leq |T|$, *such that p does not fork over* A_0.

PROOF: Let $\bar{d} \in D$ and suppose $\bar{d} \mathop{\smile\!\!\!\!\!\!\bot}_A B$ and $\bar{d} \mathop{\smile\!\!\!\!\!\!\bot}_B C$. Then $D(\bar{d}/C, \varphi, k) = D(\bar{d}/B, \varphi, k) = D(\bar{d}/A, \varphi, k)$ for all formulas φ and all $k < \omega$. Therefore $\bar{d} \mathop{\smile\!\!\!\!\!\!\bot}_A C$ for all $\bar{d} \in D$ by Proposition 2.3.9, whence $D \mathop{\smile\!\!\!\!\!\!\bot}_A C$. This proves 7($\Leftarrow$).

Next, consider $p \in S(A)$. For every formula φ and every $k < \omega$ there is a finite $\bar{a}_{\varphi,k} \in A$ with $D(p, \varphi, k) = D(p\lceil\bar{a}_{\varphi,k}, \varphi, k)$ by Remark 2.3.5. If A_0 is the union of all these $\bar{a}_{\varphi,k}$ for all formulas φ and all $k < \omega$, then $D(p, \varphi, k) = D(p\lceil A_0, \varphi, k)$ for all φ and all $k < \omega$, so p does not fork over A_0 by Proposition 2.3.9. This proves 8.

The other assertions are obvious from the equivalence of forking and dividing, or from the definitions. ■

EXAMPLE 2.3.14 Let $\mathfrak{M}$ be a dense linear order without endpoints, and consider $a < b < c$ in $\mathfrak{M}$. Since Aut($\mathfrak{M}$) acts n-transitively on $\mathfrak{M}$ for all $n < \omega$, a sequence $a < a_0 < b_0 < a_1 < b_1 < \cdots < c$ witnesses that $a < x < c$ divides over $\emptyset$, so $\operatorname{tp}(b/ac)$ divides over $\emptyset$. However, $\operatorname{tp}(ac/b)$ does not even fork over $\emptyset$ (in fact, quantifier elimination shows that forking is the same as dividing in this example). Thus independence is not symmetric in $\mathfrak{M}$.

In particular, simplicity is not implied by the equivalence of dividing and forking.

The following observation is often useful:

LEMMA 2.3.15 *If T is simple, then for every complete type* $p(\bar{x}) \in S(A)$ *and every partial type* $\Phi(\bar{x}, \bar{y})$ *there is a partial type* $\pi(\bar{y})$ *such that for any* $\bar{a}$ *there is a non-forking extension of p to* $A\bar{a}$ *containing* $\Phi(\bar{x}, \bar{a})$ *if and only if* $\models \pi(\bar{a})$.

PROOF: If $D(p, \varphi, k) = n(\varphi, k)$, then $\pi(\bar{y})$ is the partial type in $\bar{y}$ expressing $D(p(\bar{x}) \cup \Phi(\bar{x}, \bar{y}), \varphi, k) \geq n(\varphi, k)$ for all $\mathfrak{L}$-formulas φ and all $k < \omega$. ■

DEFINITION 2.3.16 A type $p \in S(A)$ is *stationary* if it has only one non-forking extension to any superset of A.

COROLLARY 2.3.17 *If $p \in S(A)$ is a stationary type in a simple theory, then for every formula $\varphi(\bar{x}, \bar{y})$ there is a formula $\psi(\bar{y})$ over A, such that $\models \psi(\bar{b})$ if and only if $\varphi(\bar{x}, \bar{b})$ is in the (unique) non-forking extension of p to $A\bar{b}$.*

PROOF: By Lemma 2.3.15 there are partial types π and π' such that $\models \pi(\bar{b})$ if and only if $\varphi(\bar{x}, \bar{b})$ is in the non-forking extension of p to $A\bar{b}$, and $\models \pi'(\bar{b})$ if and only if $\neg\varphi(\bar{x}, \bar{b})$ is in the non-forking extension of p to $A\bar{b}$. Since exactly one of the two cases must hold, compactness implies that we can replace π and π' by finite subtypes, i.e. formulas. ■

DEFINITION 2.3.18 Let $\varphi(\bar{x}, \bar{y})$ be a formula. A type $p(\bar{x}) \in S(A)$ is φ-*definable* over B if there is a formula $d_p\varphi(\bar{y})$ over B such that for any $\bar{a} \in A$ we have $\models d_p\varphi(\bar{a})$ if and only if $\varphi(\bar{x}, \bar{a}) \in p$. We call $d_p\varphi$ a *φ-definition* for p over B; it is also denoted as $d_p\bar{x}\,\varphi(\bar{x}, \bar{y})$. Finally, p is *definable* if it is φ-definable over A for all formulas φ; a *defining scheme* for p is the collection of its φ-definitions, for all formulas φ.

REMARK 2.3.19 A stationary type in a simple theory has a unique canonical defining scheme (up to equivalence), namely the defining scheme for all its non-forking extensions given by Corollary 2.3.17.

THEOREM 2.3.20 *Let $\mathfrak{M}$ be a model of a simple theory, $A \subseteq \mathfrak{M}$ and $p \in S(\mathfrak{M})$. If p is definable over A, then p does not fork over A.*

PROOF: Suppose for every formula φ there is a φ-definition $d_p\varphi$ over A. For a formula $\varphi(\bar{x}, \bar{b}) \in p$ let $(\bar{b}_i : i < n)$ be a sequence of type $\mathrm{tp}(\bar{b}/A)$ in $\mathfrak{M}$. Then $\mathfrak{M} \models d_p\varphi(\bar{b}_i)$ for all $i < n$, so $\bigcup_{i<n} \varphi(\bar{x}, \bar{b}_i)$ is a subset of p and thus consistent. It follows that $\varphi(\bar{x}, \bar{b})$ does not fork over A, and neither does p. ■

The condition that p be over a model (or some sufficiently large set) is necessary:

EXAMPLE 2.3.21 Let $\mathfrak{M}$ be a pure set without any structure. Then $\mathrm{Th}(\mathfrak{M})$ is simple, with elimination of quantifiers, and $m \not\perp m$ for any $m \in \mathfrak{M}$. But $\mathrm{tp}(m/m)$ is definable over $\emptyset$: given $\varphi(x, \bar{y})$, we can choose $d_p\varphi$ to be the true formula if $\mathfrak{M} \models \varphi(m, m, \ldots, m)$, and the false formula otherwise. In fact, the same argument works for any simple theory.

2.4 MORLEY SEQUENCES

The original definition by Shelah of a simple theory was in terms of finiteness of the local $D(., \varphi, k)$-ranks. In this section, we shall prove that symmetry of independence, and hence simplicity of a theory, is equivalent to a number of conditions. To this end we shall now introduce an important technical tool.

DEFINITION 2.4.1 A sequence $(\bar{a}_i : i \in I)$ is *independent* over A, or *A-independent*, if $\bar{a}_i \mathop{\smile\!\!\!\!\!\!\,|}_A (\bar{a}_j : j < i)$ for all $i \in I$.

DEFINITION 2.4.2 Let $A \subseteq B$ and $p \in S(B)$. A *Morley sequence in p over A* is a B-indiscernible sequence $(a_i : i < \omega)$ of realizations of p, such that $\mathrm{tp}(a_i/Ba_0 \ldots a_{i-1})$ does not fork over A for all $i < \omega$.
If $A = B$, a Morley sequence in p over A is simply called a Morley sequence.

Thus a Morley sequence in $p \in S(A)$ is an A-indiscernible A-independent sequence of realizations of p. We may occasionally index Morley sequences by infinite ordered sets other than ω.

LEMMA 2.4.3 *If p does not fork over A, then there is a Morley sequence in p.*

PROOF: Suppose for some ordinal α we have found a sequence $(\bar{a}_i : i < \alpha)$ of realizations of p, such that $\mathrm{tp}(\bar{a}_i/B, \bar{a}_j : j < i)$ does not fork over A for all $i < \alpha$, and $\mathrm{tp}(\bar{a}_i/B, \bar{a}_j : j < i) = \mathrm{tp}(\bar{a}_k/B, \bar{a}_j : j < i)$ whenever $i \leq k < \alpha$. By Lemma 2.2.3 there is some $\bar{a}_\alpha \models \bigcup_{i<\alpha} \mathrm{tp}(\bar{a}_i/B, \bar{a}_j : j < i)$ such that $\mathrm{tp}(\bar{a}_\alpha/B, \bar{a}_i : i < \alpha)$ does not fork over A. So we find such a sequence of arbitrary length. By the Erdös-Rado Theorem 1.2.13 we can extract a descending chain of n-indiscernible subsequences for $n < \omega$; by compactness and the finite character of forking (even in a theory which is not simple) we obtain a Morley sequence in p. ■

REMARK 2.4.4 In particular, every type in a simple theory has a Morley sequence.

PROOF: As $p \in S(A)$ does not divide over A, it does not fork over A. ■
The following construction yields Morley sequences in any theory.

DEFINITION 2.4.5 Let $\mathfrak{M}$ be a model, $\mathfrak{M} \subseteq A \subseteq B$, and $p \in S(B)$ a coheir of $p\lceil\mathfrak{M}$. Suppose $(\bar{a}_i : i < \omega)$ is a sequence of tuples of B such that $\bar{a}_i \models p\lceil A \cup (\bar{a}_j : j < i)$. Then $(\bar{a}_i : i < \omega)$ is called a *coheir sequence in p over* $(\mathfrak{M}, A)$. If $A = \mathfrak{M}$, it is omitted.

LEMMA 2.4.6 *Let $\mathfrak{M}$, A, B, p be as in Definition 2.4.5, and suppose $I = (\bar{a}_i : i < \omega)$ is a coheir sequence in p over $(\mathfrak{M}, A)$. Then I is a Morley sequence in $p\lceil A$ over $\mathfrak{M}$.*

PROOF: As $\mathrm{tp}(\bar{a}_i/A\bar{a}_j : j < i)$ is finitely satisfiable in $\mathfrak{M}$ for all $i < \omega$, it does not fork over $\mathfrak{M}$.

For indiscernibility, we show by induction on $n < \omega$ that $\mathrm{tp}(\bar{a}_0 \ldots \bar{a}_n/A) = \mathrm{tp}(\bar{a}_{i_0} \ldots \bar{a}_{i_n}/A)$ for all $i_0 < \cdots < i_n$. This is trivial for $n = 0$; suppose it holds for $n - 1$. Then $\mathrm{tp}(\bar{a}_j : j < n/A) = \mathrm{tp}(\bar{a}_{i_j} : j < n/A)$, whence $\text{tp}(A, \bar{a}_0 \ldots \bar{a}_{n-1}\bar{a}_{i_n}/\mathfrak{M}) = \text{tp}(A, \bar{a}_{i_0} \ldots \bar{a}_{i_{n-1}}\bar{a}_{i_n}/\mathfrak{M})$ by Lemma 2.2.6. Therefore $\text{tp}(\bar{a}_0 \ldots \bar{a}_{n-1}\bar{a}_{i_n}/A) = \text{tp}(\bar{a}_{i_0} \ldots \bar{a}_{i_{n-1}}\bar{a}_{i_n}/A)$; as clearly $\text{tp}(\bar{a}_0 \ldots \bar{a}_{n-1}\bar{a}_n/A) = \text{tp}(\bar{a}_0 \ldots \bar{a}_{n-1}\bar{a}_{i_n}/A)$, the assertion follows. ■

THEOREM 2.4.7 *The following are equivalent:*

1. *T is simple.*
2. *Forking (dividing) satisfies symmetry.*
3. *Forking (dividing) satisfies transitivity.*
4. *Forking (dividing) satisfies local character.*
5. $D(., \varphi, k) < \omega$ *for all formulas* φ *and all* $k < \omega$.
6. *A formula* $\varphi(\bar{x}, \bar{a})$ *does not fork (divide) over* A *if and only if for some Morley sequence* I *in* $\text{tp}(\bar{a}/A)$ *the set* $\{\varphi(\bar{x}, \bar{a}') : \bar{a}' \in I\}$ *is consistent.*

PROOF: 1. ⇒ 2.–5. is Theorem 2.3.13; the implication 2. ⇒ 5. is Proposition 2.3.8.

3. ⇒ 5. If $D(., \varphi, k) \not< \omega$, then by Lemma 2.3.7 there is an indiscernible sequence $(\bar{b}_i\bar{a}_i : i \in \{\pm(1 + 1/n) : 0 < n < \omega\} \cup \{\pm 1\})$, such that $\models \varphi(\bar{b}_i, \bar{a}_{-2})$ for all $i \geq -2$, and $\varphi(\bar{x}, \bar{a}_i)$ k-divides over $\{\bar{b}_j\bar{a}_j : j < i\}$ for all i. Put $I = \{\bar{b}_i : i \in \{-1 - 1/n : 0 < n < \omega\}\}$ and $J = \{\bar{b}_i : i \in \{1 + 1/n : 0 < n < \omega\}\}$. Since $\text{tp}(\bar{b}_1/IJ)$ is finitely satisfiable in I, it does not fork over I. Similarly, $\text{tp}(\bar{b}_1/IJ\bar{a}_0)$ is finitely satisfiable in J and does not fork over IJ. However, $\varphi(\bar{b}_1, \bar{a}_0)$ witnesses that $\text{tp}(\bar{b}_1/IJ\bar{a}_0)$ divides over I. Therefore transitivity of forking and of dividing fails.

4. ⇒ 5. If $D(., \varphi, k) \not< \omega$, then by Lemma 2.3.7 there is a tuple $\bar{b}$ and a $\bar{b}$-indiscernible sequence $(\bar{a}_i : i < |T|^+)$ such that $\models \varphi(\bar{b}, \bar{a}_0)$, and $\varphi(\bar{x}, \bar{a}_i)$ k-divides over $\{\bar{a}_j : j < i\}$ for all $i < |T|^+$. If $A = \{\bar{a}_i : i < |T|^+\}$ and $A_0 \subset A$ with $|A| \leq |T|$, then there is $i < |T|^+$ such that $A_0 \subseteq \{\bar{a}_j : j < i\}$. However, $\varphi(\bar{x}, \bar{a}_i) \in \text{tp}(\bar{c}/A)$ witnesses that this type divides, and hence forks, over A_0.

6. ⇒ 5. If $D(., \varphi, k) \not< \omega$, then by Lemma 2.3.7 there is a tuple $\bar{b}$ and a $\bar{b}$-indiscernible sequence $(\bar{a}_i : i \in \{\pm 1/n : 0 < n < \omega\})$ such that $\models \varphi(\bar{b}, \bar{a}_0)$, and $\varphi(\bar{x}, \bar{a}_i)$ k-divides over $\{\bar{a}_j : j < i\}$ for all $i \in \{\pm 1/n : 0 < n < \omega\}$. Put $A = \{\bar{a}_i : i \in \{-1/n : 0 < n < \omega\}\}$. As $\text{tp}(\bar{a}_{1/n+1}/A, \bar{a}_{1/i} : 0 < i \leq n\}$ is finitely satisfiable in A and thus does not fork over A, the sequence $(\bar{a}_{1/n} : 0 < n < \omega)$ is a Morley sequence in $\text{tp}(\bar{a}_1/A)$. However, $\varphi(\bar{x}, \bar{a}_1)$ divides (and hence forks) over A, while $\bigwedge_{0<n<\omega} \varphi(\bar{x}, \bar{a}_{1/i})$ is consistent.

5. ⇒ 6. Assume $D(., \varphi, k) < \omega$ for all formulas φ and all $k < \omega$, and suppose $\varphi(\bar{x}, \bar{a})$ divides over A. Let I be a Morley sequence in $\text{tp}(\bar{a}/A)$. As $\varphi(\bar{x}, \bar{a})$ divides over A, there is an A-indiscernible sequence $(\bar{a}_i : i < \omega)$ such

that $\bigwedge_{i<\omega}\varphi(\bar{x},\bar{a}_i)$ is k-inconsistent for some $k<\omega$. By Corollary 2.2.8 we may assume that $\bar{a}_j\hat{\ }I$ is A-conjugate to I for all $j<\omega$, and $(\bar{a}_j : j<\omega)$ is indiscernible over AI.

Put $p(\bar{x})=\bigwedge_{\bar{a}'\in I}\varphi(\bar{x},\bar{a}')$, and suppose p is consistent. Then the sequence $(\bar{a}_i : i<\omega)$ witnesses that $D(p(\bar{x})\cup\{\varphi(\bar{x},\bar{a}_0)\},\varphi,k)<D(p,\varphi,k)$. But these two partial types are A-conjugate and must have the same $D(.,\varphi,k)$-rank, a contradiction. Therefore p is not consistent.

For the converse, note that since $D(\bar{a}/A,\varphi,k)<\omega$ for all formulas φ and all $k<\omega$, Remark 2.3.10 implies that $\mathrm{tp}(\bar{a}/A)$ does not fork over A. By Lemma 2.4.3 there is a Morley sequence I in $\mathrm{tp}(\bar{a}/A)$. By definition, if $\varphi(\bar{x},\bar{a})$ does not divide over A, then $\{\varphi(\bar{x},\bar{a}') : \bar{a}'\in I\}$ is consistent.

Next, we show that forking is the same as dividing. So let $\varphi(\bar{x},\bar{a})$ be a formula which forks over A. So there are $n<\omega$ and $\psi_i(\bar{x},\bar{b}_i)$ for $i<n$ such that $\varphi(\bar{x},\bar{a})\vdash\bigvee_{i<n}\psi_i(\bar{x},\bar{b}_i)$ and each $\psi_i(\bar{x},\bar{b}_i)$ divides over A; adding dummy variables, we may assume $\bar{a}=\bar{b}_0=\cdots=\bar{b}_{n-1}$. Let $(\bar{a}_i : i<\omega)$ be a Morley sequence in $\mathrm{tp}(\bar{a}/A)$. If $\varphi(\bar{x},\bar{a})$ does not divide over A, then $\bigwedge_{i<\omega}\varphi(\bar{x},\bar{a}_i)$ is consistent, say it is realized by $\bar{b}$. But then there is $i_0<n$ and an infinite subset $I\subseteq\omega$ such that $\bar{b}$ realizes $\psi_{i_0}(\bar{x},\bar{a}_i)$ for all $i\in I$. As $(a_i : i\in I)$ is also a Morley sequence in $\mathrm{tp}(\bar{a}/A)$, the formula $\psi_{i_0}(\bar{x},\bar{a})$ cannot divide over A, a contradiction.

5. $\Rightarrow$ 1. By the last part, forking is the same as dividing, and 6. holds. Suppose $\mathrm{tp}(\bar{a}/A\bar{b})$ does not fork over A. By Lemma 2.2.3 there is a Morley sequence $(\bar{a}_i : i<\omega)$ in $\mathrm{tp}(\bar{a}/A\bar{b})$ over A; by partial transitivity, Lemma 2.2.2.8, this is also a Morley sequence in $\mathrm{tp}(\bar{a}/A)$. But if $\varphi(\bar{x},\bar{a},A)\in\mathrm{tp}(\bar{b}/A\bar{a})$, then by indiscernibility $\models\varphi(\bar{b},\bar{a}_i,A)$ for all $i<\omega$. Therefore $\varphi(\bar{x},\bar{a},A)$ does not fork over A by 6. Thus $\mathrm{tp}(\bar{b}/A\bar{a})$ does not fork over A, and T is simple. ∎

REMARK 2.4.8 The characterization of non-forking in a simple theory given by Theorem 2.4.7.6 is of particular importance; we shall often use it in the sequel.

2.5 THE INDEPENDENCE THEOREM

In this section, T will be a simple theory.

LEMMA 2.5.1 *Let I be an ordered index set and $(\bar{a}_i : i\in I)$ an A-independent sequence. Then $\bar{a}_i\mathop{\smile\!\!\!\!\!\!\mid}_A(\bar{a}_j : j\neq i)$ for any $i\in I$.*

PROOF: Suppose $\bar{a}_i\not\mathop{\smile\!\!\!\!\!\!\mid}_A(\bar{a}_j : j\neq i)$. By symmetry $(\bar{a}_j : j\neq i)\not\mathop{\smile\!\!\!\!\!\!\mid}_A\bar{a}_i$, so by Finite Character there is a minimal finite set $J\subset I-\{i\}$ such that $(a_j : j\in J)\not\mathop{\smile\!\!\!\!\!\!\mid}_A\bar{a}_i$. If $j<i$ for all $i\in J$, symmetry yields a contradiction to our assumption; otherwise let k be the maximal element of J. So $(\bar{a}_j : j\in J, j\neq k)\mathop{\smile\!\!\!\!\!\!\mid}_A\bar{a}_i$ by minimality of J; as $\bar{a}_k\mathop{\smile\!\!\!\!\!\!\mid}_A(\bar{a}_i,\bar{a}_j : j\in J, j\neq k)$ and thus

$\bar{a}_k \mathop{\smile\!\!\!\!\!\!\,|}_{A(\bar{a}_j:j\in J, j\neq k)} \bar{a}_i$, symmetry and transitivity yield $\bar{a}_i \mathop{\smile\!\!\!\!\!\!\,|}_A (\bar{a}_j : j \in J)$, a contradiction. ■

LEMMA 2.5.2 *Let $(\bar{a}_i : i < \omega + \omega)$ be an A-indiscernible sequence, and put $I = (\bar{a}_i : i < \omega)$ and $I' = (\bar{a}_{\omega+i} : i < \omega)$. Then I' is a Morley sequence in* $\mathrm{tp}(\bar{a}_\omega/AI)$.

PROOF: I' is clearly an AI-indiscernible sequence of realizations of $\mathrm{tp}(\bar{a}_\omega/AI)$; we have to check that it is independent. By Lemma 2.5.1 (with the order reversed) it is sufficient to check that $\mathrm{tp}(a_{\omega+i}/AI\bar{a}_j : j > i)$ does not fork over AI for all $i < \omega$. But this is obvious, since it is finitely satisfiable in I. ■

PROPOSITION 2.5.3 *Let $\pi(\bar{x}, \bar{a})$ be a partial type over $A\bar{a}$ which does not fork over A. If $(\bar{a}_i : i < \omega)$ is a Morley sequence over A in* $\mathrm{tp}(\bar{a}/A)$, *then $\pi := \bigcup_{i<\omega} p(\bar{x}, \bar{a}_i)$ is consistent and does not fork over A.*

PROOF: Note that π is consistent by Lemma 2.2.7. Suppose $\bar{b} \models \pi$, and $\pi(\bar{x}) \vdash \varphi(\bar{x}, \bar{a}_i : i < n)$ (where we have suppressed possible parameters from A). Put $\bar{b}_j = (\bar{a}_{jn+i} : i < n)$, then $(\bar{b}_j : j < \omega)$ is also an infinite A-indiscernible A-independent sequence, and $\bigwedge_{j<\omega} \varphi(\bar{x}, \bar{b}_j)$ is realized by $\bar{b}$, whence consistent. Hence $\varphi(\bar{x}, \bar{b}_0) = \varphi(\bar{x}, \bar{a}_i : i < n)$ does not fork over A, and neither does π. ■

THEOREM 2.5.4 *Let $\pi(\bar{x}, \bar{a})$ be a partial type over $A\bar{a}$ which does not fork over A. If $(\bar{a}_i : i < \omega)$ is indiscernible over A with* $\mathrm{tp}(\bar{a}_0/A) = \mathrm{tp}(\bar{a}/A)$, *then $\pi := \bigcup_{i<\omega} \pi(\bar{x}, \bar{a}_i)$ is consistent and does not fork over A.*

PROOF: Let I be a sequence of order type ω such that $I^\frown(\bar{a}_i : i < \omega)$ is indiscernible over A, and let $p(\bar{x}, \bar{a}_0)$ be a completion of $\pi(\bar{x}, \bar{a}_0)$ to $AI\bar{a}_0$ which does not fork over A. Then $(\bar{a}_i : i < \omega)$ is a Morley sequence in $\mathrm{tp}(\bar{a}_0/AI)$ by Lemma 2.5.2; since $p(\bar{x}, \bar{a}_0)$ does not fork over AI, the set $\bigcup_{i<\omega} p(\bar{x}, \bar{a}_i)$ is consistent by Proposition 2.5.3 and does not fork over AI. Its restriction to AI is equal to $p(\bar{x}, \bar{a}_0)\lceil_{AI}$ which does not fork over A; it follows that $\bigcup_{i<\omega} \pi(\bar{x}, \bar{a}_i)$ does not fork over A. ■

DEFINITION 2.5.5 A relation $R(\bar{x}, \bar{y})$ (which need not be definable) is A-*invariant* if it is invariant under all automorphisms fixing A pointwise.
An A-invariant relation $R(\bar{x}, \bar{y})$ is *stable* if there is no A-indiscernible sequence $(\bar{a}_i\bar{b}_i : i < \omega)$ such that $R(\bar{a}_i, \bar{b}_j)$ holds if and only if $i \leq j$.

REMARK 2.5.6 Instead of the condition $i \leq j$ we could equally well have put $i < j$, $i > j$, or $i \geq j$.

LEMMA 2.5.7 *Let $\Phi(\bar{x}, \bar{y})$ and $\Psi(\bar{x}, \bar{z})$ be partial types. Then the relation "$\Phi(\bar{x}, \bar{a}) \wedge \Psi(\bar{x}, \bar{b})$ does not fork over A" is a stable relation.*

PROOF: Let $(\bar{a}_i\bar{b}_i : i < \omega)$ be an A-indiscernible sequence such that $\Phi(\bar{x}, \bar{a}_i) \wedge \Psi(\bar{x}, \bar{b}_j)$ forks over A if and only if $i > j$. In particular $\Phi(\bar{x}, \bar{a}_0) \wedge \Psi(\bar{y}, \bar{b}_0)$ does not fork over A. But then by Theorem 2.5.4 the whole set $\bigcup_{i<\omega} \Phi(\bar{x}, \bar{a}_i) \wedge \Psi(\bar{x}, \bar{b}_i)$ does not fork over A, and neither does $\Phi(\bar{x}, \bar{a}_i) \wedge \Psi(\bar{x}, \bar{b}_j)$, for any $i, j < \omega$. ■

LEMMA 2.5.8 *Let R be an A-invariant relation. Suppose there is an A-indiscernible sequence $(\bar{b}_i : i \in I)$, some $i_0 \in I$ and some $\bar{a}$, such that*

1. *$\text{tp}(\bar{a}\bar{b}_i/A)$ is constant for all $i \leq i_0$ and for all $i > i_0$,*
2. *$R(\bar{a}, \bar{b}_i)$ holds if and only if $i \leq i_0$, and*
3. *both $\{i \in I : i \leq i_0\}$ and $\{i \in I : i > i_0\}$ are infinite.*

Then R is unstable.

PROOF: By compactness, we may assume that $I = \mathbb{Z}$ and $i_0 = 0$. By indiscernibility, we find $\bar{a}_i$ for $i \in I$ such that $\text{tp}(\bar{a}_i, \bar{b}_{j-i} : j \in I/A) = \text{tp}(\bar{a}, \bar{b}_j : j \in I/A)$; by Ramsey's Theorem we may assume that the sequence $(\bar{a}_i\bar{b}_i : i \in I)$ is indiscernible over A. But then $R(\bar{a}_i, \bar{b}_j)$ holds if and only if $i < j$, so R is unstable. ■

THEOREM 2.5.9 *Suppose $\mathfrak{M}$ is a model of T and R is an $\mathfrak{M}$-invariant stable relation such that $R(\bar{a}, \bar{b})$ holds for some $\bar{a} \perp_{\mathfrak{M}} \bar{b}$. Then $R(\bar{a}', \bar{b}')$ holds for all $\bar{a}' \models \text{tp}(\bar{a}/\mathfrak{M})$ and $\bar{b}' \models \text{tp}(\bar{b}/\mathfrak{M})$ with $\bar{a} \perp_{\mathfrak{M}} \bar{b}$.*

PROOF: Suppose $R(\bar{a}', \bar{b}')$ fails for some $\bar{a}' \models \text{tp}(\bar{a}/\mathfrak{M})$ and $\bar{b}' \models \text{tp}(\bar{b}/\mathfrak{M})$ with $\bar{a}' \perp_{\mathfrak{M}} \bar{b}'$. By $\mathfrak{M}$-invariance of R we may assume $\bar{a} = \bar{a}'$. Let $\mathfrak{N}$ be an $|\mathfrak{M}|^+$-saturated and -homogeneous elementary extension of $\mathfrak{M}$, and let p be a coheir of $\text{tp}(\bar{b}/\mathfrak{M})$ to $\mathfrak{N}$. Let $(\bar{b}_i : i < \omega)$ and $(\bar{b}'_i : i < \omega)$ be coheir sequences in p over $\mathfrak{M}$, with $\bar{b} = \bar{b}_0$ and $\bar{b}' = \bar{b}'_0$. By Lemma 2.4.6 both sequences are $\mathfrak{M}$-independent and $\mathfrak{M}$-indiscernible. By Lemma 2.2.7 we may assume that both $(\bar{b}_i : i < \omega)$ and $(\bar{b}'_i : i < \omega)$ are indiscernible over $\mathfrak{M} \cup \{\bar{a}\}$. So $R(\bar{a}, \bar{b}_i)$ holds for all $i < \omega$, and $R(\bar{a}, \bar{b}'_i)$ fails for all $i < \omega$.

Now let $(c_i : i < \omega)$ be a coheir sequence in p over $\mathfrak{M} \cup \{\bar{b}_i\bar{b}'_i : i < \omega\}$. Then $(\bar{b}_i : i < \omega)\hat{}(\bar{c}_i : i < \omega)$ and $(\bar{b}'_i : i < \omega)\hat{}(\bar{c}_i : i < \omega)$ are coheir sequences in p over $\mathfrak{M}$, and thus $\mathfrak{M}$-independent and $\mathfrak{M}$-indiscernible. By Ramsey's Theorem, we may assume that $(\bar{c}_i : i < \omega)$ is in fact indiscernible over $\mathfrak{M} \cup \{\bar{a}\}$.

Now either $R(\bar{a}, \bar{c}_i)$ holds for all $i < \omega$, in which case the sequence $(\bar{b}'_i : i < \omega)\hat{}(\bar{c}_i : i < \omega)$ contradicts Lemma 2.5.8, or $R(\bar{a}, \bar{c}_i)$ fails for all $i < \omega$ and $(\bar{b}_i : i < \omega)\hat{}(\bar{c}_i : i < \omega)$ contradicts Lemma 2.5.8. ■

COROLLARY 2.5.10 INDEPENDENCE THEOREM OVER A MODEL *Let $\mathfrak{M}$ be a model, $p \in S(\mathfrak{M})$, A and B supersets of $\mathfrak{M}$ with $A \downarrow_{\mathfrak{M}} B$, and $p_A \in S(A)$ and $p_B \in S(B)$ non-forking extensions of p. Then $p_A \cup p_B$ does not fork over $\mathfrak{M}$.*

PROOF: Let $\bar{a} \models p_A$ and $\bar{a}' \models p_B$. There is an $\mathfrak{M}$-automorphism mapping $\bar{a}'$ to $\bar{a}$ and B to B', and an $\mathfrak{M} \cup \{\bar{a}\}$-automorphism mapping B' to some B'' with $B'' \downarrow_{\mathfrak{M}\bar{a}} A$. As $A \downarrow_{\mathfrak{M}} \bar{a}$ we get $A \downarrow_{\mathfrak{M}} B''\bar{a}$ and hence $A \downarrow_{B''} \bar{a}$; since $\bar{a}' \downarrow_{\mathfrak{M}} B$ implies $\bar{a} \downarrow_{\mathfrak{M}} B'$, whence $\bar{a} \downarrow_{\mathfrak{M}} B''$, we obtain $\bar{a} \downarrow_{\mathfrak{M}} AB''$. Therefore $p_A \cup p_{B''}$ does not fork over $\mathfrak{M}$. But "$p_X \cup p_Y$ does not fork over $\mathfrak{M}$" is a stable $\mathfrak{M}$-invariant relation by Lemma 2.5.7, so $p_A \cup p_B$ does not fork over $\mathfrak{M}$ by Theorem 2.5.9. ■

COROLLARY 2.5.11 *Let $\mathfrak{M}$ be a model, $p \in S(\mathfrak{M})$, $(\bar{A}_i : i \in I)$ an independent sequence over $\mathfrak{M}$, and for each $i \in I$ consider a non-forking extension p_i of p to A_i. Then $\bigcup_{i \in I} p_i$ is consistent and does not fork over $\mathfrak{M}$.*

PROOF: Since consistency and non-forking are local properties, we only have to check the assertion for finite I. But here it follows by induction on $|I|$ from the Independence Theorem 2.5.10. ■

COROLLARY 2.5.12 *Let $n < \omega$ and complete types $p_i \in S(\mathfrak{M})$ be given, for $i < n$. Suppose for all $i < j < n$ there are partial types $p_{ij}(\bar{x}_i, \bar{x}_j)$ over a model $\mathfrak{M}$ extending $p_i(\bar{x}_i) \cup p_j(\bar{x}_j)$, such that $\bar{a}_i\bar{a}_j \models p_{ij}$ implies $\bar{a}_i \downarrow_{\mathfrak{M}} \bar{a}_j$. Then there is a type $p(\bar{x}_i : i < n) \in S(\mathfrak{M})$ extending $\bigcup_{i<j<n} p_{ij}$, such that for any $(\bar{a}_i : i < n) \models p$ the sequence $(\bar{a}_i : i < n)$ is independent over $\mathfrak{M}$.*

PROOF: We use induction on n, the case $n = 2$ being trivial and the case $n = 3$ being the Independence Theorem. So suppose the assertion holds for n, and let $p_{ij}(\bar{x}_i, \bar{x}_j)$ for $i < j \leq n$ be types over $\mathfrak{M}$ which imply the independence of $\bar{x}_i$ and $\bar{x}_j$ over $\mathfrak{M}$. By inductive hypothesis there is an $\mathfrak{M}$-independent sequence $(\bar{a}_i : i < n)$ realizing $\bigcup_{i<j<n} p_{ij}$. Now for all $i < n$ the type $p_{in}(\bar{a}_i, \bar{x}_n)$ is a non-forking extension of $p_n \in S(\mathfrak{M})$. By Corollary 2.5.11 there is a realization $\bar{a}_n \models \bigcup_{i<n} p_{in}(\bar{a}_i, \bar{x}_n)$ independent of $(\bar{a}_i : i < n)$ over $\mathfrak{M}$. We may take $p = \mathrm{tp}(\bar{a}_i : i \leq n/\mathfrak{M})$. ■

REMARK 2.5.13 It is easy to see, by the local character of forking, that all the results in this section also hold for types in infinitely many variables.

EXAMPLE 2.5.14 Here is an example to show that the independence of A and B over $\mathfrak{M}$ in the Independence Theorem 2.5.10 is necessary. Let $\mathfrak{M}$ be the random graph, with edge relation $R(x, y)$. Then $\mathfrak{M}$ has quantifier elimination and is simple; the only forking formulas are of the form $x = a$. Let p be a non-algebraic type over $\mathfrak{M}$, and consider a new vertex a (in some elementary extension $\mathfrak{C}$ of $\mathfrak{M}$). Then p has two non-forking extensions to a, namely p^+ where $R(x, a)$ holds and p^- where $R(x, a)$ fails. Obviously these two cannot be amalgamated.

In particular we see that in a simple theory a type over a model may have more than one non-forking extension to a given superset. In fact:

LEMMA 2.5.15 *Suppose $p \in S(\mathfrak{M})$ has only a bounded number of non-forking extensions. Then p is stationary.*

PROOF: Suppose not, and let $A \supseteq \mathfrak{M}$ be such that there are two distinct non-forking extensions $p_1(\bar{x}, A)$ and $p_2(\bar{x}, A)$ over A. Let $(A_i : i < \alpha)$ be a long Morley sequence in $\mathrm{tp}(A/\mathfrak{M})$. By Corollary 2.5.11 for every $I \subseteq \alpha$ there is a non-forking extension q_I of p extending $\bigcup_{i \in I} p_1(\bar{x}, A_i) \cup \bigcup_{j \notin I} p_2(\bar{x}, A_j)$. Clearly $q_I \neq q_J$ for $I \neq J$, so p cannot have a bounded number of non-forking extensions. ■

PROPOSITION 2.5.16 *Suppose $A = \mathrm{acl}(A)$ and $p \in S(A)$ has only a bounded number of non-forking extensions. Then p is stationary.*

PROOF: Let $\mathfrak{M}$ be a model containing A. It is clearly sufficient to show that p has a unique non-forking extension to $\mathfrak{M}$. So suppose p_1 and p_2 are two non-forking extensions of p to $\mathfrak{M}$.

CLAIM. p_1 is definable over A.

PROOF OF CLAIM: By Corollary 2.3.17, for every formula $\varphi(\bar{x}, \bar{y}) \in \mathfrak{L}$ there is a φ-definition $d_1\varphi(\bar{y})$ over $\mathfrak{M}$. Now if $d_1\varphi$ were not over A, then by compactness it would have arbitrarily many A-conjugates, all φ-definitions for distinct non-forking extensions of p. But this contradicts boundedness. ■

Similarly, p_2 is definable over A; let $d_2\varphi$ be the φ-definition for p_2, and consider some $\bar{b} \in \mathfrak{M}$. By Lemma 2.3.15 there is a partial type $\pi(\bar{x})$ such that $\models \pi(\bar{a}')$ if and only if $\varphi(\bar{a}', \bar{y})$ is in some non-forking extension of $\mathrm{tp}(\bar{b}/A)$ to $A\bar{a}'$.

Now suppose $\varphi(\bar{x}, \bar{b}) \in p_1$. If $\bar{a}_1 \models p_1$, then $\models \varphi(\bar{a}_1, \bar{b})$, whence $\models \pi(\bar{a}_1)$ and $\pi \subseteq p$. In particular, $\models \pi(\bar{a}_2)$ for any $\bar{a}_2 \models p_2$. But p_1 and p_2 are conjugate over A, so $\models \pi(\bar{a}')$ if and only if $\varphi(\bar{a}', \bar{y})$ is in some non-forking extension of $\mathrm{tp}(\bar{b}/A)$ to $A\bar{a}'$. Hence there is $\bar{b}' \models \mathrm{tp}(\bar{b}/A)$ such that $\bar{b}' \mathop{\smile\!\!\!\!|}_A \bar{a}_2$ and $\models \varphi(\bar{a}_2, \bar{b}')$. Therefore $\models d_2\varphi(\bar{b}')$; as this is a formula over A, we get $\models d_2\varphi(\bar{b})$ and $\varphi(\bar{x}, \bar{b}) \in p_2$.

This shows that $p_1 = p_2$, and p is stationary. ■

We shall now generalize the Independence Theorem by weakening the assumption that the base set $\mathfrak{M}$ (over which we consider non-forking extensions of a fixed type to independent parameters) is a model.

DEFINITION 2.5.17 Let A be a set of parameters. The group of *strong* automorphisms of $\mathfrak{C}$ over A is the subgroup of $\mathrm{Aut}_A(\mathfrak{C})$ generated by all automorphisms fixing some model $\mathfrak{M} \supseteq A$. We denote it by $\mathrm{Autf}_A(\mathfrak{C})$.
Two tuples $\bar{a}$ and $\bar{b}$ have the same *Lascar strong type* over A if they are conjugate by a strong automorphism over A.

It is clear that equality of Lascar strong type over A is an equivalence relation; we shall denote the equivalence class of $\bar{a}$ by $\mathrm{lstp}(\bar{a}/A)$. In other words, $\mathrm{lstp}(\bar{a}/A)$ represents the orbit of $\bar{a}$ under $\mathrm{Autf}_A(\mathfrak{C})$. Trivially, if A is a model, then $\mathrm{Autf}_A(\mathfrak{C}) = \mathrm{Aut}_A(\mathfrak{C})$, so types over a model are the same as Lascar strong types.

LEMMA 2.5.18 *Suppose $A \subseteq \mathfrak{M}$ and $\mathrm{tp}(\bar{a}/\mathfrak{M}) = \mathrm{tp}(\bar{b}/\mathfrak{M})$. Then there is a model $\mathfrak{M}'$ with $\bar{a}\bar{b} \mathop{\smile\!\!\!\!\!\!|}_A \mathfrak{M}'$ and $\mathrm{tp}(\bar{a}/\mathfrak{M}') = \mathrm{tp}(\bar{b}/\mathfrak{M}')$.*

PROOF: By Extension there is a model $\mathfrak{M}_0 \supseteq A$ with $\mathfrak{M}_0 \mathop{\smile\!\!\!\!\!\!|}_A \mathfrak{M}$. Let $\mathfrak{M}'$ realize a coheir of $\mathrm{tp}(\mathfrak{M}_0/\mathfrak{M})$ to $\mathfrak{M} \cup \{\bar{a}, \bar{b}\}$. Then $\mathfrak{M}'$ contains A, and $\mathfrak{M}' \mathop{\smile\!\!\!\!\!\!|}_A \mathfrak{M}\bar{a}\bar{b}$. Furthermore, $\mathrm{tp}(\mathfrak{M}'\bar{a}/\mathfrak{M}) = \mathrm{tp}(\mathfrak{M}'\bar{b}/\mathfrak{M})$ by Lemma 2.2.6. This yields the assertion. ■

THEOREM 2.5.19 *Suppose $\bar{a} \mathop{\smile\!\!\!\!\!\!|}_A \bar{b}$ and $\mathrm{lstp}(\bar{a}/A) = \mathrm{lstp}(\bar{b}/A)$. Then there is a model $\mathfrak{M}$ containing A with $\bar{a}\bar{b} \mathop{\smile\!\!\!\!\!\!|}_A \mathfrak{M}$ and $\mathrm{tp}(\bar{a}/\mathfrak{M}) = \mathrm{tp}(\bar{b}/\mathfrak{M})$.*

PROOF: Since $\mathrm{lstp}(\bar{a}/A) = \mathrm{lstp}(\bar{b}/A)$, there are $n < \omega$ and a sequence $\bar{a} = \bar{a}_0, \bar{a}_1, \ldots, \bar{a}_n = \bar{b}$ and models $\mathfrak{M}_0, \ldots, \mathfrak{M}_{n-1}$ containing A, such that $\mathrm{tp}(\bar{a}_i/\mathfrak{M}_i) = \mathrm{tp}(\bar{a}_{i+1}/\mathfrak{M}_i)$ for all $i < n$. Put $C = \bigcup_{i<n} \mathfrak{M}_i$.
CLAIM. We may assume $(\bar{a}_i : i \leq n) \mathop{\smile\!\!\!\!\!\!|}_A C$.
PROOF OF CLAIM: By Lemma 2.5.18 we may assume that $\bar{a}_i\bar{a}_{i+1} \mathop{\smile\!\!\!\!\!\!|}_A \mathfrak{M}_i$ for all $i < n$. Conjugating $\mathfrak{M}_i$ over $A\bar{a}_i\bar{a}_{i+1}$, we may also assume inductively that $\mathfrak{M}_i \mathop{\smile\!\!\!\!\!\!|}_{A\bar{a}_i\bar{a}_{i+1}} \bar{a}_0 \ldots \bar{a}_n \mathfrak{M}_0 \ldots \mathfrak{M}_{i-1}$ for all $i < n$. Then $\mathfrak{M}_i \mathop{\smile\!\!\!\!\!\!|}_A \bar{a}_0 \ldots \bar{a}_n \mathfrak{M}_0 \ldots \mathfrak{M}_{i-1}$ by transitivity, and $\bar{a}_0 \ldots \bar{a}_n \mathop{\smile\!\!\!\!\!\!|}_{A\mathfrak{M}_0 \ldots \mathfrak{M}_{i-1}} \mathfrak{M}_i$ for all $i < n$. Again by transitivity, $\bar{a}_0 \ldots \bar{a}_n \mathop{\smile\!\!\!\!\!\!|}_A \mathfrak{M}_0 \ldots \mathfrak{M}_{n-1}$. ■
Note that $\bar{a}\bar{b} \mathop{\smile\!\!\!\!\!\!|}_A C$ implies $\bar{a} \mathop{\smile\!\!\!\!\!\!|}_C \bar{b}$.
CLAIM. We may assume that $(\bar{a}_i : i \leq n)$ is independent over C.
PROOF OF CLAIM: For $0 < i < n$ let a_i' be the non-forking extension of $\mathrm{tp}(\bar{a}_i/C)$ to $C\bar{a}_0\bar{a}_n\bar{a}_1 \ldots \bar{a}_{i-1}$. Then $(\bar{a}_i : i \leq n)$ is independent over C, and since $\mathrm{tp}(\bar{a}_i'/C) = \mathrm{tp}(\bar{a}_i/C)$ and C contains all $\mathfrak{M}_i$ (for $i < n$), the new sequence still witnesses the equality of Lascar strong type. ■
We now use induction on n. If $n = 1$, the assertion is already shown (and follows from Lemma 2.5.18). So assume it is true for $n-1$. Since $\bar{a}_0 \mathop{\smile\!\!\!\!\!\!|}_A \bar{a}_{n-1}$, there is a model $\mathfrak{M}$ containing A such that $\mathrm{tp}(\bar{a}_0/\mathfrak{M}) = \mathrm{tp}(\bar{a}_{n-1}/\mathfrak{M})$. This means we may assume that $n = 2$.

Since $\bar{a}_0\bar{a}_1\bar{a}_2 \mathop{\smile\!\!\!\!\!|}_A \mathfrak{M}_0\mathfrak{M}_1$, we get $\bar{a}_0\bar{a}_1\bar{a}_2 \mathop{\smile\!\!\!\!\!|}_{\mathfrak{M}_0} \mathfrak{M}_1$. Furthermore, independence of $\{\bar{a}_0, \bar{a}_1, \bar{a}_2\}$ over $\mathfrak{M}_0\mathfrak{M}_1$ implies $\bar{a}_0 \mathop{\smile\!\!\!\!\!|}_{\mathfrak{M}_0} \bar{a}_1\bar{a}_2$. Let $p = \mathrm{tp}(\mathfrak{M}_1/\mathfrak{M}_0\bar{a}_1\bar{a}_2)$, and p' be the conjugate over $\mathfrak{M}_0\bar{a}_0$ of $\mathrm{tp}(\mathfrak{M}_1/\mathfrak{M}_0\bar{a}_1)$, which exists since $\mathrm{tp}(\bar{a}_0/\mathfrak{M}_0) = \mathrm{tp}(\bar{a}_1/\mathfrak{M}_0)$. Then both p and p' are non-forking extensions of $\mathrm{tp}(\mathfrak{M}_1/\mathfrak{M}_0)$. By the Independence Theorem 2.5.10 over $\mathfrak{M}_0$ they have a common realization $\mathfrak{M}$, such that $\mathfrak{M} \mathop{\smile\!\!\!\!\!|}_{\mathfrak{M}_0} \bar{a}_0\bar{a}_1\bar{a}_2$. Since $\mathrm{tp}(\bar{a}_0\mathfrak{M}) = \mathrm{tp}(\bar{a}_1\mathfrak{M}_1) = \mathrm{tp}(\bar{a}_2\mathfrak{M}_1) = \mathrm{tp}(\bar{a}_2\mathfrak{M})$, the assertion is shown. ■

THEOREM 2.5.20 INDEPENDENCE THEOREM *If* $B \mathop{\smile\!\!\!\!\!|}_A C$, $\mathrm{tp}(\bar{b}/AB)$ *and* $\mathrm{tp}(\bar{c}/AC)$ *do not fork over* A, *and* $\mathrm{lstp}(\bar{b}/A) = \mathrm{lstp}(\bar{c}/A)$, *then there is* $\bar{a} \models \mathrm{lstp}(\bar{b}/A) \cup \mathrm{tp}(\bar{b}/AB) \cup \mathrm{tp}(\bar{c}/AC)$, *with* $\bar{a} \mathop{\smile\!\!\!\!\!|}_A BC$.

PROOF: Let $\mathfrak{M}$ be a model containing A with $\mathfrak{M} \mathop{\smile\!\!\!\!\!|}_A BC\bar{b}\bar{c}$, and let $\bar{b}'$ and $\bar{c}'$ realize non-forking extensions of $\mathrm{tp}(\bar{b}/\mathfrak{M}B)$ to $\mathfrak{M}BC$ and $\mathrm{tp}(\bar{c}/\mathfrak{M}C)$ to $\mathfrak{M}BC\bar{b}'$, respectively. Replacing $\bar{b}$ by $\bar{b}'$ and $\bar{c}$ by $\bar{c}'$ then preserves the types and Lascar strong types in question; we may thus assume $B\bar{b} \mathop{\smile\!\!\!\!\!|}_A C\bar{c}$.

By Theorem 2.5.19 there is a model $\mathfrak{M}'$ containing A, such that $\bar{b}\bar{c} \mathop{\smile\!\!\!\!\!|}_A \mathfrak{M}'$ and $\mathrm{tp}(\bar{b}/\mathfrak{M}') = \mathrm{tp}(\bar{c}/\mathfrak{M}')$. Conjugating $\mathfrak{M}'$ over $A\bar{b}\bar{c}$, we may assume $\mathfrak{M}' \mathop{\smile\!\!\!\!\!|}_{A\bar{b}\bar{c}} BC$, whence $\mathfrak{M}' \mathop{\smile\!\!\!\!\!|}_A BC\bar{b}\bar{c}$. But then $B \mathop{\smile\!\!\!\!\!|}_{\mathfrak{M}'} C$, and $\mathrm{tp}(\bar{b}/\mathfrak{M}'B)$ and $\mathrm{tp}(\bar{c}/\mathfrak{M}'C)$ both do not fork over $\mathfrak{M}'$ and have the same restriction to $\mathfrak{M}'$. By the Independence Theorem over $\mathfrak{M}'$ they have a common realization $\bar{a}$ with $\bar{a} \mathop{\smile\!\!\!\!\!|}_{\mathfrak{M}'} BC$. So $\bar{a} \mathop{\smile\!\!\!\!\!|}_A BC$, as $\mathrm{tp}(\bar{a}/\mathfrak{M}') = \mathrm{tp}(\bar{b}/\mathfrak{M}')$ does not fork over A. This proves the theorem. ■

2.6 SIMPLICITY AND INDEPENDENCE

In this section, we shall show a remarkable converse to the Independence Theorem 2.5.10, characterizing simple theories in terms of an abstract independence relation.

THEOREM 2.6.1 *Suppose in a complete theory* T *there is an abstract independence relation* $\mathop{\smile\!\!\!\!\!|}^0$, *invariant under automorphisms, such that*

1. SYMMETRY $\bar{a} \mathop{\smile\!\!\!\!\!|}^0_A \bar{b}$ *if and only if* $\bar{b} \mathop{\smile\!\!\!\!\!|}^0_A \bar{a}$.

2. TRANSITIVITY $\bar{a} \mathop{\smile\!\!\!\!\!|}^0_A BC$ *if and only if* $\bar{a} \mathop{\smile\!\!\!\!\!|}^0_A B$ *and* $\bar{a} \mathop{\smile\!\!\!\!\!|}^0_{AB} C$.

3. EXTENSION *For any* $\bar{a}, A, B$ *there is* $\bar{a}' \models \mathrm{tp}(\bar{a}/A)$ *with* $\bar{a}' \mathop{\smile\!\!\!\!\!|}^0_A B$.

4. LOCAL CHARACTER *For any* $\bar{a}, A$ *there is* $A_0 \subseteq A$ *with* $|A_0| \leq |T|$ *and* $\bar{a} \mathop{\smile\!\!\!\!\!|}^0_{A_0} A$.

5. FINITE CHARACTER $\bar{a} \mathop{\smile\!\!\!\!\!|}^0_A B$ *if and only if for all finite* $\bar{b} \in B$ *we have* $\bar{a} \mathop{\smile\!\!\!\!\!|}^0_A \bar{b}$.

6. INDEPENDENCE THEOREM *If* $\bar a \mathop{\downarrow}^0_{\mathfrak{M}} \bar b$ *for some model* $\mathfrak{M}$, $\bar x \mathop{\downarrow}^0_{\mathfrak{M}} \bar a$, $\bar y \mathop{\downarrow}^0_{\mathfrak{M}} \bar b$ *and* $\mathrm{tp}(\bar x/\mathfrak{M}) = \mathrm{tp}(\bar y/\mathfrak{M})$, *then there is* $\bar z \models \mathrm{tp}(\bar x/\mathfrak{M}\bar a) \cup \mathrm{tp}(\bar y/\mathfrak{M}\bar b)$ *with* $\bar z \mathop{\downarrow}^0_{\mathfrak{M}} \bar a\bar b$.

Then T *is simple and* $\mathop{\downarrow}^0$ *is non-forking independence.*

PROOF: We shall show first that if $\bar a \mathop{\downarrow}^0_A \bar b$ and $(\bar b_i : i < \omega)$ is an A-indiscernible sequence in $\mathrm{tp}(\bar b/A)$, then there is $\bar a' \mathop{\downarrow}^0_A (\bar b_i : i < \omega)$ such that $\mathrm{tp}(\bar a'\bar b_i/A) = \mathrm{tp}(\bar a\bar b/A)$ for all $i < \omega$.

First we extend $(\bar b_i : i < \omega)$ to an A-indiscernible sequence $(\bar b'_i : i \leq |T|^+)$. For all $i \leq |T|^+$ we then successively find models $\mathfrak{M}_i$ of cardinality $|T|$, such that $A \cup \bigcup_{j<i} \mathfrak{M}_j \bar b'_j \subseteq \mathfrak{M}_i$, and $(\bar b'_j : i \leq j \leq |T|^+)$ is indiscernible over $\mathfrak{M}_i$, in the following way: Suppose $\mathfrak{M}_j$ have been found for $j < i$. Let $\mathfrak{M}'_i$ be any model of size $|T|$ containing $A \cup \bigcup_{j<i} \mathfrak{M}_j \bar b'_j$. By Ramsey's Theorem and compactness we obtain an $\mathfrak{M}'_i$-indiscernible sequence $(\bar b''_j : i \leq j \leq |T|^+)$ whose type over $\bigcup_{j<i} \mathfrak{M}_j$ is the same as that of $(\bar b'_j : i \leq j \leq |T|^+)$. So we choose $\mathfrak{M}_i$ to be the image of $\mathfrak{M}'_i$ under a $\bigcup_{j<i} \mathfrak{M}_j$-isomorphism mapping $(\bar b''_j : i \leq j \leq |T|^+)$ to $(\bar b'_j : i \leq j \leq |T|^+)$.

By local character and transitivity of $\mathop{\downarrow}^0$ there is some $i < |T|^+$ such that $\bar b'_{|T|^+} \mathop{\downarrow}^0_{\mathfrak{M}_i} \bigcup_{j<|T|^+} \mathfrak{M}_j$. So by transitivity again, $\bar b'_{|T|^+} \mathop{\downarrow}^0_{\mathfrak{M}_i} (\bar b'_j : i \leq j < |T|^+)$; by transitivity and $\mathfrak{M}_i$-indiscernibility, we obtain $\bar b'_j \mathop{\downarrow}^0_{\mathfrak{M}_i} (\bar b'_k : k < j)$ for all $i \leq j < |T|^+$.

Clearly, we may assume $\bar b_j = \bar b'_{i+j}$ for all $j < \omega$, and $\bar b = \bar b_0 = \bar b'_i$. By extension, there is $\bar a' \models \mathrm{tp}(\bar a/A\bar b)$ with $\bar a' \mathop{\downarrow}^0_{A\bar b} \mathfrak{M}_i$, whence $\bar a' \mathop{\downarrow}^0_A \mathfrak{M}_i \bar b_0$ by transitivity. So repeated applications of the Independence Theorem yield $\bar a'' \mathop{\downarrow}^0_{\mathfrak{M}_i} (\bar b_j : j < \omega)$ such that $\mathrm{tp}(\bar a''\bar b_j/\mathfrak{M}_i) = \mathrm{tp}(\bar a'\bar b_0/\mathfrak{M}_i)$ for all $j < \omega$. This clearly implies $\mathrm{tp}(\bar a''\bar b_j/A) = \mathrm{tp}(\bar a\bar b/A)$ for all $j < \omega$. And by transitivity, $\bar a'' \mathop{\downarrow}^0_A (\bar b_j : j < \omega)$.

Next, we shall show that $\bar a \mathop{\downarrow}^0_A B$ if and only if $\mathrm{tp}(\bar a/AB)$ does not divide over A. This will imply symmetry (and all the other properties) for non-dividing, and thus simplicity of T. Clearly, if $\bar a \mathop{\downarrow}^0_A B$, then (using Finite Character) we have just seen that $\mathrm{tp}(\bar a/A\bar b)$ does not divide over A for all finite $\bar b \in B$, so $\mathrm{tp}(\bar a/AB)$ does not divide over A.

Conversely, assume that $\mathrm{tp}(\bar a/A\bar b)$ does not divide over A. Extension and Finite Character (for the limit stages) for $\mathop{\downarrow}^0$ imply that we can find an infinite sequence $(\bar b_i : i \leq \alpha)$ in $\mathrm{tp}(\bar b/A)$ for big α, such that $\bar b_i \mathop{\downarrow}^0_A (\bar b_j : j < i)$ for all $i \leq \alpha$. By the Erdős-Rado Theorem 1.2.13 we obtain a descending sequence of n-indiscernible subsequences for $n < \omega$; Finite Character and compactness now imply that we may assume that $(\bar b_i : i \leq \alpha)$ is in fact A-indiscernible.

As $\mathrm{tp}(\bar{a}/A\bar{b})$ does not divide over A, there is $\bar{a}'$ such that $\mathrm{tp}(\bar{a}'\bar{b}_i/A) = \mathrm{tp}(\bar{a}\bar{b}/A)$ for all $i \leq \alpha$. By Local Character for $\mathop{\smile\!\!\!\!\!\!|}^0$ there is some $i < \alpha$ such that $\bar{a}' \mathop{\smile\!\!\!\!\!\!|}^0_{A\cup(\bar{b}_j:j<i)} \bar{b}_\alpha$. As $\bar{b}_\alpha \mathop{\smile\!\!\!\!\!\!|}^0_A (\bar{b}_j : j < i)$, we get $\bar{b}_\alpha \mathop{\smile\!\!\!\!\!\!|}^0_A \{\bar{a}', \bar{b}_j : j < i\}$, whence $\bar{b}_\alpha \mathop{\smile\!\!\!\!\!\!|}^0_A \bar{a}'$. Therefore $\bar{a} \mathop{\smile\!\!\!\!\!\!|}^0_A \bar{b}$. ■

2.7 BOUNDED EQUIVALENCE RELATIONS

In this section, T is not necessarily simple. Recall that κ is the ordinal such that the monster model $\mathfrak{C}$ is κ-homogeneous and κ-saturated.

DEFINITION 2.7.1 An A-invariant equivalence relation E on $\mathfrak{C}^n$ (not necessarily definable) is *bounded* if it has less than κ classes on $\mathfrak{C}^n$.
An A-invariant relation R is *type-definable* if the set of tuples realizing R is given by a partial type.

REMARK 2.7.2 If a type $p \in S(A)$ has a realization satisfying some A-invariant relation R, then every realization of p must satisfy R. It follows that a relation is A-invariant if and only of it is given by an infinite disjunction of partial types over A.

LEMMA 2.7.3 *If R is A-invariant and type-definable, then it is type-definable over A.*

PROOF: If R is given by the partial type $\Phi(\bar{x}, B)$ for some parameters B, then it is also given by $\exists Y\, [\Phi(\bar{x}, Y) \wedge Y \models \mathrm{tp}(B/A)]$, which is equivalent to a partial type over A. ■

LEMMA 2.7.4 *If E is a bounded A-invariant equivalence relation and $(\bar{a}_i : i \in I)$ is an infinite A-indiscernible sequence, then $E(\bar{a}_i, \bar{a}_j)$ for all $i \neq j$ in I.*

PROOF: By compactness, we can find an A-indiscernible sequence $(\bar{a}'_i : i < \kappa)$ with $\mathrm{tp}(\bar{a}_i, \bar{a}_j/A) = \mathrm{tp}(\bar{a}'_0, \bar{a}'_1/A)$ for any $i < j$ in I. So if $\neg E(\bar{a}_i, \bar{a}_j)$ holds, then $\neg E(\bar{a}'_\alpha, \bar{a}'_\beta)$ holds for all $\alpha < \beta < \kappa$, and E has κ classes, a contradiction. ■

PROPOSITION 2.7.5 *Equality of Lascar strong type over A is the finest bounded A-invariant equivalence relation.*

PROOF: Equality of Lascar strong type over A is clearly an A-invariant equivalence relation. Let $\mathfrak{M}$ be a model containing A with $|\mathfrak{M}| = |T(A)|$. Then two tuples which have the same type over $\mathfrak{M}$ have the same Lascar strong type over A. So there are at most $2^{|T(A)|}$ Lascar strong types over A, and the relation is bounded.

Now let E be an A-invariant bounded equivalence relation, and consider two tuples $\bar{a}$ and $\bar{b}$ with the same Lascar strong type over A. By transitivity of E it is enough to show that if $\bar{a}$ and $\bar{b}$ have the same type over some model $\mathfrak{M}$ containing A, then $E(\bar{a}, \bar{b})$ holds. Let $\mathfrak{N}$ be an $|\mathfrak{M}|^+$-saturated elementary extension of $\mathfrak{M}$ and p a coheir of $\text{tp}(\bar{a}/\mathfrak{M})$ over $\mathfrak{N}$. Let $(\bar{c}_i : i < \omega) \subset \mathfrak{N}$ be a coheir sequence in p over $\mathfrak{M} \cup \{\bar{a}, \bar{b}\}$. Then both $\bar{a}\hat{\ }(\bar{c}_i : i < \omega)$ and $\bar{b}\hat{\ }(\bar{c}_i : i < \omega)$ are coheir sequences over $\mathfrak{M}$ and therefore $\mathfrak{M}$-indiscernible. By Lemma 2.7.4 we have $E(\bar{a}, \bar{c}_0)$ and $E(\bar{b}, \bar{c}_0)$, whence $E(\bar{a}, \bar{b})$. ■

COROLLARY 2.7.6 *An A-invariant bounded equivalence relation has at most* $2^{|T(A)|}$ *classes.* ■

PROPOSITION 2.7.7 *Let T be simple, and* $\bar{a}_0 \mathop{\smile\!\!\!\!\!\!\,|}_A \bar{a}_1$. *Then the following are equivalent:*

1. $\text{lstp}(\bar{a}_0/A) = \text{lstp}(\bar{a}_1/A)$.
2. *There is a Morley sequence* $(\bar{a}_i : i < \omega)$ *over A.*
3. *There is an A-indiscernible sequence* $(\bar{a}_i : i < \omega)$.

PROOF: 2. ⇒ 3. is trivial, and 3. ⇒ 1. follows from Lemma 2.7.4 and Proposition 2.7.5. So assume $\text{lstp}(\bar{a}_0/\mathfrak{A}) = \text{lstp}(\bar{a}_1/A)$. By Theorem 2.5.19 there is a model $\mathfrak{M}$ containing A such that $\text{tp}(\bar{a}_0/\mathfrak{M}) = \text{tp}(\bar{a}_1/\mathfrak{M})$ and $\bar{a}_0\bar{a}_1 \mathop{\smile\!\!\!\!\!\!\,|}_A \mathfrak{M}$. We now construct an $\mathfrak{M}$-independent sequence $(\bar{a}_i : i < \omega)$ such that $\text{tp}(\bar{a}_i\bar{a}_j/\mathfrak{M}) = \text{tp}(\bar{a}_0\bar{a}_1/\mathfrak{M})$ for all $i < j < \omega$. We can clearly start with $\bar{a}_0\bar{a}_1$.

Suppose we have already found $(\bar{a}_i : i < n)$ for some $n < \omega$, and let $p_i(\bar{x})$ be the conjugate of $\text{tp}(\bar{a}_1/\mathfrak{M}\bar{a}_0)$ over $\mathfrak{M}\bar{a}_i$ for all $i < n$. By Corollary 2.5.11 there is $\bar{a}_n$ realizing $\bigcup_{i<n} p_i$ with $\bar{a}_n \mathop{\smile\!\!\!\!\!\!\,|}_{\mathfrak{M}} (\bar{a}_i : i < n)$. Then $\text{tp}(\bar{a}_i\bar{a}_n/\mathfrak{M}) = \text{tp}(\bar{a}_0\bar{a}_1/\mathfrak{M})$, and we are done.

Since $\mathfrak{M}$-independence is type-definable as "$D(\bar{a}_i/\mathfrak{M}\bar{a}_j : j < i) \geq D(\bar{a}_0/A)$ for all formulas φ and all $k < \omega$", we can apply Ramsey's Theorem and compactness to obtain a Morley sequence $(\bar{a}'_i : i < \omega)$ in $\text{tp}(\bar{a}_0/\mathfrak{M})$, with $\text{tp}(\bar{a}'_0\bar{a}'_1/\mathfrak{M}) = \text{tp}(\bar{a}_0\bar{a}_1/\mathfrak{M})$. Conjugating $\bar{a}'_0\bar{a}'_1$ to $\bar{a}_0\bar{a}_1$, we are done. ■

THEOREM 2.7.8 *Let T be simple. Then the following are equivalent:*

1. $\text{lstp}(\bar{a}/A) = \text{lstp}(\bar{b}/A)$.
2. *There is* $\bar{c}$ *with* $\bar{c} \mathop{\smile\!\!\!\!\!\!\,|}_A \bar{a}\bar{b}$ *and models* $\mathfrak{M}, \mathfrak{M}'$ *containing A, such that* $\text{tp}(\bar{a}/\mathfrak{M}) = \text{tp}(\bar{c}/\mathfrak{M})$ *and* $\text{tp}(\bar{c}/\mathfrak{M}') = \text{tp}(\bar{b}/\mathfrak{M}')$.
3. *There is* $\bar{c}$ *and sequences I and I′, such that* $\bar{a}\hat{\ }I$, $\bar{c}\hat{\ }I$, $\bar{c}\hat{\ }I'$ *and* $\bar{b}\hat{\ }I'$ *are all A-indiscernible.*

PROOF: 1. $\Rightarrow$ 2. Suppose $\mathrm{lstp}(\bar{a}/A) = \mathrm{lstp}(\bar{b}/A)$, and let $\mathfrak{M}$ be a model containing A with $\mathfrak{M} \mathop{\smile\!\!\!\!\!\!|}_A \bar{a}\bar{b}$. Choose a realization $\bar{c}$ of $\mathrm{tp}(\bar{a}/\mathfrak{M})$ with $\bar{c} \mathop{\smile\!\!\!\!\!\!|}_{\mathfrak{M}} \bar{a}\bar{b}$. Then $\bar{c} \mathop{\smile\!\!\!\!\!\!|}_A \mathfrak{M}$, whence $\bar{c} \mathop{\smile\!\!\!\!\!\!|}_A \mathfrak{M}\bar{a}\bar{b}$. Clearly $\mathrm{lstp}(\bar{c}/A) = \mathrm{lstp}(\bar{a}/A) = \mathrm{lstp}(\bar{b}/A)$. By Theorem 2.5.19 there is a model $\mathfrak{M}'$ with $\mathrm{tp}(\bar{c}/\mathfrak{M}') = \mathrm{tp}(\bar{b}/\mathfrak{M}')$.

2. $\Rightarrow$ 3. Assume there are $\mathfrak{M}, \mathfrak{M}', \bar{c}$ as in 2. We can then construct I as a coheir sequence in some coheir of $\mathrm{tp}(\bar{c}/\mathfrak{M})$ over $\mathfrak{M}\bar{a}\bar{c}$, and I' as a coheir sequence in some coheir of $\mathrm{tp}(\bar{c}/\mathfrak{M}')$ over $\mathfrak{M}\bar{c}\bar{b}$ (as in the proof of Proposition 2.7.5).

3. $\Rightarrow$ 1. Obvious, by Lemma 2.7.4 and transitivity. ■

COROLLARY 2.7.9 *Equality of Lascar strong type is type-definable in a simple theory. More precisely, for every set A and tuples $\bar{x}$ and $\bar{y}$ of the same length* $\mathrm{lstp}(\bar{x}/A) = \mathrm{lstp}(\bar{y}/A)$ *is given by a partial type* $r_A(\bar{x}, \bar{y})$. *Furthermore,* $r_A = \bigcup_{\bar{a} \in A \text{ finite}} r_{\bar{a}}$, *so* $\mathrm{lstp}(\bar{x}/A) = \mathrm{lstp}(\bar{y}/A)$ *if and only if* $\mathrm{lstp}(\bar{x}/\bar{a}) = \mathrm{lstp}(\bar{y}/\bar{a})$ *for all finite* $\bar{a} \in A$.

PROOF: By Theorem 2.7.8 we can take r_A to say

> There are $(\bar{x}_i : i < \omega)$, $(\bar{y}_i : i < \omega)$ and $\bar{z}$ such that $\bar{x}\hat{\ }(\bar{x}_i : i < \omega)$, $\bar{z}\hat{\ }(\bar{x}_i : i < \omega)$, $\bar{z}\hat{\ }(\bar{y}_i : i < \omega)$ and $\bar{y}\hat{\ }(\bar{y}_i : i < \omega)$ are indiscernible over A.

Clearly, $r_A = \bigcup_{\bar{a} \in A \text{ finite}} r_{\bar{a}}$. ■

REMARK 2.7.10 So in a simple theory equality of Lascar strong type over A is the finest bounded equivalence relation type-definable over A.

The following example shows that independence of $\bar{a}$ and $\bar{b}$ over A is necessary in Theorem 2.5.19.

EXAMPLE 2.7.11 Let $\mathfrak{L} = \{P, Q, R\}$, where P and Q are unary predicates and R is a binary relation. Let $\mathfrak{M}$ be the disjoint union of two countable sets $P^{\mathfrak{M}}$ and $Q^{\mathfrak{M}}$. Suppose $P^{\mathfrak{M}}$ is the disjoint union of two sets U and V, and let f be a bijection from U to V (but U, V and f are not in the language). Let R be a binary relation such that R forms the random bipartite graph on U and $Q^{\mathfrak{M}}$, and $R(u, x)$ holds if and only if $\neg R(f(u), x)$ holds, for all $u \in U$ and $x \models Q$. Then $\mathfrak{M}$ is interpretable in the random bipartite graph, which is simple, so $\mathfrak{M}$ is simple.

Let $u \in U$ and $v = f(u)$. If $\mathfrak{M}'$ and $\mathfrak{M}''$ are two disjoint elementary substructures of $\mathfrak{M}$ (both containing neither u nor v), then it is consistent to say for some w that it is related to the same points in $\mathfrak{M}'$ as u and to the same points in $\mathfrak{M}''$ as v. It is then easy to see that $\mathrm{tp}(u/\mathfrak{M}') = \mathrm{tp}(w/\mathfrak{M}')$ and $\mathrm{tp}(w/\mathfrak{M}'') = \mathrm{tp}(v/\mathfrak{M}'')$, so $\mathrm{lstp}(u) = \mathrm{lstp}(v)$. But any model $\mathfrak{N}$ must contain a point $x \models Q$, and $R(u, x)$ holds if and only if $\neg R(v, x)$ holds. Therefore $\mathrm{tp}(u/\mathfrak{N}) \neq \mathrm{tp}(v/\mathfrak{N})$.

Nevertheless, if $U - \{u\} = \{u_i : i < \omega\}$, then $(u, u_0, u_1, \ldots)$ and $(v, u_0, u_1, \ldots)$ are both indiscernible over $\emptyset$.

2.8 TYPES

In this section, we shall characterize simplicity in terms of the number of types over sets. So the ambient theory is not necessarily simple.

DEFINITION 2.8.1 Let $\varphi(\bar{x}, \bar{y})$ be an $\mathfrak{L}$-formula. A *φ-formula* is a formula of the form $\varphi(\bar{x}, \bar{a})$ or $\neg\varphi(\bar{x}, \bar{a})$ for some $\bar{a}$; in the first case it is called a *positive* φ-formula. A (partial) *φ-type* is a type which consists only of φ-formulas; it is *positive* if it only consists of positive φ-formulas.
Let λ', λ be infinite cardinal numbers. $NT_\varphi(\lambda', \lambda)$ is the supremum of the numbers of pairwise incompatible φ-types of size λ' over sets of size λ; similarly $NT_\varphi^+(\lambda', \lambda)$ is the supremum of the numbers of pairwise incompatible positive φ-types over sets of size λ. Finally, $NT(\lambda', \lambda)$ is the supremum of the numbers of pairwise incompatible partial types of size λ' over sets of size λ.

REMARK 2.8.2 In any theory we have $NT_\varphi^+(\lambda', \lambda) \leq NT_\varphi(\lambda', \lambda) \leq \lambda^{\lambda'}$, and $NT(\lambda', \lambda) \leq (|T| + \lambda)^{\lambda'}$.

We shall always assume that $\lambda' \leq \lambda + |T|$, as there are no types of size greater than $\lambda + |T|$ over a set of size λ.

In the following definition, ${}^{<\alpha}\omega$ will denote sequences of natural numbers of length $< \alpha$, excluding the empty sequence.

DEFINITION 2.8.3 Let $\varphi(\bar{x}, \bar{y})$ be an $\mathfrak{L}$-formula, $k < \omega$, and α an ordinal. A *(φ, k)-tree* of *heigth* α is a sequence $(\bar{a}_\mu : \mu \in {}^{<\alpha}\omega)$ such that for all $\mu \in {}^{<\alpha^-}\omega$ the set $\{\varphi(\bar{x}, \bar{a}_{\mu^\frown i}) : i < \omega\}$ is k-inconsistent (where α^- is the least ordinal whose successor is $\geq \alpha$). We say that φ has the *k-tree property* if there is a (φ, k)-tree of height ω. Finally, φ has the *tree property* if φ has the k-tree property for some $k < \omega$.
Let α be an ordinal. A *dividing chain* of length α for φ of length α is a sequence $(\bar{a}_i : i < \alpha)$ such that $\bigcup_{i<\alpha} \varphi(\bar{x}, \bar{a}_i)$ is consistent and $\varphi(\bar{x}, \bar{a}_i)$ divides over $\{\bar{a}_j : j < i\}$ for all $i < \alpha$. It is a *k-dividing chain* if $\varphi(\bar{x}, \bar{a}_i)$ k-divides over $\{\bar{a}_j : j < i\}$ for all $i < \alpha$. We say that φ *k-divides α times* if there is a k-dividing chain of length α; it *divides α times* if there is a dividing chain of length α.

REMARK 2.8.4 Shelah and Kim call our k-tree property the *tree property with respect to k*; their k-tree property means that there is a φ-tree such that every node has k successors (rather than ω).

PROPOSITION 2.8.5 *Let $\varphi(\bar{x}, \bar{y})$ be a formula. Then the following are equivalent:*

1. *There is a (φ, k)-tree of height n, for all $n < \omega$.*

2. *There is a (φ, k)-tree of height α, for all ordinals α.*

3. *φ k-divides n times, for all $n < \omega$.*

4. *φ k-divides α times, for all ordinals α.*

5. *$D(., \varphi, k) \geq n$ for all $n < \omega$.*

6. *$D(., \varphi, k) \geq \alpha$, for all ordinals α.*

PROOF: 1. $\Rightarrow$ 2. and 3. $\Rightarrow$ 4. are true by compactness, 6. $\Rightarrow$ 5. is trivial.

2. $\Rightarrow$ 3. Let $(\bar{a}_\mu : \mu \in {}^{<n}\omega)$ be a (φ, k)-tree of height n. By compactness and Ramsey's Theorem, we may assume that $(\bar{a}_{\mu\frown i} : i < \omega)$ is indiscernible over $(\bar{a}_\nu : \nu \leq \mu)$ for all $\mu \in {}^{<n-1}\omega$. But that means that $(\bar{a}_\mu : \mu \in {}^{<n}\{0\})$ is a k-dividing chain for φ of length n.

4. $\Rightarrow$ 6. Let $(\bar{a}_i : i < \omega)$ be a k-dividing chain of length ω. Let $\bar{c} \models \bigwedge_{i<\omega} \varphi(\bar{x}, \bar{a}_i)$, and suppose $D(., \varphi, k) \leq \alpha$ for some ordinal α. Since $\varphi(\bar{x}, \bar{a}_i)$ k-divides over $(\bar{a}_j : j < i)$ for all $i < \omega$, we get $D(\bar{c}/(\bar{a}_j : j \leq i), \varphi, k) < D(\bar{c}/(\bar{a}_j : j < i), \varphi, k)$ for all $i < \omega$. But this yields an infinite descending sequence of ordinals below α, a contradiction.

5. $\Rightarrow$ 1. We show by induction on n that if $D(\pi(\bar{x}), \varphi, k) \geq n$, then there is a (φ, k)-tree of height n consistent with π. This is clearly true for $n = 0$, as both conditions merely say that π is consistent. So suppose it is true for n, and $D(\pi, \varphi, k) \geq n + 1$, where π is a partial type over A. So there is some $\bar{a}$ such that $\varphi(\bar{x}, \bar{a})$ k-divides over A, and $D(\pi(\bar{x}) \cup \{\varphi(\bar{x}, \bar{a})\}, \varphi, k) \geq n$. By inductive hypothesis, there is a (φ, k)-tree $(\bar{a}_\mu : \mu \in {}^{<n}\omega)$ consistent with $\pi(\bar{x}) \cup \{\varphi(\bar{x}, \bar{a})\}$. By the definition of dividing, there is an A-indiscernible sequence $(\bar{a}'_i : i < \omega)$ of type $\mathrm{tp}(\bar{a}/A)$ such that $\{\varphi(\bar{x}, \bar{a}_i) : i < \omega\}$ is k-inconsistent; let $(\bar{a}'_{i\frown\mu} : \mu \in {}^{<n}\omega)$ denote the image of $(\bar{a}_\mu : \mu \in {}^{<n}\omega)$ under an A-automorphism mapping $\bar{a}$ to $\bar{a}'_i$. Then $(\bar{a}'_\mu : \mu \in {}^{<n+1}\omega)$ is a (φ, k)-tree of height $n + 1$ consistent with π. ■

PROPOSITION 2.8.6 *Let $\varphi(\bar{x}, \bar{y})$ be a formula. Then the following are equivalent:*

1. *φ has the tree property.*

2. *φ divides ω_1 times.*

3. *φ divides α times for all ordinals α.*

PROOF: 1. $\Rightarrow$ 3. follows from Proposition 2.8.5, 3. $\Rightarrow$ 2. is obvious. So assume that φ divides ω_1 times. Hence there is some $k < \omega$ such that φ k-divides ω_1 times, and 1. follows. ■

DEFINITION 2.8.7 Let $\varphi(\bar{x}, \bar{y})$ be a formula. The *n-th conjunction* of φ, denoted $\bigwedge_n \varphi(\bar{x}, \bar{y}_i : i < n)$, is the formula $\bigwedge_{i<n} \varphi(\bar{x}, \bar{y}_i)$.

PROPOSITION 2.8.8 *Let $\varphi(\bar{x}, \bar{y})$ be a formula. Then the following are equivalent:*

1. *$\bigwedge_n \varphi$ does not have the tree property, for all $n < \omega$.*

2. *$\bigwedge_n \varphi$ does not have the 2-tree property, for all $n < \omega$.*

3. *$NT^+_\varphi(\lambda', \lambda) \leq \lambda^\omega + 2^{\lambda'}$, for all infinite cardinals λ, λ'.*

4. *$NT^+_\varphi(\lambda', \lambda) \leq \lambda^\omega + 2^{\lambda'}$, for some regular $\lambda' \geq \omega_1$ and all cardinals λ.*

5. *There are cardinals λ and λ' with $\lambda = \lambda^{<\lambda'}$ and $NT^+_\varphi(\lambda', \lambda) < \lambda^{\lambda'}$.*

PROOF: 1. $\Rightarrow$ 2. and 3. $\Rightarrow$ 4. are trivial.

2. $\Rightarrow$ 3. We assume $\neg$3.; in order to show $\neg$2., we may clearly assume that the language is countable. Suppose A is a set of size λ such that there is a family P of pairwise incompatible positive φ-types of size λ' over A, with $|P| > \lambda^\omega + 2^{\lambda'}$. Write $P = \{p_i : i < \lambda''\}$, and $p_i = \{\varphi(\bar{x}, \bar{a}^j_i) : j < \lambda'\}$. Since T is countable and $\lambda'' > 2^{\lambda'}$, we may assume that $\text{tp}(\bar{a}^j_i : j < \lambda') = \text{tp}(\bar{a}^j_{i'} : j < \lambda')$ for all $i, i' < \lambda''$. We shall construct inductively a sequence $(S_t : t < \omega_1)$ of finite subsets of λ', such that for every $t < \omega_1$ the formula $\bigwedge_{j \in S_t} \varphi(\bar{x}, \bar{a}^j_i)$ 2-divides over $\bigcup_{s<t}\{\bar{a}^j_i : j \in S_s\}$.

Suppose $t < \omega_1$ and we have obtained S_s for all $s < t$. Since $|A|^\omega = \lambda^\omega < \lambda''$ and $t < \omega_1$, we may assume that there is a subset $I_t \subseteq \lambda''$ with $|I_t| = \lambda''$, such that $(\bar{a}^j_i : j \in \bigcup_{s<t} S_s)$ is constant for $i \in I_t$. Let $i, i' \in I_t$ be distinct. Since $p_i \cup p_{i'}$ is inconsistent, there is a finite subset $S(i, i') \subset \lambda'$ such that

$$\bigwedge_{s \in S(i,i')} \varphi(\bar{x}, \bar{a}^s_i) \wedge \varphi(\bar{x}, \bar{a}^s_{i'})$$

is inconsistent. As $\lambda'' \geq (2^{\lambda'})^+$, we may apply the Erdös-Rado Theorem 1.2.13 to obtain an infinite set $J \subseteq I_t$ (in fact, with $|J| \geq (\lambda')^+$), such that $S(i, i') = S(j, j') =: S_t$ for all pairs $\{i, i'\}$ and $\{j, j'\}$ in J. Then $\bigwedge_{s \in S_t} \varphi(\bar{x}, \bar{a}^s_i)$ 2-divides over $\bigcup_{s<t}\{\bar{a}^j_i : j \in S_s\}$, for all $i \in J$. In fact, this holds for all $i < \lambda''$, as all types in P are conjugates of one another.

There must be an infinite subset $I \subseteq \omega_1$ and $n < \omega$ such that $|S_t| = n$ for all $t \in I$. But then $(\bar{a}^j_i : j \in S_t)_{t \in I}$ is an infinite 2-dividing chain for $\bigwedge_n \varphi$, so $\bigwedge_n \varphi$ has the 2-tree property by Proposition 2.8.5.

3. $\Rightarrow$ 1. Suppose $\bigwedge_n \varphi =: \psi$ has the k-tree property, choose a regular ordinal $\lambda' \geq \omega_1$, and put $\lambda = \beth_{\lambda'}(\lambda')$. Then $\text{cf}(\lambda) = \lambda'$, whence $\lambda^\omega \leq \lambda^{<\lambda'} = \lambda$, $2^{2^{\lambda'}} \leq \lambda$, and $\lambda^{\lambda'} > \lambda$ (an easy diagonalisation argument).

By compactness, there is a (ψ, k)-tree $(\bar{a}_\mu : \mu \in {}^{<\lambda'}\lambda)$ of height λ', such that every node has not only ω but in fact λ successors. Consider the branches of the tree, which induce positive ψ- and hence positive φ-types p_i, for $i < \lambda^{\lambda'}$. For a subset I of $\lambda^{\lambda'}$, put $p_I = \bigcup_{i \in I} p_i$. Suppose p_I is consistent and consider $\varphi(\bar{x}, \bar{a}_\mu) \in p_I$. Then there at most $k - 1$ different $i < \omega$ such that $\varphi(\bar{x}, \bar{a}_{\mu\hat{}i} \in p_I$, so $|p_I| \leq 2^{\lambda'}$. Now let P be the set of maximally consistent p_I, for $I \subset \lambda^{\lambda'}$. Since $\lambda^{\lambda'} > 2^{\lambda'}$ we get $|P| = \lambda^{\lambda'}$; as the size of the tree is $\lambda^{<\lambda'} = \lambda$, we get $NT^+_\varphi(2^{\lambda'}, \lambda) \geq \lambda^{\lambda'} > \lambda = \lambda^\omega + 2^{2^{\lambda'}}$.

4. $\Rightarrow$ 5. Take a suitable regular $\lambda' \geq \omega_1$, and choose $\lambda = \beth_{\lambda'}(\lambda')$. Then

$$NT^+_\varphi(\lambda', \lambda) \leq \lambda^\omega + 2^{\lambda'} < \lambda^{\lambda'}.$$

5. $\Rightarrow$ 2. Suppose $\bigwedge_n \varphi =: \psi$ has the 2-tree property, and consider λ and λ' with $\lambda = \lambda^{<\lambda'}$. As in the proof of 3. $\Rightarrow$ 1. we obtain a λ-splitting $(\psi, 2)$-tree A of height λ', whose $\lambda^{\lambda'}$ branches induce pairwise contradictory positive φ-types of size λ'. As $|A| = \lambda^{<\lambda'} = \lambda$, we get $NT^+_\varphi(\lambda', \lambda) \geq \lambda^{\lambda'}$. ■

COROLLARY 2.8.9 *The following are equivalent:*

1. *T is simple.*
2. *There is no tuple $\bar{b}$ and a sequence $(\bar{a}_i : i < |T|^+)$ such that* $\text{tp}(\bar{b}/\bar{a}_j : j \leq i)$ *divides over $(\bar{a}_j : j < i)$ for all $i < \omega_1$.*
3. *No formula in T has the tree property.*
4. *No formula in T has the 2-tree property.*
5. *$NT(\lambda', \lambda) \leq \lambda^{|T|} + 2^{\lambda'}$, for all infinite ordinals λ, λ'.*
6. *$NT(\lambda', \lambda) \leq \lambda^{|T|} + 2^{\lambda'}$ for some regular $\lambda' \geq |T|^+$ and all λ.*
7. *$NT(\lambda', \lambda) < \lambda^{\lambda'}$ for some λ', λ with $\lambda^{<\lambda'} = \lambda$.*

PROOF: 1. $\Rightarrow$ 2. follows from local character of forking, 2. $\Rightarrow$ 3. from the fact that out of $(\bar{a}_i : i < |T|^+)$ we can extract a dividing chain of length $|T|^+$ for some formula $\varphi \in \mathfrak{L}$, and hence an infinite k-dividing chain. The rest follows from Propositions 2.8.5 and 2.8.8, noting that

$$NT(\lambda', \lambda) \leq \prod_{\varphi \in \mathfrak{L}} NT_\varphi(\lambda', \lambda) \leq \max\{NT_\varphi(\lambda', \lambda)\}^{|T|}. \quad ■$$

REMARK 2.8.10 In fact, we may restrict ourselves to formulas $\varphi(x, \bar{y})$ (i.e. where x is a single variable) and to counting 1-types in Corollary 2.8.9.

PROOF: We shall use induction on n to show that if there is a tuple $\bar{b} = (b_0, \ldots, b_n)$ and a sequence $(\bar{a}_i : i < |T|^+)$ such that $\text{tp}(\bar{b}/\bar{a}_j : j \leq i)$ divides over $(\bar{a}_j : j < i)$ for all $i < |T|^+$, then we can find an element b and a sequence $(\bar{a}'_i : i < |T|^+)$ such that $\text{tp}(b/\bar{a}'_j : j \leq i)$ divides over $(\bar{a}'_j : j < i)$ for all $i < |T|^+$. This is trivial for $n = 0$.

So suppose the assertion holds for n and let $\bar{b} = (b_0, \ldots, b_n, b_{n+1})$ and a sequence $(\bar{a}_i : i < |T|^+)$ be given, with $\text{tp}(\bar{b}/\bar{a}_j : j \leq i)$ dividing over $(\bar{a}_j : j < i)$ for all $i < |T|^+$. For every $i < |T|^+$ either $\text{tp}(b_{n+1}/(\bar{a}_j : j \leq i) \cup (b_0, \ldots, b_n))$ divides over $(\bar{a}_j : j < i) \cup (b_0, \ldots, b_n)$, or $\text{tp}(b_0, \ldots, b_n/\bar{a}_j : j \leq i)$ divides over $(\bar{a}_j : j < i)$ by Proposition 2.2.9. If the first case happens $|T|^+$ times we are done, otherwise we use the induction hypothesis.

Finally, note that all the equivalences in Corollary 2.8.9 (except for 1. obviously) work without affecting the arity of the first variable tuple. ■

COROLLARY 2.8.11 *A theory interpretable in a simple theory is simple. In particular, T is simple if and only if T^{eq} is simple.*

PROOF: We use the characterization in terms of counting types. ■

DEFINITION 2.8.12 A simple theory T is called *supersimple* if no type divides over all finite subsets of its domain.

We shall treat supersimple theories in Chapter 5.

PROPOSITION 2.8.13 *Let φ be a formula. The following are equivalent:*

1. *T is supersimple.*

2. *There are no $\bar{b}$ and a sequence $(\bar{a}_i : i < \omega)$ such that* $\text{tp}(\bar{b}/\bar{a}_j : j \leq i)$ *divides over $(\bar{a}_j : j < i)$ for all $i < \omega$.*

3. *There are no $\bar{b}$ and a sequence $(\bar{a}_i : i < \omega)$ such that* $\text{tp}(\bar{b}/\bar{a}_j : j \leq i)$ *2-divides over $(\bar{a}_j : j < i)$ for all $i < \omega$.*

4. *$NT(\lambda', \lambda) \leq \lambda + 2^{|T|+\lambda'}$, for all infinite cardinals λ, λ'.*

5. *$NT(\lambda', \lambda) \leq \lambda + 2^{\lambda'}$, for some $\lambda' \geq |T|$ and all cardinals λ.*

6. *There are infinite cardinals λ and λ' with $NT(\lambda', \lambda) < \lambda^\omega$.*

PROOF: 1. ⇒ 2., 2. ⇒ 3., 4. ⇒ 5. and 5. ⇒ 6. is obvious. The other arguments are similar to those in the proof of Proposition 2.8.9, so we shall be short.

3. ⇒ 4. Suppose A is a set of cardinality λ, and P is a family of pairwise incompatible partial types over A of size λ', with $|P| > \lambda + 2^{|T|+\lambda'}$. As in the proof of 2. ⇒ 3. of Proposition 2.8.8 (but with ω replacing ω_1) we obtain a sequence $(\varphi_i(\bar{x}, \bar{a}_i) : i < \omega)$ with $\bigwedge_{i<\omega} \pi(\bar{x}, \bar{a}_i)$ consistent, such that $\varphi_i(\bar{x}, \bar{a}_i)$

2-divides over $(\bar{a}_j : j < i)$. Choosing a realization $\bar{b}$ of $\bigwedge_{i<\omega} \varphi_i(\bar{x}, \bar{a}_i)$, we contradict 3..

6. $\Rightarrow$ 3. If $\bar{b}$ is a tuple and $(\bar{a}_i : i < \omega)$ is a sequence such that $\mathrm{tp}(\bar{b}/\bar{a}_j : j \leq i)$ 2-divides over $(\bar{a}_j : j < i)$ for all $i < \omega$, then by compactness for every λ we get a tree $(\bar{a}_\mu : \mu \in {}^{<\omega}\lambda)$ and a family $\{p_i : i \in \lambda^\omega\}$ of pairwise inconsistent types, such that p_i is a type over $\{\bar{a}_\mu : \mu < i\}$ for all $i \in \lambda^\omega$. So p_i is countable for all $i < \lambda^\omega$, and $NT(\omega, \lambda) \geq \lambda^\omega$.

4. $\Rightarrow$ 1. Let $\lambda > 2^{|T|+2^\omega}$ be a cardinal with $\lambda < \lambda^\omega$. Suppose p is a type which divides over all finite subsets of its domain. If $\bar{b} \models p$, there must be a sequence of finite tuples $(\bar{a}_i : i < \omega)$ such that $\mathrm{tp}(\bar{b}/\bar{a}_j : j \leq i)$ divides over $(\bar{a}_j : j < i)$. By compactness we get a tree $(\bar{a}_\mu : \mu \in {}^{<\omega}\lambda)$ and a family $\{p_i : i \in \lambda^\omega\}$ of types induced by the branches of the tree. For $I \subset \lambda^\omega$ put $p_I = \bigcup_{i \in I} p_i$. If p_I is consistent, then it must come from a finitely splitting subtree, and $|I| \leq \omega^\omega = 2^\omega$. Hence there are λ^ω pairwise incompatible maximal consistent types p_I of size 2^ω, and

$$NT(2^\omega, \lambda) \geq \lambda^\omega > \lambda = \lambda + 2^{|T|+2^\omega}. \quad \blacksquare$$

REMARK 2.8.14 Again, a theory interpretable in a supersimple theory is supersimple. So T is supersimple if and only if T^{eq} is supersimple.

REMARK 2.8.15 *A fortiori*, in a supersimple theory $NT_\varphi^+(\lambda', \lambda) \leq \lambda + 2^{\lambda'}$ for every formula φ. However, there are simple non-supersimple theories which satisfy this bound; in fact, the *stable* theories defined in the next section will even satisfy $NT_\varphi^+(\lambda', \lambda) \leq \lambda$.

PROPOSITION 2.8.16 *Let $\varphi(\bar{x}, \bar{y})$ be a formula, where the length of $\bar{y}$ is n. The following are equivalent:*

1. *φ does not divide ω times.*

2. *There is no sequence $(\bar{a}_i : i < \omega)$ (where for all $i < \omega$ the length of $\bar{a}_i$ is nn_i for some $n_i < \omega$) such that $\bigwedge_{n_i} \varphi(\bar{x}, \bar{a}_i)$ divides over $(\bar{a}_j : j < i)$ for all $i < \omega$.*

3. *There is no sequence $(\bar{a}_i : i < \omega)$ (where for all $i < \omega$ the length of $\bar{a}_i$ is nn_i for some $n_i < \omega$) such that $\bigwedge_{n_i} \varphi(\bar{x}, \bar{a}_i)$ 2-divides over $(\bar{a}_j : j < i)$ for all $i < \omega$.*

4. *$NT_\varphi^+(\lambda', \lambda) \leq \lambda + 2^{\lambda'}$, for all infinite cardinals λ, λ'.*

5. *$NT_\varphi^+(\lambda', \lambda) \leq \lambda + 2^{\lambda'}$, for some infinite λ' and all cardinals λ.*

6. *There are infinite cardinals λ and λ' with $NT_\varphi^+(\lambda', \lambda) < \lambda^\omega$.*

PROOF: Similar to the proof of Proposition 2.8.13 (which can be localized to a single formula). ■

DEFINITION 2.8.17 A simple theory is *low* if for every formula φ there is some $n_\varphi < \omega$ such that there is no dividing φ-chain of length n_φ.

We shall study low theories in Section 6.4.

REMARK 2.8.18 A low theory satisfies the equivalent conditions of Proposition 2.8.16 for all formulas φ. However, we cannot use compactness to deduce from the failure of lowness the existence of a dividing φ-chain of length ω, since the dividing numbers in the dividing chain can be different. In fact, Casanovas and Kim [21] have constructed a supersimple non-low theory.

2.9 STABILITY

Historically, simple theories are only a recent extension of the narrower class of *stable* theories, which were developped by Shelah in order to classify all models of a complete first-order theory. In a stable theory a type over a model has a unique non-forking extension to any superset of the model; in general a type over a set A has only a bounded number (called the *multiplicity* of the type) of non-forking extensions to any superset, which in addition is determined by its class modulo definable *finite equivalence relations* (equivalence relations with only finitely many classes). Multiplicity and simplicity (the behaviour of non-forking) were originally seen as closely intertwined; only the advent of simplicity theory has shown that one can have a reasonable forking theory without multiplicity considerations.

DEFINITION 2.9.1 A theory is *stable in* λ, or λ*-stable*, if $|S(A)| \leq \lambda$ for all A of size λ. A theory is *stable* if it is stable for some infinite λ.

LEMMA 2.9.2 *A stable theory is simple.*

PROOF: Suppose T is not simple. Fix a cardinal λ and let λ' be minimal with $\lambda^{\lambda'} > \lambda$. By Proposition 2.8.9 and compactness there is a formula φ and a $(\varphi, 2)$-tree A of height λ', such that every node has λ successors. But that means that over A there are $\lambda^{\lambda'}$ pairwise incompatible positive φ-types, and $|A| = \lambda^{<\lambda'} = \lambda$. So T is not stable in λ. ■

THEOREM 2.9.3 *In a stable theory, a type over a model has a unique non-forking extension to any superset.*

PROOF: Suppose not, and let $\mathfrak{M}$ be a model, $p \in S(M)$, and $p_1(\bar{x}, B)$ and $p_2(\bar{x}, B)$ be two non-forking extensions of p to a set $B \supseteq \mathfrak{M}$. Clearly we may assume that the language, $\mathfrak{M}$ and B are all countable. Let $(B_i : i < \lambda)$ be a

Morley sequence in $\mathrm{tp}(B/\mathfrak{M})$. By Corollary 2.5.11 for every $I \subset \lambda$ the partial type

$$p_I := \bigcup_{i\in I} p_1(\bar{x}, B_i) \cup \bigcup_{i\notin I} p_2(\bar{x}, B_i)$$

is consistent. As the family $\{p_I : I < 2^\lambda\}$ is pairwise incompatible, there are $2^\lambda > \lambda$ types over the λ parameters $\mathfrak{M} \cup \bigcup_{i<\lambda} B_i$. So T is not λ-stable. ■

REMARK 2.9.4 In particular, if $\mathfrak{M}$ is a model, A and B are arbitrary supersets of $\mathfrak{M}$, and $p \in S(A)$ and $q \in S(B)$ have the same restriction to $\mathfrak{M}$, then $p \cup q$ is consistent and does not fork over $\mathfrak{M}$, as it is part of the non-forking extension of $p\lceil\mathfrak{M}$ to $A \cup B$. So a very strong form of the Independence Theorem (without independence) holds in stable theories.

COROLLARY 2.9.5 *A stable theory is λ-stable for all λ with $\lambda^{|T|} = \lambda$.*

PROOF: Let A be a set of size λ, contained in a model $\mathfrak{M}$ of size λ. Clearly it is sufficient to count types over $\mathfrak{M}$. But for every $p \in S(\mathfrak{M})$ there is a submodel $\mathfrak{M}_p \prec \mathfrak{M}$ with $|\mathfrak{M}_p| \leq |T|$, such that p does not fork over $\mathfrak{M}_p$. As p is uniquely determined as the non-forking extension of $p\lceil\mathfrak{M}_p$, we get $|S(\mathfrak{M})| \leq \lambda^{|T|} + 2^{|T|} = \lambda$. ■

REMARK 2.9.6 In fact, the proof shows that a simple theory is stable if and only if every type over a model has a unique non-forking extension.

REMARK 2.9.7 As $|S(\mathfrak{M})| \leq |S(\mathfrak{M}^{eq})| \leq |S(\mathfrak{M})| + |T|$, a theory T is stable if and only if T^{eq} is stable. More generally, a theory interpretable in a stable theory is stable.

COROLLARY 2.9.8 *A type over an arbitrary set in a stable theory has at most $2^{|T|}$ non-forking extensions to any given superset.*

PROOF: Suppose $p \in S(A)$ and $A \subseteq B$. Let $A_0 \subseteq A$ be such that p does not fork over A_0 and $|A_0| \leq |T|$, and $\mathfrak{M} \supseteq A_0$ be a model of size $|T|$. Then any non-forking extension q of p over B can be extended non-forkingly to some $q' \in S(\mathfrak{M} \cup B)$, and does not fork over $\mathfrak{M}$. So it is uniquely determined by $q'\lceil\mathfrak{M}$. As there are only $2^{|T|}$ types over $\mathfrak{M}$, the assertion follows. ■

COROLLARY 2.9.9 *In a stable theory, every type over a model or over an algebraically closed set (in T^{eq}) is stationary, and hence definable.*

PROOF: This follows immediately from Corollary 2.3.17 and Corollary 2.9.8. ■

DEFINITION 2.9.10 The *strong type* of a tuple $\bar{a}$ over A, denoted by $\operatorname{stp}(\bar{a}/A)$, is the type of $\bar{a}$ over $\operatorname{acl}^{eq}(A)$.

A strong type is stationary, and a stationary type extends to a unique strong type.

THEOREM 2.9.11 *Every type in a stable theory is definable.*

PROOF: Consider $p \in S(A)$, and let q be a non-forking extension of p to $\operatorname{acl}^{eq}(A)$. Then q is definable over $\operatorname{acl}^{eq}(A)$; clearly for any formula $\varphi \in \mathfrak{L}$ the disjunction or the conjunction of the finitely many A-conjugates of $d_q\varphi$ will be a φ-definition for p over A. ■

REMARK 2.9.12 This is a considerable improvement on Lemma 2.3.15: if we take the disjunction in the proof of Theorem 2.9.11, then we get a *formula* $\psi(\bar{y})$ over A such that $\psi(\bar{a})$ holds for any $\bar{a}$ if and only if some non-forking extension of p to $A\bar{a}$ contains $\varphi(\bar{x}, \bar{a})$. Similarly, if we take the conjunction, we get a *formula* $\psi'(\bar{y})$ over A such that $\psi(\bar{a})$ holds for any $\bar{a}$ if and only if all non-forking extensions of p to $A\bar{a}$ contain $\varphi(\bar{x}, \bar{a})$.

COROLLARY 2.9.13 FINITE EQUIVALENCE RELATION THEOREM *Suppose T is stable and* $\operatorname{tp}(\bar{a}/A) = \operatorname{tp}(\bar{a}'/A)$. *Then $\bar{a}$ and $\bar{a}'$ have the same strong type over A if and only if they agree on all A-definable equivalence relations with a finite number of classes.*

PROOF: Suppose there are $\bar{b} \in \operatorname{acl}(A)$ and some formula φ with $\models \varphi(\bar{a}, \bar{b}) \wedge \neg\varphi(\bar{a}', \bar{b})$, where $\bar{b}$ lies in the finite A-definable set B. Consider the relation E defined by $E(\bar{x}, \bar{y})$ if and only if

$$\forall \bar{z} \in B\, [\varphi(\bar{x}, \bar{z}) \leftrightarrow \varphi(\bar{y}, \bar{z})].$$

This clearly is an A-definable equivalence relation with a finite number of classes, and $\neg E(\bar{a}, \bar{a}')$.

The assertion now follows from the fact that types over algebraically closed sets (in T^{eq}) are stationary. ■

DEFINITION 2.9.14 The *canonical base* of a strong type $p \in S(\operatorname{acl}^{eq}(A))$ is the definable closure of $\{\operatorname{Cb}(d_p x\, \varphi) : \varphi \in \mathfrak{L}\}$.

We shall write $\operatorname{Cb}(a/A)$ for $\operatorname{Cb}(\operatorname{stp}(a/A))$.

THEOREM 2.9.15 *Let T be stable, and p a strong type over A. Then p does not fork over some A_0 if and only if* $\operatorname{Cb}(p) \subseteq \operatorname{acl}^{eq}(A_0)$. *Furthermore, $p\lceil\operatorname{Cb}(p)$ is stationary, and if p does not fork over A_0 and $p\lceil A_0$ is stationary, then* $\operatorname{Cb}(p) \in \operatorname{dcl}^{eq}(A_0)$.

PROOF: Theorem 2.3.20 says that p does not fork over $\mathrm{Cb}(p)$. Hence if $\mathrm{Cb}(p) \subseteq \mathrm{acl}^{eq}(A_0)$, then p cannot fork over A_0. Conversely, if p does not fork over A_0, then it is definable over $\mathrm{acl}^{eq}(A_0)$; as the canonical parameters for the φ-definitions are unique up to interdefinability, we get $\mathrm{Cb}(p) \subseteq \mathrm{acl}^{eq}(A_0)$.

For the furthermore, let q be a non-forking extension of $p\lceil\mathrm{Cb}(p)$ to some model. Then q can be conjugated over $\mathrm{Cb}(p)$ to a type whose φ-definitions are $d_p\varphi$, for all formulas φ; it follows that $d_q\varphi = d_p\varphi$ and p is stationary. Finally, if p does not fork over A_0 and $p\lceil A_0$ is stationary, then $\mathrm{Cb}(d_p\varphi)$ is in $\mathrm{dcl}^{eq}(A_0)$ for all formulas φ, so $\mathrm{Cb}(p) \subseteq \mathrm{dcl}^{eq}(A_0)$. ■

2.10 BIBLIOGRAPHICAL REMARKS

The definitions and results of Section 2.2 are due to Shelah [147, 149], those of Sections 2.3 and 2.4 mostly to Kim [64, 66]; the quick proof of symmetry from Theorem 2.4.7 is due to Shelah and can be found in [38]. Lemma 2.3.7, Proposition 2.3.8 and Theorem 2.4.7, which allow us to define simplicity in terms of symmetry of independence, come from Kim [68]. Sections 2.5 and 2.6 were shown by Kim and Pillay in [70]. Lascar strong type was introduced by Lascar in [82, 87]; type-definability of Lascar strong type in simple theories is again due to Kim [67], who also gives Example 2.7.11; Example 2.2.4 was found by Poizat. Section 2.8 is taken from Casanovas [20].

Chapter 3

HYPERIMAGINARIES

3.1 HYPERIMAGINARIES

Recall that if $\bar{x}$ and $\bar{y}$ are tuples of the same (possibly infinite) length α, an equivalence relation $E(\bar{x}, \bar{y})$ on $\mathfrak{C}^\alpha$ is type-definable over A if it is given by a partial type $\pi(\bar{x}, \bar{y})$ over A. We shall usually assume that E is closed under finite conjunctions.

REMARK 3.1.1 If E is a type-definable equivalence relation on a partial type π, we may extend E to a type-definable equivalence relation E' defined on the whole of $\mathfrak{C}$, by putting

$$E'(\bar{x}, \bar{y}) = [E(\bar{x}, \bar{y}) \wedge \pi(\bar{x}) \wedge \pi(\bar{y})] \vee \bar{x} = \bar{y}.$$

(By compactness, this can be rewritten as an infinite conjunction.)

DEFINITION 3.1.2 A type-definable equivalence relation E is *countable* if α and the partial type defining E are countable; E is *finitary* if α is finite.

LEMMA 3.1.3 *If $E(\bar{x}, \bar{y})$ is a type-definable equivalence relation, then for some index set I there are type-definable countable equivalence relations $\{E_i(\bar{x}_i, \bar{y}_i) : i \in I\}$, with $\bar{x}_i \subseteq \bar{x}$ and $\bar{y}_i \subseteq \bar{y}$ for all $i \in I$, such that $\models E(\bar{x}, \bar{y})$ if and only if $\models E_i(\bar{x}_i, \bar{y}_i)$ for all $i \in I$.*

PROOF: Since E is a type-definable equivalence relation, and hence symmetric and transitive, compactness implies that for every formula $\varphi(\bar{x}, \bar{y}) \in E$ there is a formula $\varphi'(\bar{x}, \bar{y}) \in E$ such that $\varphi'(\bar{x}, \bar{y}) \wedge \varphi'(\bar{y}, \bar{z})$ implies $\varphi(\bar{x}, \bar{z}) \wedge \varphi(\bar{y}, \bar{x})$. For every formula $\varphi_i \in E$ let E_i be the closure of φ_i under the operation $\varphi \mapsto \varphi'$. It is easy to see that E_i is reflexive, symmetric and transitive, and thus a type-definable countable equivalence relation. Clearly $\bigwedge_{i \in I} E_i$ is equivalent to E. ■

DEFINITION 3.1.4 Let $E(\bar{x}, \bar{y})$ be a type-definable equivalence relation. If $\bar{a}$ is a tuple of the right length, we denote the class of $\bar{a}$ modulo E by $\bar{a}_E$ and call such a quantity a *hyperimaginary* element of *type* E. A hyperimaginary $\bar{a}_E$ is *countable* or *finitary* if E is countable or finitary, respectively.
$\mathfrak{M}^{heq}$ is $\mathfrak{M}$ together with the collection of all countable hyperimaginaries modulo type-definable equivalence relations over $\emptyset$.

In particular, every sequence, and indeed every set, of real or imaginary elements is a hyperimaginary element. Note that every automorphism of a model $\mathfrak{M}$ extends uniquely to an automorphism of $\mathfrak{M}^{heq}$.

REMARK 3.1.5 If $E_A(\bar{x}, \bar{y})$ is type-definable over a set A, we can type-define an equivalence relation F over $\emptyset$ by

> $F(\bar{x}X, \bar{y}Y)$ if and only if $X = Y \models \text{tp}(A)$ and $E_X(\bar{x}, \bar{y})$, or $X = Y$ and $\bar{x} = \bar{y}$.

Then every class modulo E corresponds to a class modulo F coming from a tuple with second co-ordinate A. Furthermore, if E is countable, so is F.

In contrast to what happens for definable equivalence relations, where the *eq*-construction allows us to treat imaginary elements as real elements of the structure, this is not possible for hyperimaginaries.

EXAMPLE 3.1.6 Let $\mathfrak{M}$ be the unit circle, with binary relations $R_n(x, y)$ for each $i < \omega$, such that $\mathfrak{M} \models R_n(x, y)$ if the distance between x and y is less than $1/n$. It is easy to see that $\bigwedge R_n(x, y)$ is a type-definable equivalence relation E, saying "x is infinitely close to y"; in any elementary extension $\mathfrak{N}$ of $\mathfrak{M}$ the E-class of a point $n \in \mathfrak{N}$ is represented by the standard part $\text{std}(n) \in \mathfrak{M}$. So E has only continuum many classes in E. Now if we could add a predicate for $\mathfrak{M}/E$ and a definable map $\pi_E : x \mapsto x_E$, then compactness would imply that there are only finitely many E-classes! (Note, though, that this example is not simple.)

If E is a type-definable equivalence relation, and F is a type-definable equivalence relation coarsening E, then we may view F as a type-definable equivalence relation on hyperimaginaries modulo E.

We shall now try to give meaning to the logic of hyperimaginaries.

DEFINITION 3.1.7 Let A be a set (possibly containing hyperimaginaries).

1. The (hyperimaginary) *definable closure* of A, denoted $\text{dcl}(A)$, is the set of all countable hyperimaginaries which are fixed under all A-automorphisms.
2. The (hyperimaginary) *algebraic closure* of A, denoted $\text{acl}(A)$, is the set of all countable hyperimaginaries which have only finitely many images under A-automorphisms.

3. The (hyperimaginary) *bounded closure* of A, denoted $\text{bdd}(A)$, is the set of all countable hyperimaginaries which have only boundedly many (i.e. less than κ) images under A-automorphisms.

If we want to emphasize that we take the hyperimaginary definable or algebraic closure, we may indicate this by a superscript ^{heq}.

We shall call two sets A and B *interdefinable* (resp. *interalgebraic* or *interbounded*) if $\text{dcl}(A) = \text{dcl}(B)$ (resp. $\text{acl}(A) = \text{acl}(B)$ or $\text{bdd}(A) = \text{bdd}(B)$).

REMARK 3.1.8 ■ By Lemma 3.1.3 every hyperimaginary element is interdefinable with a sequence of countable hyperimaginaries.

- If a is an imaginary element in $\text{bdd}(A)$, compactness implies that it is in $\text{acl}(A)$.
- If a is an uncountable hyperimaginary fixed under all A-automorphisms, we shall still say that $a \in \text{dcl}(A)$ (and similarly for the algebraic and bounded closures). In this sense, $A \subseteq \text{dcl}(A) \subseteq \text{acl}(A) \subseteq \text{bdd}(A)$.
- If $\bar{a}_E \in \text{bdd}(A)$, then by compactness for every formula $\varphi(\bar{x}, \bar{y}) \in E(\bar{x}, \bar{y})$ there are some $\psi(\bar{x}) \in \text{tp}(\bar{a}/A)$ and some $n_\varphi < \omega$ such that $\bigwedge_{0 \leq i < j < n_\varphi} [\psi(\bar{x}_i) \wedge \neg\varphi(\bar{x}_i, \bar{x}_j)]$ is inconsistent.

REMARK 3.1.9 A *closure operator* $\text{cl}(.)$ is a map from subsets of $\mathfrak{C}$ to subsets of $\mathfrak{C}$ satisfying:

- $A \subseteq \text{cl}(A)$,
- If $A \subseteq B$, then $\text{cl}(A) \subseteq \text{cl}(B)$,
- $\text{cl}(\text{cl}(A)) = \text{cl}(A)$;

we say that $\text{cl}(.)$ is *finitary* if $\text{cl}(A) = \bigcup\{\text{cl}(A_0) : A_0 \subseteq A \text{ finite}\}$.

It is easy to see that definable, algebraic and bounded closures are closure operators. However, they need not be finitary (although their restriction to imaginary elements is finitary).

DEFINITION 3.1.10 If π and π' are two partial types with $\pi \vdash \pi'$ and $\pi' \vdash \pi$, we shall say that π and π' are *equivalent*, and denote this by $\pi \equiv \pi'$.

DEFINITION 3.1.11 Let E be a type-definable equivalence relation. A *partial E-type* is a type $\pi(\bar{x})$ which is invariant under E, i.e. whenever $E(\bar{a}, \bar{a}')$ holds, then $\models \pi(\bar{a})$ if and only if $\models \pi(\bar{a}')$.
A partial type $\pi(\bar{x})$ over a hyperimaginary $\bar{a}_E$ is a partial type over some parameters A, such that for any $\bar{a}_E$-automorphism σ we have $\pi(\bar{x}) \equiv \sigma(\pi(\bar{x}))$.

A set X of hyperimaginaries of type E is *hyperdefinable* if it is given by a partial E-type.

In particular, a type over a hyperimaginary $\bar{a}_E$ may contain parameters not in $\bar{a}_E$. In fact, no single *formula* in a type over $\bar{a}_E$ need be invariant under all $\bar{a}_E$-automorphisms.

REMARK 3.1.12 If π and π' are two equivalent partial types over the same set of real or imaginary elements, and both are closed under implication, then $\pi = \pi'$. However, two equivalent types over the same hyperimaginary set are only truly equal (contain the same formulas) if we represent them over the same set of real or imaginary parameters.

LEMMA 3.1.13 *If $\pi(\bar{x})$ is a partial type over $\bar{a}_E$ and $\bar{a}_E \in \mathrm{dcl}(B)$, then there is a partial type π' over B equivalent to π.*

PROOF: Suppose $\pi = \pi(\bar{x}, A)$ is a partial type with parameters A. Define $\pi'(\bar{x})$ as

$$\exists X \models \mathrm{tp}(A/B)\ \pi(\bar{x}, X).$$

Clearly $\pi \vdash \pi'$, as A witnesses the existential quantifier. Conversely, if $\bar{c} \models \pi'$, there is $A' \models \mathrm{tp}(A/B)$ with $\models \pi(\bar{c}, A')$. So there is a B-automorphism mapping A' to A; since $\bar{a}_E \in \mathrm{dcl}(B)$, this is in fact an a_E-automorphism. Hence $\pi(\bar{x}, A') \vdash \pi(\bar{x}, A) = \pi$, and we are done. ■

DEFINITION 3.1.14 Let E and F be type-definable equivalence relations over $\emptyset$. Then $\mathrm{tp}(\bar{a}_E/\bar{b}_F)$ is the partial type over $\bar{b}$ which consists of the union of all partial types

$$\exists \bar{x}'\bar{y}' \left[E(\bar{x}', \bar{x}) \wedge F(\bar{y}', \bar{b}) \wedge \varphi(\bar{x}', \bar{y}')\right]$$

which are true of $\bar{a}$. (Note that these can indeed be re-written as partial types.) Let π be a partial E-type. We say that a hyperimaginary $\bar{a}_E$ realizes π if $\models \pi(\bar{a})$; this is invariant under the choice of representative $\bar{a}$ for $\bar{a}_E$.

So for a hyperimaginary $\bar{a}_E$ to satisfy a partial type π, it is first of all necessary that π is an E-type. Clearly, $\mathrm{tp}(\bar{a}_E/\bar{b}_F)$ is a partial E-type over $\bar{b}_F$.

LEMMA 3.1.15 *If $\bar{a}_E$ and $\bar{b}_F$ are hyperimaginaries, then $\bar{a}^* \models \mathrm{tp}(\bar{a}_E/\bar{b}_F)$ if and only if there is an automorphism of $\mathfrak{C}$ fixing the F-class of $\bar{b}$ and mapping the E-class of $\bar{a}$ to the E-class of $\bar{a}^*$.*

PROOF: If $\bar{a}^* \models \mathrm{tp}(\bar{a}_E/\bar{b}_F)$, then by compactness there are $\bar{a}'$ and $\bar{b}'$ with $\mathrm{tp}(\bar{a}'\bar{b}') = \mathrm{tp}(\bar{a}\bar{b})$ and $E(\bar{a}', \bar{a}^*)$ and $F(\bar{b}', \bar{b})$. So there is an automorphism σ mapping $\bar{a}\bar{b}$ to $\bar{a}'\bar{b}'$; since σ fixes $\bar{b}_F$ and maps $\bar{a}_E$ to $\bar{a}^*_E$, it is as required.

Conversely, suppose there is such an automorphism σ. If

$$\exists\bar{x}'\bar{y}'\,[E(\bar{x}',\bar{x})\wedge F(\bar{y}',\bar{b})\wedge\varphi(\bar{x}',\bar{y}')]$$

is satisfied by $\bar{a}$, then

$$\exists\bar{x}'\bar{y}'\,[E(\bar{x}',\bar{x})\wedge F(\bar{y}',\sigma(\bar{b}))\wedge\varphi(\bar{x}',\bar{y}')]$$

is satisfied by $\sigma(\bar{a})$. But as $F(\bar{b},\sigma(\bar{b}))$ holds, this partial type is clearly equivalent to the first one. Moreover, $E(\bar{a}^*,\sigma(\bar{a}))$ implies that it is also satisfied by $\bar{a}^*$, whence $\bar{a}^*\models \mathrm{tp}(\bar{a}_E/\bar{b}_F)$. ∎

The following proposition shows that in the same way a finite set of imaginary elements can be considered as a single imaginary element, a bounded set of hyperimaginaries can be considered as a single hyperimaginary.

PROPOSITION 3.1.16 *Let a be a hyperimaginary element and suppose $a\in \mathrm{bdd}(B)$. Then there is a type-definable equivalence relation F' such that the set of B-conjugates of a over B is interdefinable with $B_{F'}$. In particular, it is in* $\mathrm{dcl}(B)$.

PROOF: Suppose E is a type-definable equivalence relation (over $\emptyset$) and $\bar{a}$ is a real sequence such that $a=\bar{a}_E$. For every symmetric formula $\varphi(\bar{x},\bar{y})\in E$, since $a\in\mathrm{bdd}(B)$, there is a maximal $n_\varphi<\omega$ such that there are n_φ conjugates $(\bar{a}_i^\varphi : i<n_\varphi)$ of $\bar{a}$ over B with $\models\neg\varphi(\bar{a}_i^\varphi,\bar{a}_j^\varphi)$ whenever $i\neq j<n_\varphi$. Put $A=(\bar{a}_i^\varphi:\varphi\in E, i<n_\varphi)$.

Type-define a relation $R(Y,X)$, where $X=(\bar{x}_i^\varphi:\varphi\in E, i<n_\varphi)$ and Y is a tuple of variables of the same order type as some enumeration of B, by $\mathrm{tp}(Y,(\bar{x}_i^\varphi)_E)=\mathrm{tp}(B,\bar{a}_E)$ for all $\varphi\in E$ and $i<n_\varphi$, and $\models\neg\varphi(\bar{x}_i^\varphi,\bar{x}_j^\varphi)$ for all $\varphi\in E$ and $i\neq j<n_\varphi$. Define a relation F on $\mathrm{tp}(B)^2$ by $F(B_1,B_2)$ if and only if there is some X with $R(B_1,X)$ and $R(B_2,X)$. We claim that for an automorphism σ (of the monster model) $F(B,\sigma(B))$ holds if and only if σ fixes the set of B-conjugates of a (setwise).

Suppose first that σ fixes the set of B-conjugates of a. Clearly $R(B,A)$; since σ maps every B-conjugate $(\bar{a}_i^\varphi)_E$ of a to some other B-conjugate of a, we obtain

$$\mathrm{tp}((\bar{a}_i^\varphi)_E,\sigma(B))=\mathrm{tp}(\sigma^{-1}(\bar{a}_i^\varphi)_E,B)=\mathrm{tp}(a,B).$$

Therefore $R(\sigma(B),A)$ holds, whence $F(B,\sigma(B))$ is true. Conversely, suppose $F(B,\sigma(B))$ and consider X with $R(B,X)$ and $R(\sigma(B),X)$. By maximality, $\varphi(\sigma(\bar{a}),\bar{x}_i^\varphi)$ must hold for some $i<n_\varphi$. But then $E(\bar{x},\sigma(\bar{a}))\cup\mathrm{tp}(\bar{a}_E/B)$ is finitely satisfyable (by those $\bar{x}_i^\varphi$), so $\sigma(\bar{a})_E$ is a B-conjugate of a. As this argument does not depend on a, the set of B-conjugates of a is fixed by σ setwise.

It follows that F is a type-definable equivalence relation on $\mathrm{tp}(B)$, which can be extended to a type-definable equivalence relation F' on $\mathfrak{C}^{|B|}$ by Remark 3.1.1. And $B_{F'}$ is interdefinable with the set of B-conjugates of a. ■

COROLLARY 3.1.17 *If $a \subset \mathrm{dcl}(b)$, then there is a type-definable equivalence relation E such that a is interdefinable with b_E.* ■

DEFINITION 3.1.18 Let A be a set of parameters (possibly containing hyperimaginaries), and E a type-definable equivalence relation over A. We denote by $S_E(A)$ the space of all E-types over A, and endow $S_E(A)$ with the following topology: for every formula $\varphi(x)$ over A the set $[\varphi]_E$ of all types in $S_E(A)$ which are consistent with φ is a basic closed set.

In particular, if $\varphi(x)$ and $\psi(x)$ intersect the same E-classes, then $[\varphi]_E$ and $[\psi]_E$ are equal.

LEMMA 3.1.19 *$S_E(A)$ is complete and compact.*

PROOF: Let $(p_i : i < \alpha)$ be a sequence in $S_E(A)$, and p an accumulation point of the sequence in the space of partial types over A. If $a \models p$, then $\mathrm{tp}(a_E/A)$ is an accumulation point of $(p_i : i < \alpha)$. ■

REMARK 3.1.20 $S_E(A)$ need not be totally disconnected.

From now on, we shall no longer distinguish between real elements, imaginary elements, and hyperimaginary elements. So $a, b, c, \ldots$ will denote hyperimaginary elements (which may or may not be tuples of real elements, or imaginaries), and $A, B, C, \ldots$ will denote sets (of size less than κ, as usual) of hyperimaginaries, unless we specify that they are real or imaginary. We may still write a_E for the class of a modulo E, if we want to stress that a_E is an equivalence class. Note however that any parameters occurring in a *formula* must be imaginary).

DEFINITION 3.1.21 An ordered infinite sequence $(a_i : i \in I)$ of hyperimaginaries is *indiscernible* over a set A, or *A-indiscernible*, if whenever $n < \omega$ and $i_0 < i_1 < \cdots < i_n$ and $j_0 < j_1 < \cdots < j_n$ are two sequences from I, then $\mathrm{tp}(a_{i_0}a_{i_1}\ldots a_{i_n}/A) \equiv \mathrm{tp}(a_{j_0}a_{j_1}\ldots a_{j_n}/A)$. If $\mathrm{tp}(a_0/A) = p$, we say that $(a_i : i \in I)$ is a sequence *of type p*.

REMARK 3.1.22 1. We shall not usually use finite indiscernible sequences.

2. If A is a real or imaginary set, then this is the usual definition by Remark 3.1.12.

3. If $(a_i : i \in I)$ is an A-indiscernible sequence and J is any ordered set, then there is an A-indiscernible sequence $(a'_j : j \in J)$ with

$\text{tp}(a_{i_0}a_{i_1}\dots a_{i_n}/A) \equiv \text{tp}(a_{j_0}a_{j_1}\dots a_{j_n}/A)$ for all $i_0 < i_1 < \cdots < i_n$ from I and $j_0 < j_1 < \cdots < j_n$ from J (and all $n < \omega$).

4. If I is an infinite A-indiscernible sequence, then it is indiscernible over $\text{bdd}(A)$.

5. If I is indiscernible over $\bar{b}_E$, then there is $\bar{b}'$ with $E(\bar{b}, \bar{b}')$, such that I is indiscernible over $\bar{b}'$.

PROOF: 3. follows immediately from compactness. For 4., Ramsey's Theorem and compactness yields an infinite $\text{bdd}(A)$-indiscernible sequence which has the same type over A. So we may move it to I by an A-automorphism, and I must have been indiscernible over $\text{bdd}(A)$.

To show 5., Ramsey's Theorem and compactness yield a $\bar{b}$-indiscernible sequence J with $\text{tp}(J/\bar{b}_E) = \text{tp}(I/\bar{b}_E)$. We may then take $\bar{b}'$ to be the image of $\bar{b}$ under an $\bar{b}_E$-automorphism moving J to I. ■

3.2 FORKING FOR HYPERIMAGINARIES

The following definition is the same as for real or imaginary elements. However, we have to check that it makes sense for hyperimaginary elements.

DEFINITION 3.2.1 Let $\pi(\bar{x}, b)$ be a partial type over ab. We say that π *divides* over a if there is an a-indiscernible sequence $(b_i : i < \omega)$ of type $\text{tp}(b/a)$ such that $\bigwedge_{i<\omega} \pi(\bar{x}, b_i)$ is inconsistent.
We say that π forks over a if it implies a finite disjunction of formulas, each of which divides over a.

LEMMA 3.2.2 *Dividing and forking is well-defined for types over hyperimaginaries.*

PROOF: Consider hyperimaginaries $a \subseteq b$ and a partial type $\pi(\bar{x}, b)$ over b. Let B be an imaginary set with $ab \in \text{dcl}(B)$, and $\pi_B(\bar{x})$ a partial type over B with $\pi_B(\bar{x}) \equiv \pi(\bar{x}, b)$. If B' is another set with $ab \in \text{dcl}(B)$ and $\pi_{B'}(\bar{x})$ a partial type over B' equivalent to $\pi(\bar{x}, b)$, then π_B divides over a if and only if $\pi_{B'}$ divides over a, so our definition of dividing is independent of the particular parameter set we take π to be defined over.

As the definition of forking uses π only in the form of the formulas it implies, it is independent of choice of domain for π as well. ■

LEMMA 3.2.3 *A partial type π divides over some hyperimaginary a if and only if there is some tuple $\bar{a}$ of real elements, with $a \in \text{dcl}(\bar{a})$, such that π divides over $\bar{a}$.*

PROOF: If π divides over $a = \bar{a}_E$ (where $\bar{a}$ is a real tuple) and this is witnessed by an a-indiscernible sequence I, then by Remark 3.1.22.5 we can choose $\bar{a}$ such that I is $\bar{a}$-indiscernible. Hence π divides over $\bar{a}$; the assertion follows. ■

We can now re-constitute the basic theory of dividing and forking as in Chapter 2, Sections 2–5 and 7, but for hyperimaginary elements and sets.

LEMMA 3.2.4 *A partial type π over b does not divide over a if and only if for any a-indiscernible sequence I with $b \in I$ there is $c \models \pi$ such that I is ac-indiscernible. In particular,* $\mathrm{tp}(c/b)$ *does not divide over a if and only if for any a-indiscernible sequence I with $b \in I$ there is an ab-automorphic image J of I which is ac-indiscernible.*

PROOF: If $\pi(x, b)$ does not divide over a and I is a-indiscernible, we may first assume that b is a tuple of real elements (enlarging b and then I, if necessary). By Remark 3.1.22.5 there is a tuple $\bar{a}$ of real elements such that $a \in \mathrm{dcl}(\bar{a})$ and I is $\bar{a}$-indiscernible. As π does not divide over $\bar{a}$ either, there is c' such that $c' \models \pi(x, b')$ for all $b' \in I$. By Ramsey's Theorem and compactness we find an ac'-indiscernible sequence J realizing $\mathrm{tp}(I/a)$, such that $\models \pi(c', b')$ for all $b' \in J$. If σ is an a-automorphism taking J to I, then $c = \sigma(c')$ will do. The converse is clear from the definition of dividing.

The second assertion follows immediately from the first by an ab-automorphism argument. ■

COROLLARY 3.2.5 *If* $\mathrm{tp}(a/bcd)$ *does not divide over bc and* $\mathrm{tp}(b/cd)$ *does not divide over c, then* $\mathrm{tp}(ab/cd)$ *does not divide over c.*

PROOF: As the proof of Proposition 2.2.9. ■

REMARK 3.2.6 By the proof of Proposition 2.3.9 (3. $\Rightarrow$ 1.) and Remark 2.3.10, every partial type over a hyperimaginary a in a simple theory does not fork over a, and hence has a non-forking extension to any superset of a.

As before, a *Morley sequence* in some type π over a is an infinite a-indiscernible sequence $(b_i : i < \omega)$ of realizations of π, such that $\mathrm{tp}(b_i/a, b_j : j < i)$ does not fork over a for all $i < \omega$. If T is simple, then for every partial type π over a there is a Morley sequence in π over a by Remark 3.2.6.

PROPOSITION 3.2.7 *Let T be simple, a, b be hyperimaginaries, and $\pi(x, b)$ a partial type over b. Then the following are equivalent:*

1. *$\pi(x, b)$ does not fork over a.*

2. *$\pi(x, b)$ does not divide over a.*

3. *$\bigwedge_{i<\omega} \pi(x, b_i)$ is consistent for all Morley sequences $(b_i : i < \omega)$ in* $\mathrm{tp}(b/a)$.

4. *There is a Morley sequence* $(b_i : i < \omega)$ *in* $\mathrm{tp}(b/a)$ *such that* $\bigwedge_{i<\omega} \pi(x, b_i)$ *is consistent.*

PROOF: 1. $\Rightarrow$ 2. and 2. $\Rightarrow$ 3. are obvious, and 3. $\Rightarrow$ 4. follows from the existence of Morley sequences.

4. $\Rightarrow$ 2. Extend the Morley sequence to one indexed by ω^2, and find inductively hyperimaginaries b'_i and real tuples $\bar{b}_i$ for $i < \omega^2$, such that

- $\mathrm{tp}(b'_j : j \leq i/a) = \mathrm{tp}(b_j : j \leq i/a)$ and $\mathrm{tp}(b'_i/a, \bar{b}_j : j < i)$ does not fork over a, and
- $ab'_i \in \mathrm{dcl}(\bar{b}_i)$ and $\mathrm{tp}(\bar{b}_i/a, b'_i, \bar{b}_j : j < i)$ does not fork over ab'_i.

This is possible, since $\mathrm{tp}(b'_i/a, b'_j : j < i)$ does not fork over a and non-forking extensions (or completions) exist. By Ramsey's Theorem, compactness, and after conjugating by an a-automorphism, we may assume that $b'_i = b_i$ for all $i < \omega^2$ and $(\bar{b}_i : i < \omega^2)$ is indiscernible over a. Corollary 3.2.5 implies that $\mathrm{tp}(\bar{b}_i/a, \bar{b}_j : j < i)$ does not divide over a, and $\mathrm{tp}(\bar{b}_j : j > i/a, \bar{b}_i)$ does not divide over a, for all $i < \omega^2$. By indiscernibility for any $i < \omega$

$$D(\bar{b}_i/\{\bar{b}_j : j \geq \omega\}, \varphi, k) = D(\bar{b}_i/\{\bar{b}_j : j > i\}, \varphi, k)$$

for any formula φ and any $k < \omega$, so $(\bar{b}_i : i < \omega)$ is independent over $(\bar{b}_j : j \geq \omega)$ by Proposition 2.3.9, and forms a Morley sequence. Since $\bigwedge_{i<\omega} \pi(x, b_i)$ is consistent, $\pi(x, b_0)$ does not fork over $(\bar{b}_j : j \geq \omega)$ by Theorem 2.4.7. But since $\mathrm{tp}(\bar{b}_j : j \geq \omega/a, \bar{b}_0)$ does not divide over a, any a-indiscernible sequence I containing b_0 has a b_0a-conjugate J which is $\{a, \bar{b}_j : j \geq \omega\}$-indiscernible by Lemma 3.2.4. Now $\pi(x, \bar{b}_0)$ does not divide over $(\bar{b}_j : j \geq \omega)$, so $\bigwedge_{b' \in J} \pi(x, b')$ is consistent, as is $\bigwedge_{b' \in I} \pi(x, b')$. Hence $\pi(x, b_0)$ cannot divide over a.

2. $\Rightarrow$ 1. Suppose $\pi(x, b)$ proves a disjunction $\bigvee_{i<n} \varphi_i(x, c_i)$ of formulas dividing over a. Put $\bar{c} = (c_i : i < n)$, and consider a Morley sequence in $\bar{c}$ over a. Then the co-ordinates form Morley sequences $(c_i^j : j < \omega)$ of type $\mathrm{tp}(c_i/a)$ for all $i < n$. By 4. $\Rightarrow$ 2. the conjunctions $\bigwedge_{j<\omega} \varphi_i(x, c_i^j)$ are all inconsistent, so $\bigvee_{i<n} \varphi_i(x, c_i)$ divides over a, as does $\pi(x, b)$. ■

THEOREM 3.2.8 PROPERTIES OF INDEPENDENCE *Suppose* T *is simple, and let* $A \subseteq B \subseteq C$ *be sets (possibly of hyperimaginaries). Then dividing is equal to forking, and:*

1. EXISTENCE *If* π *is a partial type over* A*, then* π *does not fork over* A*.*
2. EXTENSION *Every partial type over* B *which does not fork over* A *has a completion which does not fork over* A*.*
3. REFLEXIVITY $B \mathop{\smile\!\!\!\!\!\!|}_A B$ *if and only if* $B \subseteq \mathrm{bdd}(A)$*.*

4. MONOTONICITY *If π and π' are partial types with $\pi \vdash \pi'$ and π does not fork over A, then π' does not fork over A.*

5. FINITE CHARACTER *$D \downfall_A B$ if and only if $\bar{d} \downfall_A B$ for every finite tuple $\bar{d} \in D$.*

6. SYMMETRY *$D \downfall_A B$ if and only if $B \downfall_A D$.*

7. TRANSITIVITY *$D \downfall_A C$ if and only if $D \downfall_A B$ and $D \downfall_B C$.*

8. LOCAL CHARACTER *For any partial type $\pi(\bar{x})$ over A there is $A_0 \subseteq A$ with $|A_0| \leq |T| + |\bar{x}|$, such that p does not fork over A_0.*

Furthermore, if a is a hyperimaginary element, then $a \downfall_A B$ if and only if $D(a/A, \varphi, k) = D(a/B, \varphi, k)$ for all formulas φ and all $k < \omega$.

PROOF: Existence, extension, and monotonicity are nothing new; finite character follows straight from the definition. For reflexivity, recall that if $B \subseteq \mathrm{bdd}(A)$ and $(B_i : i < \omega)$ is A-indiscernible, then it is $\mathrm{bdd}(A)$-indiscernible. Conversely, if $B \notin \mathrm{bdd}(A)$ and $B = \bar{b}_E$, then there are unboundedly many E-classes of the same type as B over A. Therefore the partial type $E(X, B) \subseteq \mathrm{tp}(B/B)$ divides over A.

Symmetry follows as before from the existence of Morley sequences and the characterization in Proposition 3.2.7.4; Transitivity is implied by Symmetry and Corollary 3.2.5. Local character follows from the furthermore clause.

Now let a be a hyperimaginary modulo E. We have already seen that if $D(a/A, \varphi, k) = D(a/B, \varphi, k)$ for all formulas φ and all $k < \omega$, then $a \downfall_A B$. So suppose conversely that $a \downfall_A B$. Fix a formula φ and some $k < \omega$, and let $\bar{a}$ be a real tuple (of the appropriate length) such that $\bar{a}_E \models \mathrm{tp}(a/A)$ (i.e. $\bar{a}$ realizes $\mathrm{tp}(a/A)$ considered as a partial type over A) and $D(\bar{a}/A, \varphi, k) = D(a/A, \varphi, k) = n$. Then there is a hyperimaginary $a' \models \mathrm{tp}(a/A)$ with $a' = \bar{a}_E$; by Lemma 3.1.15 we may assume $a = a'$ and $\bar{a} \downfall_A B$. Then $D(\bar{a}/B, \varphi, k) = n$; since $\bar{a} \models \mathrm{tp}(a/B)$ (considered as a partial type over B), we are done. ■

REMARK 3.2.9 It follows that Lemma 2.3.15 holds not only for complete types of real tuples, but also for complete types of hyperimaginaries: If T is simple, then for every complete E-type $p(\bar{x})$ over A and every partial type $\Phi(\bar{x}, \bar{y})$ there is a partial type $\pi(\bar{y})$ such that for any $\bar{a}$ there is a non-forking extension of p to $A\bar{a}$ containing $\exists \bar{z}\, [E(\bar{x}, \bar{z}) \wedge \Phi(\bar{z}, \bar{a})]$ if and only if $\models \pi(\bar{a})$. In other words, if we fix a (real or hyperimaginary) type p over some set A, then the condition

$$\exists x\, [x \models p \wedge x \underset{A}{\downfall} b \wedge \Phi(x, b)]$$

is a type-definable condition on b over A.

DEFINITION 3.2.10 Let A be a set of hyperimaginaries. The group of *strong A-automorphisms* is the subgroup of $\mathrm{Aut}(\mathfrak{C})$ generated by all automorphisms which fix some model $\mathfrak{M}$ with $A \subseteq \mathrm{dcl}(\mathfrak{M})$; we denote it by $\mathrm{Autf}_A(\mathfrak{C})$.
Two hyperimaginaries a and b have the *same Lascar strong type* over a set A, denoted $\mathrm{lstp}(a/A) = \mathrm{lstp}(b/A)$, if there is a strong A-automorphism mapping a to b.

As models consist of real elements, we have to replace the condition $A \subseteq \mathfrak{M}$ by the condition $A \subseteq \mathrm{dcl}(\mathfrak{M})$ in the definition of a strong A-automorphism.

REMARK 3.2.11 If $\mathrm{lstp}(a/A) = \mathrm{lstp}(b/A)$, then there are a sequence $\mathfrak{M}_0, \ldots, \mathfrak{M}_{n-1}$ of models with $A \subseteq \mathrm{dcl}(\mathfrak{M}_i)$ for all $i < n$, and hyperimaginary elements $a = a_0, a_1, \ldots, a_n = b$, such that $\mathrm{tp}(a_i/\mathfrak{M}_i) = \mathrm{tp}(a_{i+1}/\mathfrak{M}_i)$ for all $i < n$. In particular, two hyperimaginaries have the same Lascar strong type over a model if and only if they have the same type. Furthermore, if a and b are classes modulo a type-definable equivalence relation E, then $\mathrm{lstp}(a/A) = \mathrm{lstp}(b/A)$ if and only if there are real tuples $\bar{a}$ and $\bar{b}$ with $\bar{a}_E = a$, $\bar{b}_E = b$, and $\mathrm{lstp}(\bar{a}/A) = \mathrm{lstp}(\bar{b}/A)$.

PROOF: The first assertion follows from the definition of strong automorphism. For the second assertion, choose a tuple $\bar{a}$ with $a = \bar{a}_E$, and let $\bar{b}$ be the image of $\bar{a}$ under a strong automorphism mapping a to b. ■

LEMMA 3.2.12 *Equality of Lascar strong type is a type-definable equivalence relation on hyperimaginaries (modulo the same E).*

PROOF: Let F be the relation $\exists \bar{z}\,[E(\bar{x}, \bar{z}) \wedge \mathrm{lstp}(\bar{z}/A) = \mathrm{lstp}(\bar{y}/A)]$. By Corollary 2.7.9 this is a type-definable equivalence relation over A, and its classes are exactly the Lascar strong types of hyperimaginaries modulo E. ■

LEMMA 3.2.13 *Two hyperimaginaries a and b have the same Lascar strong type over a set A if and only if they have the same type over* $\mathrm{bdd}(A)$.

PROOF: Suppose $\mathrm{tp}(a/\mathrm{bdd}(A)) = \mathrm{tp}(b/\mathrm{bdd}(A))$. Since equality of Lascar strong type over A is a bounded type-definable equivalence relation E over A, we get $a_E = b_E \in \mathrm{bdd}(A)$, whence $\mathrm{lstp}(a/A) = \mathrm{lstp}(b/A)$. Conversely, suppose $\mathrm{lstp}(a/A) = \mathrm{lstp}(b/A)$. As $\mathrm{bdd}(A) \subseteq \mathrm{dcl}(\mathfrak{M})$ for every model $\mathfrak{M}$ with $A \subseteq \mathrm{dcl}(\mathfrak{M})$, any strong A-automorphism must stabilize $\mathrm{bdd}(A)$ pointwise. Therefore $\mathrm{tp}(a/\mathrm{bdd}(A)) = \mathrm{tp}(b/\mathrm{bdd}(A))$. ■

LEMMA 3.2.14 *Suppose* $\mathrm{lstp}(a/A) = \mathrm{lstp}(b/A)$ *and $a \mathop{\smile\!\!\!\!\!\!\;|}_A b$. Then there is a model $\mathfrak{M}$ with $A \subseteq \mathrm{dcl}(\mathfrak{M})$ and $\mathfrak{M} \mathop{\smile\!\!\!\!\!\!\;|}_A ab$, such that* $\mathrm{tp}(a/\mathfrak{M}) = \mathrm{tp}(b/\mathfrak{M})$.

PROOF: By Remark 3.2.11, we may assume that a and b are tuples of real elements. The proof is now exactly the same as the proof of Theorem 2.5.19. (We cannot just quote Theorem 2.5.19 since A may be hyperimaginary.) ■

THEOREM 3.2.15 *Let A, B, C be sets (of hyperimaginaries) with $B \mathop{\smile\!\!\!\!\!\!\mid}_A C$, and suppose b and c are hyperimaginaries with $b \mathop{\smile\!\!\!\!\!\!\mid}_A B$, $c \mathop{\smile\!\!\!\!\!\!\mid}_A C$ and $\mathrm{lstp}(b/A) = \mathrm{lstp}(c/A)$. Then there is $a \mathop{\smile\!\!\!\!\!\!\mid}_A BC$ realizing $\mathrm{tp}(b/B) \cup \mathrm{tp}(c/C) \cup \mathrm{lstp}(b/A)$.*

PROOF: Replacing b by a realization of a non-forking extension of $\mathrm{tp}(b/\mathrm{bdd}(A)B)$ to $\mathrm{bdd}(A)BC$, and c by a realization of a non-forking extension of $\mathrm{tp}(c/\mathrm{bdd}(A)C)$ to $\mathrm{bdd}(A)BCb$, we may assume $Bb \mathop{\smile\!\!\!\!\!\!\mid}_A Cc$. By Lemma 3.2.14 there is a model $\mathfrak{M} \mathop{\smile\!\!\!\!\!\!\mid}_A bc$ with $A \subseteq \mathrm{dcl}(\mathfrak{M})$ and $\mathrm{tp}(b/\mathfrak{M}) = \mathrm{tp}(c/\mathfrak{M})$; we may assume $\mathfrak{M} \mathop{\smile\!\!\!\!\!\!\mid}_{Abc} BC$. Then $B \mathop{\smile\!\!\!\!\!\!\mid}_{\mathfrak{M}} C$, $b \mathop{\smile\!\!\!\!\!\!\mid}_{\mathfrak{M}} B$, and $c \mathop{\smile\!\!\!\!\!\!\mid}_{\mathfrak{M}} C$.

By Remark 3.2.11 we can find real tuples $\bar{b}$ and $\bar{c}$ with $b = \bar{b}_E$, $c = \bar{c}_E$, and $\mathrm{tp}(\bar{b}/\mathfrak{M}) = \mathrm{tp}(\bar{c}/\mathfrak{M})$, where E is some type-definable equivalence relation; we may choose them in addition such that $\bar{b} \mathop{\smile\!\!\!\!\!\!\mid}_{\mathfrak{M}b} B$ and $\bar{c} \mathop{\smile\!\!\!\!\!\!\mid}_{\mathfrak{M}c} C$. Hence $\bar{b} \mathop{\smile\!\!\!\!\!\!\mid}_A \mathfrak{M}B$ and $\bar{c} \mathop{\smile\!\!\!\!\!\!\mid}_A \mathfrak{M}C$. Furthermore we may choose sets B' and C' of real elements with $B \in \mathrm{dcl}(B')$, $C \in \mathrm{dcl}(C')$, $B' \mathop{\smile\!\!\!\!\!\!\mid}_B \mathfrak{M}C\bar{b}\bar{c}$ and $C' \mathop{\smile\!\!\!\!\!\!\mid}_C \mathfrak{M}B'\bar{b}\bar{c}$. Then $B' \mathop{\smile\!\!\!\!\!\!\mid}_{\mathfrak{M}} C'$, $\bar{b} \mathop{\smile\!\!\!\!\!\!\mid}_{\mathfrak{M}} B'$, and $\bar{c} \mathop{\smile\!\!\!\!\!\!\mid}_{\mathfrak{M}} C'$. By the Independence Theorem 2.5.10, we obtain a realization $\bar{a} \models \mathrm{tp}(\bar{b}/\mathfrak{M}B) \cup \mathrm{tp}(\bar{c}/\mathfrak{M}C)$ with $\bar{a} \mathop{\smile\!\!\!\!\!\!\mid}_{\mathfrak{M}} BC$. Then $\bar{a} \mathop{\smile\!\!\!\!\!\!\mid}_A \mathfrak{M}$. Put $a = \bar{a}_E$. Then $a \mathop{\smile\!\!\!\!\!\!\mid}_A \mathfrak{M}BC$, and $a \models \mathrm{tp}(b/B) \cup \mathrm{tp}(c/C) \cup \mathrm{lstp}(b/A)$. ■

3.3 CANONICAL BASES

In this section, T will be a simple theory. We have seen in Section 2.9 that in a stable theory every strong type p has a canonical base $\mathrm{Cb}(p)$, which is characterized by two properties:

1. p does not fork over $\mathrm{Cb}(p)$ and $p\lceil\mathrm{Cb}(p)$ is stationary.

2. Whenever a non-forking extension q of p does not fork over some subset A of its domain and $q\lceil A$ is stationary, then $\mathrm{Cb}(p) \in \mathrm{dcl}(A)$.

In this section, we shall prove an analogous result for Lascar strong types in simple theories. However, their canonical bases will turn out to be hyperimaginary.

We shall start with a general lemma which reconstructs an equivalence relation from a "generically given" one.

LEMMA 3.3.1 *Suppose R is a type-definable reflexive symmetric relation on a partial type π, such that whenever $a, a', a'' \models \pi$ with $a' \mathop{\smile\!\!\!\!\!\!\mid}_a a''$ and $R(a, a')$ and $R(a, a'')$ hold, then $R(a', a'')$ holds. Then the transitive closure E of R equals the 2-step iteration of R, and is thus a type-definable equivalence relation. Furthermore, $E(a, b)$ holds for some $a, b \models \pi$ if and only if there is some $c \models \pi$ with $c \mathop{\smile\!\!\!\!\!\!\mid}_a b$ and $c \mathop{\smile\!\!\!\!\!\!\mid}_b a$, such that $R(a, c)$ and $R(c, b)$ hold.*

PROOF: Enumerate all pairs (φ, k) of a formula φ and some $k < \omega$ as $(\varphi_i, k_i)_{i<\alpha}$. Fix a completion p of π. Since R is type-definable, for every $i < \alpha$ there is n_i such that there are $a \models p$ and $a' \models \pi$ with $R(a, a')$ and $D(a'/a, \varphi_j, k_j) = n_j$ for all $j < i$, and $D(a'/a, \varphi_i, k_i) = n_i$ is maximal possible. Put $R^*(a, a')$ if $R(a, a')$ and $D(a'/a, \varphi_i, k_i) \geq n_i$ for all $i < \alpha$. This is a type-definable condition; by compactness, there are $a \models p$ and $a' \models \pi$ with $R^*(a, a')$.

CLAIM. If $a \models p$, $a' \models \pi$ and a, a' are related by an R-chain, then there is $a'' \models \pi$ with $R^*(a', a'')$.

PROOF OF CLAIM: We us induction on the length n of an R-chain, the case $n = 0$ being trivial. So suppose a and a' are R-related by a chain of length $n+1$. Hence there is $a_0 \models \pi$ which is R-related to a by a chain of length n, with $\models R(a_0, a')$. By inductive hypothesis, there is $a'' \models \pi$ with $\models R^*(a_0, a'')$; we may take $a'' \mathop{\smile\!\!\!\!\!\mid}_{a_0} a'$. Then $\models R(a', a'')$; furthermore

$$D(a''/a', \varphi_i, k_i) \geq D(a''/a_0a', \varphi_i, k_i) = D(a''/a_0, \varphi_i, k_i) \geq n_i$$

for all $i < \alpha$. Thus $R^*(a', a'')$ holds. ■

By maximality, it follows that the n_i for $i < \alpha$, and hence the definition of R^*, does not depend on the choice of p within an R-connected component: if $a \models p$ and a' is related to a by some R-chain, then starting from $\mathrm{tp}(a')$ would have yielded the same sequence $(n_i : i < \alpha)$.

CLAIM. Suppose $a_1, a_2, a_3, a_4, b \models \pi$, $R(a_1, a_2)$, $R(a_2, a_3)$, $R(a_3, a_4)$ and $R^*(a_2, b)$. Assume that a_1 (and hence all elements) are related to a realization of p by some R-chain. Suppose further that $a_2 \mathop{\smile\!\!\!\!\!\mid}_{a_1} a_3$, $a_2 \mathop{\smile\!\!\!\!\!\mid}_{a_3} a_1$, $b \mathop{\smile\!\!\!\!\!\mid}_{a_2} a_1a_3a_4$ and $a_2 \mathop{\smile\!\!\!\!\!\mid}_{a_1a_3} a_4$. Then $R^*(a_j, b)$ for all $j \leq 4$ and $b \mathop{\smile\!\!\!\!\!\mid}_{a_j} a_k$ for all $j, k \leq 4$,

PROOF OF CLAIM: By assumption on R we have $R(a_1, b)$ and $R(a_3, b)$. The independences imply $a_2 \mathop{\smile\!\!\!\!\!\mid}_{a_3} a_4$, whence $R(a_2, a_4)$ and $R(a_4, b)$. For any $j \leq 4$ and any $i < \alpha$ we have by maximality of n_i

$$n_i \geq D(b/a_j, \varphi_i, k_i) \geq D(b/a_1a_2a_3a_4, \varphi_i, k_i) = D_i(b/a_2) = n_i;$$

whence $R^*(a_j, b)$ and $b \mathop{\smile\!\!\!\!\!\mid}_{a_j} a_k$ for all $j, k \leq 4$. ■

We now prove by induction on n that if a and a' are related by an R-chain of length n, and are related to some realization of p by an R-chain, then there is $a'' \models p$ with $R^*(a, a'')$, $R^*(a', a'')$, $a'' \mathop{\smile\!\!\!\!\!\mid}_a a'$ and $a'' \mathop{\smile\!\!\!\!\!\mid}_{a'} a$. This is clearly trivial for a chain of length 0, as $R^*(a, x)$ is consistent by the first claim.

So suppose that a_1 and a_4 are related by an R-chain of length $n + 1$, and are related to some realization of p by an R-chain. Then there is $a_3 \models \pi$ such that $R(a_3, a_4)$ and a_1 and a_3 are related by a chain of length n. By inductive hypothesis, there is $a_2 \models \pi$ with $a_2 \mathop{\smile\!\!\!\!\!\mid}_{a_1} a_3$, $a_2 \mathop{\smile\!\!\!\!\!\mid}_{a_3} a_1$, $R^*(a_1, a_2)$, and $R^*(a_3, a_2)$. We may assume that $a_2 \mathop{\smile\!\!\!\!\!\!\,|}_{a_1a_3} a_4$. By the first claim, we may choose $b \models \pi$ such that $R^*(a_2, b)$ and $b \mathop{\smile\!\!\!\!\!\!\,|}_{a_2} a_1a_3a_4$. By the second claim, $R^*(a_1, b), R^*(a_4, b), b \mathop{\smile\!\!\!\!\!\!\,|}_{a_1} a_4$ and $b \mathop{\smile\!\!\!\!\!\!\,|}_{a_4} a_1$. This finishes the induction.

As we might have started with an arbitrary completion p of π, the assertion follows. ■

DEFINITION 3.3.2 Let p and q be Lascar strong types. We write $p \sim_0 q$ if p and q have a common non-forking extension over the union of their domains. Define $\sim$ to be the transitive closure of $\sim_0$.
If $p = p_A(x)$ is a (Lascar strong) type over a boundedly closed set A, define a relation $\approx_0$ on $\operatorname{tp}(A)$ by $A \approx_0 A'$ if and only if $p_A \sim_0 p_{A'}$, and let $\approx$ be the transitive closure of $\approx_0$.

REMARK 3.3.3 By Lemma 3.2.13, a Lascar strong type might just be considered a type over a boundedly closed set (e.g. a model).

Since $\sim_0$ is reflexive and symmetric, $\sim$ is an equivalence relation. Similarly, $\approx_0$ is reflexive and symmetyric, and $\approx$ is an equivalence relation. Clearly $\approx$ depends not only on $\operatorname{tp}(A)$, but also on the choice of p_A. Until further notice, we therefore fix a boundedly closed set A and a type p_A over A, and consider the corresponding relation $\approx$.

LEMMA 3.3.4 *$A \approx A'$ if and only if $p_A \sim p_{A'}$.*

PROOF: Clearly $A \approx A'$ implies $p_A \sim p_{A'}$. For the converse, it is enough to show that if $p_A \sim_0 q$ and $q \sim_0 q'$, then there is $A' \models \operatorname{tp}(A)$ such that $p_A \sim_0 p_{A'}$ and $p_{A'} \sim_0 q$; this will inductively prove the assertion. So let a realize a common non-forking extension of $p_A \cup q$, and a' realize a common non-forking extension of $q \cup q'$. Suppose B is the domain of q and B' is the domain of q', both boundedly closed.

Since $\operatorname{tp}(a/B) = \operatorname{tp}(a'/B) = q$, there is a B-automorphism σ with $\sigma(a) = a'$; we may choose it such that $\sigma(A) \mathop{\smile\!\!\!\!\!\!\,|}_{Ba'} AB'$. Put $A' = \sigma(A)$. Now $a \mathop{\smile\!\!\!\!\!\!\,|}_B A$ implies $a' \mathop{\smile\!\!\!\!\!\!\,|}_B A'$, so $A' \mathop{\smile\!\!\!\!\!\!\,|}_B AB'a'$. If q_0 is a common non-forking extension of $q \cup p_{A'}$ to BA' and q_1 is a common non-forking extension of $q \cup q'$ to BB', then by the Independence Theorem $q_0 \cup q_1$ has a common non-forking extension q_2 to $A'BB'$, which forks neither over A' nor over B'. Hence $p_{A'} \sim_0 q'$; similarly $p_A \sim_0 p_{A'}$ and we are done. ■

PROPOSITION 3.3.5 *The equivalence relation $\approx$ is type-definable on $\operatorname{tp}(A)$; in fact it is the two-step iterate of $\approx_0$. Furthermore, $A \approx A'$ for any $A' \models \operatorname{tp}(A)$ if and only if there is $A'' \models \operatorname{tp}(A)$ with $A'' \mathop{\smile\!\!\!\!\!\!\,|}_A A'$, $A'' \mathop{\smile\!\!\!\!\!\!\,|}_{A'} A$, $A'' \approx_0 A$ and $A'' \approx_0 A'$.*

PROOF: We have $A \approx_0 A'$ if and only if $D(p_A \cup p_{A'}, \varphi, k) \geq D(p_A, \varphi, k)$ for all formulas φ and all $k < \omega$, so $\approx_0$ is type-definable. Furthermore, if

$A' \mathop{\smile\!\!\!\!\!\!\!\;|}_A A''$ and $A \approx_0 A'$ and $A \approx_0 A''$, choose common non-forking extensions q of $p_A \cup p_{A'}$ and q' of $p_A \cup p_{A''}$. By the Independence Theorem $q \cup q'$ is consistent and does not fork over A, so there is a common non-forking extension q'' over $AA'A''$. This shows $A' \approx_0 A''$. Now apply Lemma 3.3.1. ■

By Remark 3.1.1 we can extend $\approx$ to a type-definable equivalence relation on the whole universe. Hence we can consider $A_\approx$ as a hyperimaginary element.

THEOREM 3.3.6 *1. p_A does not fork over $A_\approx$.*

2. *$p_A\lceil A_\approx$ is a Lascar strong type.*

3. *If q is a Lascar strong type over B such that p_A and q have a common non-forking extension, then $A_\approx \subseteq \mathrm{dcl}(B)$.*

4. *If $q \in S(B)$ and p_A and q have a common non-forking extension, then $A_\approx \subseteq \mathrm{bdd}(B)$.*

PROOF:

1. Choose $A' \models \mathrm{tp}(A/A_\approx)$ with $A \mathop{\smile\!\!\!\!\!\!\!\;|}_{A_\approx} A'$. Then $A \approx A'$, so there is $A'' \models \mathrm{tp}(A)$ with $A'' \mathop{\smile\!\!\!\!\!\!\!\;|}_{A'} A$ and $A'' \approx_0 A$. Let a realise the common non-forking extension of $p_A \cup p_{A''}$. Then $a \mathop{\smile\!\!\!\!\!\!\!\;|}_{A''} A$; since $A \mathop{\smile\!\!\!\!\!\!\!\;|}_{A_\approx} A''$, we get $a \mathop{\smile\!\!\!\!\!\!\!\;|}_{A_\approx} A$ by transitivity, and p_A does not fork over $A_\approx$.

2. Let $q_1 \in S(B_1)$ and $q_2 \in S(B_2)$ be non-forking extensions of $p_A\lceil A_\approx$, with $B_1 \mathop{\smile\!\!\!\!\!\!\!\;|}_{A_\approx} B_2$.

   CLAIM. If $A \mathop{\smile\!\!\!\!\!\!\!\;|}_{A_\approx} B$ and q is a non-forking extension of $p_A\lceil A_\approx$ to B, then $p_A \sim_0 q$.

   PROOF OF CLAIM: If $a \models p_A$ and $a' \models q$ with $a' \mathop{\smile\!\!\!\!\!\!\!\;|}_B A$, there is an $A_\approx$-automorphism σ mapping a to a'; put $A' = \sigma(A)$. We may assume that $A' \mathop{\smile\!\!\!\!\!\!\!\;|}_{A_\approx a'} BA$. Now $a \mathop{\smile\!\!\!\!\!\!\!\;|}_{A_\approx} A$ implies $a' \mathop{\smile\!\!\!\!\!\!\!\;|}_{A_\approx} A'$, whence $A' \mathop{\smile\!\!\!\!\!\!\!\;|}_{A_\approx} BAa'$ and $A' \mathop{\smile\!\!\!\!\!\!\!\;|}_B a'$; similarly $a' \mathop{\smile\!\!\!\!\!\!\!\;|}_{A_\approx} B$ yields $a' \mathop{\smile\!\!\!\!\!\!\!\;|}_{A'} B$. Hence $\mathrm{tp}(a'/A'B)$ is a common non-forking extension of $p_{A'}$ and q. Moreover $A \mathop{\smile\!\!\!\!\!\!\!\;|}_{Ba'} A'$; whence $A \mathop{\smile\!\!\!\!\!\!\!\;|}_B A'a'$ and finally $A \mathop{\smile\!\!\!\!\!\!\!\;|}_{A_\approx} BA'a'$ by transitivity.

   Since $A \approx A'$, there is $A'' \models \mathrm{tp}(A)$ with $A'' \mathop{\smile\!\!\!\!\!\!\!\;|}_A A'$, $A'' \mathop{\smile\!\!\!\!\!\!\!\;|}_{A'} A$, $A'' \approx_0 A$ and $A'' \approx_0 A'$. Then $A \mathop{\smile\!\!\!\!\!\!\!\;|}_{A_\approx} A'A''$ and $A \mathop{\smile\!\!\!\!\!\!\!\;|}_{A''} A'$. So a common non-forking extension of $p_A \cup p_{A''}$ and a common non-forking extension of $p_{A'} \cup p_{A''}$ combine by the Independence Theorem 3.2.15 to a common non-forking extension of $p_A \cup p_{A'} \cup p_{A''}$. In particular p_A and p'_A have a common non-forking extension; since $A \mathop{\smile\!\!\!\!\!\!\!\;|}_{A'} B$, another application of the

Independence Theorem yields a common non-forking extension of p_A and q (and, in fact, $p_{A'}$). ■

Now let $A' \models \text{tp}(A/A_\approx)$ with $A' \mathop{\smile\!\!\!\!\!\!\perp}_{A_\approx} B_1B_2$. Then there are common non-forking extensions p_1 of $p_{A'} \cup q_1$ and p_2 of $p_{A'} \cup q_2$; since $B_2 \mathop{\smile\!\!\!\!\!\!\perp}_{A_\approx B_1} A'$ implies $B_2 \mathop{\smile\!\!\!\!\!\!\perp}_{A_\approx} B_1A'$ and hence $B_1 \mathop{\smile\!\!\!\!\!\!\perp}_{A'} B_1$, the Independence Theorem yields a common non-forking extension of $p_1 \cup p_2$, whence of $q_1 \cup q_2$.

It follows that $p_A \lceil_{A_\approx}$ has a unique extension to $\text{bdd}(A_\approx)$, so it must be a Lascar strong type.

3. Suppose p_A and q have a common non-forking extension, where q is a Lascar strong type over B. Let σ be an automorphism fixing B. Then $p_A \sim_0 q = \sigma(q) \sim_0 p_{\sigma(A)}$, so $A \approx \sigma(A)$ and σ fixes $A_\approx$. Therefore $A_\approx \subseteq \text{dcl}(B)$.

4. Suppose p_A and $q \in S(B)$ have a common non-forking extension. Then there is an extension q' of q to $\text{bdd}(B)$ such that p_A and q' have a common non-forking extension; since q' is a Lascar strong type, part 3. yields $A_\approx \subseteq \text{bdd}(B)$. ■

It follows that $A_\approx$ is a canonical base for the $\sim$-class of p_A.

DEFINITION 3.3.7 If p_A is a Lascar strong type over A and $\approx$ is defined as in Definition 3.3.2, then the *canonical base* $\text{Cb}(p_A)$ of p_A is $\text{dcl}(A_\approx)$.

REMARK 3.3.8 In a stable theory, the two definitions of canonical base agree by Theorems 2.9.15 and 3.3.6.

DEFINITION 3.3.9 Let A be a set of hyperimaginaries, and π a partial E-type over A. Then π is *finitely satisfiable* in A if for every formula $\varphi(\bar{x}) \in \pi$ there is some hyperimaginary $\bar{a}_E$ in A (where $\bar{a}$ is a real tuple), such that $\models \varphi(\bar{a}')$ for all $\bar{a}'$ with $E(\bar{a}, \bar{a}')$.

In the case when A consists of real elements, this agrees with the usual definition. Note though that in general a formula $\varphi \in \pi$ need not be E-invariant.

REMARK 3.3.10 A partial E-type π finitely satisfiable in A does not fork over A.

PROOF: If $\varphi(\bar{x}, \bar{b}) \in \pi$ and $(\bar{b}_i : i < \omega)$ is an A-indiscernible sequence of $\text{tp}(\bar{b}/A)$, then by finite satisfiability there is $\bar{a}_E \in A$ such that $\varphi(\bar{x}, \bar{b})$ is satisfied by all $\bar{a}'$ with $E(\bar{a}', \bar{a})$. Hence $\varphi(\bar{x}, \bar{b}_i)$ is satisfied by all these $\bar{a}'$, since $\text{tp}(\bar{b}/\bar{a}_E) = \text{tp}(\bar{b}_i/\bar{a}_E)$ for all $i < \omega$. ■

LEMMA 3.3.11 *Suppose* $\text{tp}(\bar{a}_E/A)$ *is finitely satisfiable in* A, *where* $\bar{a}$ *is a real tuple. Then* $\text{tp}(\bar{a}_E/A)$ *is a Lascar strong type.*

PROOF: Let F be a type-definable equivalence relation with boundedly many classes over A which is coarser than E, and consider a symmetric formula $\varphi(\bar{x}, \bar{y})$ in F. As E and F are equivalence relations and F is coarser than E, there are symmetric formulas $\psi(\bar{x}, \bar{y})$ in E and $\varphi'(\bar{x}, \bar{y})$ in F such that

$$\psi(\bar{x}, \bar{x}') \wedge \varphi'(\bar{x}', \bar{y}') \wedge \psi(\bar{y}', \bar{y}) \vdash \varphi(\bar{x}, \bar{y}).$$

CLAIM. There is $n < \omega$ such that for any n tuples $\{\bar{a}_0, \dots, \bar{a}_{n-1}\}$ there is $i < j < n$ and $\bar{a}', \bar{a}''$ with $E(\bar{a}_i, \bar{a}')$, $E(\bar{a}_j, \bar{a}'')$ and $\models \varphi'(\bar{a}', \bar{a}'')$.

PROOF OF CLAIM: Let $\psi'' \in E$ and $\varphi'' \in F$ be symmetric formulas such that $\psi''(\bar{x}, \bar{x}') \wedge \varphi''(\bar{x}', \bar{y}') \wedge \psi''(\bar{y}', \bar{y}) \vdash \varphi'(\bar{x}, \bar{y})$. By boundedness of F there is some $n < \omega$ such that for any $\bar{a}_0, \dots, \bar{a}_{n-1}$ there is $i < j < n$ with $\models \varphi''(\bar{a}_i, \bar{a}_j)$. So if $E(\bar{a}', \bar{a}_i)$ and $E(\bar{a}_j, \bar{a}'')$ we get $\models \psi''(\bar{a}_i, \bar{a}') \wedge \psi''(\bar{a}_j, \bar{a}'')$, whence $\models \varphi'(\bar{a}', \bar{a}'')$. ■

By finite satisfiability, A cannot be empty. Let $\bar{a}_E^0, \dots, \bar{a}_E^{k-1}$ be a maximal sequence of elements in A such that for all $i < j < k$ and $\bar{a}', \bar{a}''$ with $E(\bar{a}^i, \bar{a}')$ and $E(\bar{a}^j, \bar{a}'')$ we have $\models \neg\varphi'(\bar{a}', \bar{a}'')$. This exists by the previous claim. Let ψ' be a symmetric formula in E such that $\psi'(\bar{x}, \bar{x}') \wedge \psi'(\bar{x}', \bar{x}'') \vdash \psi(\bar{x}, \bar{x}'')$.

CLAIM. There is $i < k$ and a symmetric formula $\psi^*(\bar{x}, \bar{y})$ in E with $\psi^* \vdash \psi$, such that for all $\bar{a}'$ with $\models \psi^*(\bar{a}^i, \bar{a}')$ and all $\bar{a}''$ with $\models \psi^*(\bar{a}, \bar{a}'')$ we have $\models \varphi(\bar{a}', \bar{a}'')$.

PROOF OF CLAIM: Suppose not. Then the partial type

$$\bigwedge_{i<k} \exists \bar{x}'\bar{x}'' \, [E(\bar{a}^i, \bar{x}') \wedge E(\bar{x}, \bar{x}'') \wedge \neg\varphi(\bar{x}', \bar{x}'')]$$

is implied by $\mathrm{tp}(\bar{a}_E/A)$. By finite satisfiability, there is $\bar{a}_E^k \in A$ such that the formula

$$\bigwedge_{i<k} \exists \bar{x}'\bar{x}'' \, [\psi'(\bar{a}^i, \bar{x}') \wedge \psi'(\bar{x}, \bar{x}'') \wedge \neg\varphi(\bar{x}', \bar{x}'')]$$

is satisfied by all $\bar{a}'$ with $E(\bar{a}^k, \bar{a}')$. Let $(\bar{a}_i'\bar{a}_i'' : i < k)$ witness the existential quantifier. By choice of $(\bar{a}^i : i < k)$ there must be some $i < k$ and $\bar{a}', \bar{a}''$ with $E(\bar{a}^i, \bar{a}')$ and $E(\bar{a}^k, \bar{a}'')$ such that $\models \varphi'(\bar{a}', \bar{a}'')$. As $\models \psi'(\bar{a}^i, \bar{a}_i') \wedge \psi'(\bar{a}^k, \bar{a}_i'')$, we get $\models \psi(\bar{a}_i', \bar{a}') \wedge \psi(\bar{a}'', \bar{a}_i'')$, whence $\models \varphi(\bar{a}_i', \bar{a}_i'')$, a contradiction. ■

It follows that $\mathrm{tp}(\bar{a}_E/A) \vdash \varphi(\bar{x}, \bar{a}^i)$ for some $i < k$. Thus, for every formula $\varphi(\bar{x}, \bar{y})$ in F we find $\bar{a}_\varphi$ such that $\mathrm{tp}(\bar{a}_E/A) \vdash \varphi(\bar{x}, \bar{a}_\varphi)$. It follows by compactness that any A-automorphism must fix the F-class of $\bar{a}_E$; since equality of Lascar strong type over A is a type-definable bounded equivalence relation, $\mathrm{tp}(\bar{a}_E/A)$ is a Lascar strong type. ■

THEOREM 3.3.12 *Let p be a Lascar strong type over A, and $(a_i : i \leq \omega)$ a Morley sequence in p. Then $\mathrm{tp}(a_\omega/a_i : i < \omega)$ is a Lascar strong type, and*

$p \sim_0 \mathrm{tp}(a_\omega/a_i : i < \omega)$. *In fact, the common non-forking extension is realized by* a_ω.

PROOF: As $\mathrm{tp}(a_\omega/a_i : i < \omega)$ is finitely satisfiable in $(a_i : i < \omega)$ by indiscernibility, it is a Lascar strong type by Lemma 3.3.11. As $(a_i : i \leq \omega)$ is a Morley sequence over A, we have $a_\omega \mathop{\smile\!\!\!\!\!\!|}_A (a_i : i < \omega)$. On the other hand, $\mathrm{tp}(a_\omega/A, a_i : i < \omega)$ is finitely satisfiable in $(a_i : i < \omega)$, so $a_\omega \mathop{\smile\!\!\!\!\!\!|}_{(a_i:i<\omega)} A$. ■

COROLLARY 3.3.13 *Let p be a Lascar strong type, and $(a_i : i < \omega)$ a Morley sequence in p. Then* $\mathrm{Cb}(p) \in \mathrm{dcl}(a_i : i < \omega)$. ■

REMARK 3.3.14 Let $\bar{a}$ be a countable sequence, and E a type-definable equivalence relation on countable tuples. Put $A = \mathrm{Cb}(\bar{a}/\bar{a}_E)$; clearly $A \subseteq \mathrm{bdd}(\bar{a}_E)$. Conversely, as $\bar{a} \mathop{\smile\!\!\!\!\!\!|}_A \bar{a}_E$ and $\bar{a}_E \in \mathrm{dcl}(\bar{a})$, we get $\bar{a}_E \in \mathrm{bdd}(A)$. But then $\mathrm{tp}(\bar{a}/A) \vdash \mathrm{tp}(\bar{a}/\bar{a}_E)$, whence $\bar{a}_E \in \mathrm{dcl}(A)$.

3.4 INTERNALITY AND ANALYSABILITY

Again, we shall work in a simple theory. We shall consider families $\Sigma, \Pi, \ldots$ of partial types, and study possible interrelations (dependencies) between these families. Now if E is a type-definable equivalence relation and π is a partial E-type, we may interpret π in two ways: as a partial type of real elements (which just happens to be invariant under E), or as a partial type of hyperimaginaries modulo E (and it is not obvious from π which interpretation is meant). We shall usually mean the latter, so a partial E-type will implicitly carry the "sort" E (although it is not really a sort) of hyperimaginary it is supposed to be realized by.

DEFINITION 3.4.1 A family Σ of partial types is *A-invariant* if it is invariant under all A-automorphisms.

For these families of partial types our usual restrictions on the size (to be less than κ) will not apply. In fact, if Σ is an A-invariant family containing a partial type π not over A, then π will have at least κ distinct conjugates over A.

DEFINITION 3.4.2 Let Σ be an A-invariant family of partial types. A partial type π over A is *internal* in Σ, or *Σ-internal*, if for every realization a of π there is $B \mathop{\smile\!\!\!\!\!\!|}_A a$, types $\bar{\sigma}$ from Σ based on B, and realizations $\bar{c}$ of $\bar{\sigma}$, such that $a \in \mathrm{dcl}(B\bar{c})$.
We say that π is *finitely generated over* Σ if there is some $B \supseteq A$ such that for all realizations a of π there are types $\bar{\sigma}$ from Σ based on B, and realizations $\bar{c}$ of $\bar{\sigma}$, such that $a \in \mathrm{dcl}(B\bar{c})$.
Finally, π is *almost* Σ-internal, or *almost* finitely generated over Σ, if we replace $\mathrm{dcl}(B\bar{c})$ by $\mathrm{bdd}(B\bar{c})$ in the above definitions.

REMARK 3.4.3 If a is an imaginary element, compactness implies that we can replace the bounded by the algebraic closure, and need only a finite tuple $\bar{\sigma}$ of types in Σ and a finite tuple $\bar{c}$.

In a stable theory, internality and finite generation agree. This is longer be the case in a simple structure:

EXAMPLE 3.4.4 Consider a structure consisting of three disjoint sets P, Q and R, together with a surjection $\pi : Q \to P$ whose fibres have size 2 (so Q is a 2-cover of P), and the random bipartite graph on $Q \times R$ added generically. By Proposition 6.3.15 this is supersimple; any $q \in Q$ is definable over $\pi(q)$ and some (necessarily independent) $r \in R$ such that there is exactly one edge between r and $\pi^{-1}(\pi(q))$. But we cannot add some fixed set A of parameters such that $q \in \mathrm{dcl}(A, \pi(q))$ for all $q \in Q$. So Q is P-internal, but not finitely generated over P.

LEMMA 3.4.5 *If Σ is A-invariant, π is a partial type over A and $B \supseteq A$, then π is Σ-internal over A if and only if it is Σ-internal over B. The same assertion holds for almost internality.*

PROOF: Suppose π is Σ-internal over A, and consider $a \models \pi$. There is $B' \mathop{\smile\!\!\!\!\!\!\mid}_A a$ and $\bar{c}$ realizing types in Σ, with $a \in \mathrm{dcl}(B', \bar{c})$. By A-invariance, we may assume $B', \bar{c} \mathop{\smile\!\!\!\!\!\!\mid}_{Aa} B$, whence $B' \mathop{\smile\!\!\!\!\!\!\mid}_A Ba$ and hence $B' \mathop{\smile\!\!\!\!\!\!\mid}_B a$. Hence π is Σ-internal over B.

Conversely, suppose π is Σ-internal over B, and consider $a \models \pi$. There is some $a' \models \mathrm{tp}(a/A)$ with $a' \mathop{\smile\!\!\!\!\!\!\mid}_A B$, some $B' \mathop{\smile\!\!\!\!\!\!\mid}_B a'$ and realizations $\bar{c}$ of types on Σ over B', such that $a' \in \mathrm{dcl}(B'\bar{c})$. Note that $a' \mathop{\smile\!\!\!\!\!\!\mid}_A BB'$ by transitivity; conjugating a' to a yields the result by A-invariance of Σ.

The case of almost internality is similar. ■

REMARK 3.4.6 It is obvious that the above Lemma holds also for (almost) finite generation.

COROLLARY 3.4.7 *Suppose $A \subseteq B$ and $\mathrm{tp}(a/B)$ does not fork over A. Then $\mathrm{tp}(a/B)$ is Σ-internal if and only if $\mathrm{tp}(a/A)$ is Σ-internal.*

PROOF: Right to left follows from Lemma 3.4.5. Conversely, a is Σ-internal over A as in the proof of Lemma 3.4.5. Since any $a' \models \mathrm{tp}(a/A)$ is A-conjugate to a, it is also Σ-internal.

LEMMA 3.4.8 *Suppose Π and Σ are A-invariant families of partial types such that every type in Π is Σ-internal, and π is a Π-internal partial type over A. Then π is Σ-internal. The same assertion holds with* almost internal, *or* (almost) finitely generated *instead of* internal.

PROOF: Given $a \models \pi$, find a set $B \perp_A a$, a sequence $(\pi_i : i < \alpha)$ of types in Π over B, and realizations $c_i \models \pi_i$ for $i < \alpha$ with $a \in \mathrm{dcl}(B, c_i : i < \alpha)$. For each $i < \alpha$ choose a set $C_i \perp_{AB} c_i$, a sequence $\bar\sigma_i$ of types in Σ over C_i, and realizations $\bar d_i \models \bar\sigma_i$ with $c_i \in \mathrm{dcl}(C_i, \bar d_i)$. By A-invariance of Σ, we may choose the $(C_i : i < \alpha)$ such that $C_i \perp_{ABc_i} (a, C_j : j < i)$, whence $C_i \perp_{AB} (a, C_j : j < i)$ and $a \perp_A (B, C_i : i < \alpha)$. Then $a \in \mathrm{dcl}(B, C_i, \bar d_i : i < \alpha)$; as the realization a of π was arbitrary, π is Σ-internal. The other cases are similar. ■

Although internality and finite generation need not be the same in a simple theory, they cannot differ by much:

PROPOSITION 3.4.9 *Suppose π is a partial type over A which is almost Σ-internal for some family Σ of types over A. Then π is almost finitely generated over Σ.*

PROOF: As there are only boundedly many Lascar strong types over A extending π, we may assume that $A = \mathrm{bdd}(A)$ and $\pi = p$ is a complete type. Suppose now that $a \models p$, $B \supseteq A$ with $B \perp_A a$, $\bar\sigma$ is a tuple of types in Σ, $\bar c \models \bar\sigma$ and $a \in \mathrm{bdd}(B\bar c)$. Let $(a_i, \bar c_i : i < \omega)$ be a Morley sequence in $\mathrm{lstp}(a, \bar c/B)$. We may suppose that $(a, \bar c)$ extends this sequence. Since $a \perp_A B$, we get $a \perp_A (a_i : i < \omega)$; as $a, \bar c \perp_{(a_i, \bar a_i : i<\omega)} B$, we obtain $a \perp_{(\bar c, a_i, \bar c_i : i<\omega)} B$, and $a \in \mathrm{bdd}(B, \bar c)$ implies $a \in \mathrm{bdd}(\bar c, a_i, \bar c_i : i < \omega)$.

For any $a' \models p$ there are $(\bar c', a_i', \bar c_i' : i < \omega)$ such that

$$\mathrm{tp}(a, \bar c, a_i, \bar c_i : i < \omega/A) = \mathrm{tp}(a', \bar c', a_i', \bar c_i' : i < \omega/A).$$

If $a' \perp_A B$, then since $(a_i' : i < \omega)$ forms an independent sequence over A independent of a', by the Independence Theorem we may assume that $\mathrm{tp}(a_i'/B) = \mathrm{tp}(a/B)$. In particular there are realizations $\bar d_i$ of $\bar\sigma$ with $a_i' \in \mathrm{bdd}(B, \bar d_i)$. It follows that $a' \in \mathrm{bdd}(B, \bar c', \bar d_i, \bar c_i' : i < \omega)$.

Finally, if $a' \not\perp_A B$, we may choose those $(a_i' : i < \omega)$ independent of B over A, since they are independent of a' over A. By the previous paragraph, every a_i' is bounded over B and realizations of types in Σ, and so is a'. ■

PROBLEM 3.4.10 Does Proposition 3.4.9 also hold if Σ is merely an A-invariant family of types?

DEFINITION 3.4.11 Let π be a partial type over A and Σ an A-invariant family of partial types. We say that π is *analysable* in Σ, or *Σ-analysable*, if for any $a \models \pi$ there are $(a_i : i < \alpha) \in \mathrm{dcl}(A, a)$ such that $\mathrm{tp}(a_i/A, a_j : j < i)$ is Σ-internal for all $i < \alpha$, and $a \in \mathrm{bdd}(A, a_i : i < \alpha)$.

A complete type $p \in S(A)$ is *foreign* to Σ if for all $a \models p$, $B \perp_A a$, and realizations $\bar{c}$ of possibly forking extensions of types in Σ over B, we always have $a \mathop{\smile\!\!\!\!\!\mid}_{AB} \bar{c}$.

The connection between foreignness and internality is given by the following:

PROPOSITION 3.4.12 *If Σ is an A-invariant family of partial types and $\mathrm{tp}(a/A)$ is not foreign to Σ, then there is $a_0 \in \mathrm{dcl}(Aa) - \mathrm{bdd}(A)$ such that $\mathrm{tp}(a_0/A)$ is Σ-internal.*

PROOF: We take $B \mathop{\smile\!\!\!\!\!\mid}_A a$ and $\bar{c}$ realizing types in Σ with $a \not\mathop{\smile\!\!\!\!\!\mid}_{AB} \bar{c}$. Now put $a_1 = \mathrm{Cb}(B\bar{c}/Aa)$ and let a_0 be the set of Aa-conjugates of a_1. If $(B_i, \bar{c}_i : i < \omega)$ is a Morley sequence in $\mathrm{lstp}(B\bar{c}/Aa)$, then $a_1 \in \mathrm{dcl}(B_i\bar{c}_i : i < \omega)$. By A-invariance of Σ, every c_i realizes a type in Σ over B_i; as $a \mathop{\smile\!\!\!\!\!\mid}_A B$, we get $a \mathop{\smile\!\!\!\!\!\mid}_A (B_i : i < \omega)$. Since $a_1 \in \mathrm{bdd}(Aa)$, we get $a_1 \mathop{\smile\!\!\!\!\!\mid}_A (B_i : i < \omega)$, and $\mathrm{tp}(a_1/A)$ is Σ-internal. Furthermore $a_1 \notin \mathrm{bdd}(A)$, as otherwise $a \mathop{\smile\!\!\!\!\!\mid}_A B\bar{c}$.

As a_0 is a set of boundedly many hyperimaginaries, $a_1 \in \mathrm{bdd}(a_0)$. This shows $a_0 \notin \mathrm{bdd}(A)$; by definition $a_0 \in \mathrm{dcl}(Aa)$. Finally, as any of the Aa-conjugates of a_1 in a_0 is Σ-internal, so is a_0 itself. ■

LEMMA 3.4.13 *Let p be a type over A, and Σ an A-invariant family of partial types. Then p is foreign to Σ if and only if it is foreign to all Σ-analysable types.*

PROOF: We only have to show left to right. So suppose $a \models p$, and consider a superset B of A with $B \mathop{\smile\!\!\!\!\!\mid}_A a$ and some b such that $\mathrm{tp}(b/B)$ is Σ-analysable. Let $(b_i : i < \alpha)$ be a Σ-analysis of b over B. Then $b \in \mathrm{bdd}(B, b_i : i < \alpha)$, so if $a \not\mathop{\smile\!\!\!\!\!\mid}_B b$, then there is a minimal $i < \alpha$ with $a \not\mathop{\smile\!\!\!\!\!\mid}_{B\cup\{b_j:j<i\}} b_i$. Note that $a \mathop{\smile\!\!\!\!\!\mid}_A B \cup \{b_j : j < i\}$.

Since $\mathrm{tp}(b_i/B, b_j : j < i)$ is Σ-internal, there is $C \mathop{\smile\!\!\!\!\!\mid}_{B\cup\{b_j:j<i\}} b_i$ and a tuple $\bar{c}$ of realizations of types in Σ with $b_i \in \mathrm{dcl}(B, \bar{c}, b_j : j < i)$. By A-invariance, we can choose this with $C \mathop{\smile\!\!\!\!\!\mid}_{B\cup\{b_j:j\le i\}} a$, whence $C \mathop{\smile\!\!\!\!\!\mid}_{B\cup\{b_j:j<i\}} ab_i$ and $a \mathop{\smile\!\!\!\!\!\mid}_A BC \cup \{b_j : j < i\}$. Moreover, $a \not\mathop{\smile\!\!\!\!\!\mid}_{BC\cup\{b_j:j<i\}} \bar{c}$, so $\mathrm{tp}(a/A)$ is not foreign to Σ. ■

LEMMA 3.4.14 *Let Σ be an A-invariant family of partial types. Then $\mathrm{lstp}(a/A)$ is Σ-analysable if and only if for every $B \supseteq A$ either $\mathrm{lstp}(a/B)$ is bounded, or not foreign to Σ.*

PROOF: Suppose $\mathrm{lstp}(a/A)$ is Σ-analysable, and $B \supseteq A$. Let $(a_i : i < \alpha)$ be a Σ-analysis of a over A. If $a_i \in \mathrm{bdd}(B)$ for all $i \le \alpha$, then $a \in \mathrm{bdd}(B)$ and $\mathrm{tp}(a/B)$ is bounded. Otherwise there is a minimal $i \le \alpha$ such that $a_i \notin \mathrm{bdd}(B)$. Hence $\{a_j : j < i\} \subseteq \mathrm{bdd}(B)$; since $\mathrm{tp}(a_i/A, a_j : j < i)$ is

Σ-internal, so is $\operatorname{tp}(a_i/B)$. But $a_i \in \operatorname{bdd}(Aa)$ implies $a_i \not\perp_B a$, so $\operatorname{tp}(a/B)$ is not foreign to $\operatorname{tp}(a_i/B)$; by Lemma 3.4.13 it is not foreign to Σ either.

Conversely, suppose $\operatorname{tp}(a/B)$ is either bounded or not foreign to Σ, for all $B \supseteq A$. Put $\lambda = (|T| + |a|)^+$. By Lemma 3.4.12 we can construct inductively a sequence $(a_i : i < \lambda)$ of elements in $\operatorname{dcl}(Aa)$ such that for all $i < \lambda$ either $\operatorname{tp}(a/A, a_j : j < i)$ is bounded, or $\operatorname{tp}(a_i/A, a_j : j < i)$ is unbounded and Σ-internal. We cannot have the second case for all $i < \lambda$, since then $a \not\perp_{A\cup\{a_j:j<i\}} a_i$ for all $i < \lambda$, contradicting local character. So there is $i < \lambda$ with $a \in \operatorname{bdd}(A, a_j : j < i)$; we put $\alpha = i$ and are done. ■

PROPOSITION 3.4.15 *Let Σ, Σ' be A-invariant families of partial types, and π a partial type over A.*

1. *If $B \supseteq A$ and π is Σ-analysable over A, then π is Σ-analysable over B.*

2. *If π is Σ-analysable and every type in Σ is Σ'-analysable, then π is Σ'-analysable.*

3. *π is Σ-analysable over A if and only if for every $a \models \pi$ there is a sequence $(a_i : i < \alpha)$ such that $a \in \operatorname{bdd}(A, a_i : i < \alpha)$ and $\operatorname{tp}(a_i/A, a_j : j < i)$ is almost Σ-internal for all $i < \alpha$.*

PROOF:

1. This follows immediately from Lemma 3.4.5.

2. Consider $a \models \pi$ and $B \supseteq A$ with $a \notin \operatorname{bdd}(B)$. We have to show that $\operatorname{tp}(a/B)$ is not foreign to Σ'. As $\operatorname{tp}(a/B)$ is Σ-analysable, there is Σ-internal $a_0 \in \operatorname{dcl}(Ba) - \operatorname{bdd}(B)$, a set $C \perp_B a_0$ and realizations $(c_i : i \in I)$ of types in Σ, such that $a_0 \in \operatorname{dcl}(C, c_i : i \in I)$. Since all types in Σ are Σ'-analysable, $\operatorname{tp}(a_0/BC)$ is not foreign to Σ' by Lemma 3.4.13. Therefore $\operatorname{tp}(a/BC)$ is not foreign to Σ'. By A-invariance of Σ', we may actually assume $C \perp_{Ba_0} a$, whence $C \perp_B a$. It follows that $\operatorname{tp}(a/B)$ is not foreign to Σ'.

3. If π is Σ-analysable, then certainly for every $a \models \pi$ there is a sequence as required. Conversely, suppose $a \models \pi$ and $(a_i : i < \alpha)$ is a sequence with $a \in \operatorname{bdd}(A, a_i : i < \alpha)$ and $\operatorname{tp}(a_i/A, a_j : j < \alpha)$ almost Σ-internal for all $i < \alpha$. We have to show that for every superset B of A the type $\operatorname{tp}(a/B)$ is either bounded or not foreign to Σ. Since $a \in \operatorname{bdd}(A, a_i : i < \alpha)$, if $\operatorname{tp}(a/B)$ is not bounded, there is a minimal $i < \alpha$ such that $a \not\perp_{B\cup\{a_j:j<i\}} a_i$. Since $\operatorname{tp}(a_i/B, a_j : j < i)$ must be almost Σ-internal and not bounded, we finish by Lemma 3.4.13. ■

If π is a partial type internal in a family Σ of partial types, it need not be the case that there is a finite or even just bounded subset Σ_0 of Σ such that π is Σ_0-internal. However, if Σ consists of formulas, we can say more.

DEFINITION 3.4.16 A *multi-function* from X to Y is a binary relation R on $X \times Y$, such that there is a bound $n < \omega$ on the size of $\{y \in Y : (x, y) \in R\}$ for varying $x \in X$.

LEMMA 3.4.17 *Suppose Σ is an A-invariant family of formulas, and π is a partial imaginary type over A which is (almost) finitely generated over Σ. Then there are a finite fragment φ of π, a finite subset Σ_0 of Σ, and a definable (multi-)function f, such that φ is (almost) finitely generated over Σ_0 via $f : \left(\bigcup \Sigma_0\right)^n \to \varphi$.*

PROOF: Let A be a set such that every realization a of π is definable over A and realizations of Σ. By compactness there must be finitely many finite subsets Σ_i of Σ and A-definable functions f_i such that every realization a of π lies in the image $f_i((\bigcup \Sigma_i)^n)$ for some i, and we can assemble these functions in a single function f (of greater arity). Now the image of f must contain a finite fragment φ of π.

If every realization of π is only algebraic over A and realizations of Σ, we use multi-functions instead of functions. ■

LEMMA 3.4.18 *Let Σ be a set of formulas over A and π a partial type over A, such that every $a \models \pi$ has a Σ-analysis consisting of imaginary elements. Then there are a finite fragment φ of π and a finite subset Σ_0 of Σ, such that φ is Σ_0-analysable in finitely many steps, with imaginary analyses.*

We should note that this lemma may fail if Σ is a family of size κ.

PROOF: By compactness, we may assume that π is a complete type: if every completion p of π over A is covered by a Σ_p-analysable formula φ_p for some finite Σ_p, then finitely many of these formulas will cover π, and we take the union of the corresponding Σ_p as our Σ_0.

We shall use transfinite induction and assume that the assertion is true for all types which are analysable by imaginary analyses of length less than α. Suppose that a realization a of π has an analysis $(a_i : i \leq \alpha)$ of length α over A. Since $\mathrm{tp}(a_\alpha/A \cup \{a_i : i < \alpha\})$ is Σ-internal, it is almost finitely generated over Σ by Proposition 3.4.9. By Lemma 3.4.17 there are some finite fragment φ', a finite subset Σ_1 of Σ and a B-definable multi-function f_B (for some finite set B of parameters) such that φ' is almost finitely generated over Σ_1 via f_B. But φ' mentions only finitely many of the $\{a_i : i < \alpha\}$, say $\bar{a}$. By the inductive hypothesis there are a finite fragment φ'' of $\mathrm{tp}(\bar{a}/A)$ and a finite subset Σ_2 of Σ such that φ'' is Σ_2-analysable in finitely many steps. Finally, there is a formula $\psi(a, a_\alpha, \bar{a})$ over A which says that a is algebraic over a_α and $(a_\alpha, \bar{a})$ is algebraic over a. Putting this together, we get a finite fragment φ in $\operatorname{tp}(a/A)$ such that for all realizations a' of φ there are a'_α and $\bar{a}'$ such that $\psi(a', a'_\alpha, \bar{a}')$ holds, $\operatorname{tp}(a'_\alpha/A\bar{a}')$ is almost finitely generated over Σ_1 via $f_{B'}$ for some B', and $\operatorname{tp}(\bar{a}'/A')$ is Σ_2-analysable in finitely many steps, namely

$$\exists x', \bar{x} \left[\psi(x, x', \bar{x}) \wedge \varphi''(\bar{x}) \wedge \varphi'(x, \bar{x}) \wedge \right.$$
$$\left. \exists Z\, \forall y\, [\varphi'(y, \bar{x}) \to \exists\, \bar{y} \in \Sigma_1\; y = f_Z(\bar{y})]\right];$$

replacing a'_α and $\bar{a}'$ by the finite set of their Aa'-conjugates, we see that $\operatorname{tp}(a'/A)$ is $(\Sigma_1 \cup \Sigma_2)$-analysable in finitely many steps. We put $\Sigma_0 = \Sigma_1 \cup \Sigma_2$ and are done. ∎

Note that this proof also works if Σ is a definable family of formulas (all formulas of the form $\varphi(x, \bar{a})$ with $\models \vartheta(\bar{a})$, say); this condition is stronger than analysability in the formula $\exists \bar{y}\, [\vartheta(\bar{y}) \wedge \varphi(x, \bar{y})]$.

Even if Σ is a big family, imaginary analyses in formulas can still be chosen finite.

LEMMA 3.4.19 *Let Σ be a class of formulas and suppose a type $p \in S(A)$ has a Σ-analysis consisting of imaginary elements. Then p is Σ-analysable in finitely many steps.*

PROOF: Again we shall use induction on the length of an analysis. If a is a realization of p and $(a_i)_{i \leq \alpha}$ a Σ-analysis of a over A, then as above there is a finite $\bar{a}$ in $(a_i)_{i<\alpha}$ such that $\operatorname{tp}(a_\alpha/A\bar{a})$ is almost finitely generated over Σ. By the inductive assumption $\operatorname{tp}(\bar{a}/A)$ is Σ-analysable in finitely many steps, and so is $\operatorname{tp}(a/A) = p$. ∎

3.5 P-CLOSURE AND LOCAL MODULARITY

In this section, the ambient theory will again be simple, and P will be an $\emptyset$-invariant family of types. We shall study a condition which ensures that independence cannot become too complicated for realizations of types in P (or more generally P-internal types).

DEFINITION 3.5.1 A partial type π is *co-foreign* to P if every type in P is foreign to π.
The *P-closure* $\operatorname{cl}_P(A)$ of a set A is the collection of all countable hyperimaginaries a such that $\operatorname{tp}(a/A)$ is P-analysable and co-foreign to P.
A type p is *P-minimal* if it is foreign to all types which are co-foreign to P.

We write cl_p instead of $\operatorname{cl}_{\{p\}}$. Note that if π is a partial type analysable in a family Σ of partial types and every $\sigma \in \Sigma$ is co-foreign to P, then π is co-foreign to P by Lemma 3.4.13.

REMARK 3.5.2 The P-analysability assumption could be modified or even omitted, resulting in a larger P-closure. If P is the family of all types, then $\mathrm{cl}_P(A) = \mathrm{bdd}(A)$, and only the bounded types are co-foreign to P. More generally $\mathrm{bdd}(A) \subseteq \mathrm{cl}_P(A)$; if the inequality is strict, then $\mathrm{cl}_P(A)$ has cardinality κ.

LEMMA 3.5.3 *The following are equivalent:*

1. $\mathrm{tp}(a/A)$ *is foreign to all P-analysable types which are co-foreign to P.*
2. $a \mathop{\smile\!\!\!\!\!\!|}_A \mathrm{cl}_P(A)$.
3. $a \mathop{\smile\!\!\!\!\!\!|}_A \mathrm{dcl}(aA) \cap \mathrm{cl}_P(A)$.

If $\mathrm{tp}(a/A)$ *is P-analysable, then* $a \mathop{\smile\!\!\!\!\!\!|}_A \mathrm{cl}_P(A)$ *if and only if* $\mathrm{tp}(a/A)$ *is P-minimal.*

PROOF: 1. $\Rightarrow$ 2. $\Rightarrow$ 3. is trivial. So suppose $\mathrm{tp}(a/A)$ is not foreign to some type p which is co-foreign to P. Assume either p or $\mathrm{tp}(a/A)$ is P-analysable. By Proposition 3.4.12 there is some $a_0 \in \mathrm{dcl}(aA) - \mathrm{bdd}(A)$ such that $\mathrm{tp}(a_0/A)$ is p-internal. Therefore $\mathrm{tp}(a_0/A)$ is P-analysable, and co-foreign to P (since p is). Hence $a_0 \in [\mathrm{dcl}(aA) \cap \mathrm{cl}_P(A)] - \mathrm{bdd}(A)$, so $a \not\mathop{\smile\!\!\!\!\!\!|}_A a_0$. This also proves the last assertion. ■

COROLLARY 3.5.4 *P-closure is a closure operator, except that in general* $\mathrm{cl}_P(A)$ *has cardinality κ (and thus violates our conventions on subsets of $\mathfrak{C}$).*

PROOF: $A \subseteq \mathrm{cl}_P(A)$ and $A \subseteq B \Rightarrow \mathrm{cl}_P(A) \subseteq \mathrm{cl}_P(B)$ are obvious from the definition. So consider $a \in \mathrm{cl}_P(\mathrm{cl}_P(A))$. Then $\mathrm{tp}(a/\mathrm{cl}_P(A))$ is P-analysable; if B is a subset of $\mathrm{cl}_P(A)$ such that $a \mathop{\smile\!\!\!\!\!\!|}_B \mathrm{cl}_P(A)$, then $\mathrm{tp}(a/B)$ and $\mathrm{tp}(B/A)$ are both P-analysable. Hence $\mathrm{tp}(a/A)$ is P-analysable by Proposition 3.4.15.2.

Now take any $p \in P$. Then p is foreign to $\mathrm{tp}(B/A)$ and to $\mathrm{tp}(a/B)$, whence to $\mathrm{tp}(a/A)$. Hence $a \in \mathrm{cl}_P(A)$, and $\mathrm{cl}_P(\mathrm{cl}_P(A)) = \mathrm{cl}_P(A)$. ■

In particular $\mathrm{bdd}(\mathrm{cl}_P(A)) = \mathrm{cl}_P(A)$, and the intersection of two P-closed sets is again P-closed. Moreover, $\mathrm{tp}(a/\mathrm{dcl}(aA) \cap \mathrm{cl}_P(A))$ is foreign to all P-analysable types which are co-foreign to P by Lemma 3.5.3.

We shall now assemble the basic properties of P-closure.

LEMMA 3.5.5 *Suppose* $B \mathop{\smile\!\!\!\!\!\!|}_A C$. *Then* $\mathrm{cl}_P(B) \mathop{\smile\!\!\!\!\!\!|}_{\mathrm{cl}_P(A)} \mathrm{cl}_P(C)$.

PROOF: Let $A' = \mathrm{dcl}(BA) \cap \mathrm{cl}_P(A)$. Then $B \mathop{\smile\!\!\!\!\!\!|}_{A'} C$; since $\mathrm{cl}_P(A) = \mathrm{cl}_P(A')$, Lemma 3.5.3 implies that $\mathrm{tp}(B/A')$ is foreign to all P-analysable types co-foreign to P.

Suppose $c \in \mathrm{cl}_P(AC)$. Then $\mathrm{tp}(c/A'C)$ is P-analysable and co-foreign to P; since $B \mathop{\smile\!\!\!\!\!\!|}_{A'} C$, $\mathrm{tp}(B/A'C)$ is foreign to $\mathrm{tp}(c/A'C)$. By transitivity, $B \mathop{\smile\!\!\!\!\!\!|}_{A'} cC$, whence $B \mathop{\smile\!\!\!\!\!\!|}_{A'} \mathrm{cl}_P(AC)$, and $B \mathop{\smile\!\!\!\!\!\!|}_{\mathrm{cl}_P(A)} \mathrm{cl}_P(AC)$.

Since $\mathrm{cl}_P(\mathrm{cl}_P(A)) = \mathrm{cl}_P(A)$, symmetry yields $\mathrm{cl}_P(B) \mathop{\smile\!\!\!\!\!\mid}_{\mathrm{cl}_P(A)} \mathrm{cl}_P(C)$. ■

LEMMA 3.5.6 *Suppose* $A \subseteq B \cap C$ *satisfies* $\mathrm{cl}_P(B) \cap \mathrm{cl}_P(C) = \mathrm{cl}_P(A)$. *If* $D \mathop{\smile\!\!\!\!\!\mid}_A BC$, *then* $\mathrm{cl}_P(BD) \cap \mathrm{cl}_P(CD) = \mathrm{cl}_P(AD)$.

PROOF: The assumptions imply $BD \mathop{\smile\!\!\!\!\!\mid}_B BC$ and $CD \mathop{\smile\!\!\!\!\!\mid}_C BC$. By Lemma 3.5.5 we have $\mathrm{cl}_P(BD) \mathop{\smile\!\!\!\!\!\mid}_{\mathrm{cl}_P(B)} \mathrm{cl}_P(BC)$ and $\mathrm{cl}_P(CD) \mathop{\smile\!\!\!\!\!\mid}_{\mathrm{cl}_P(C)} \mathrm{cl}_P(BC)$. Let $e \in \mathrm{cl}_P(CD) \cap \mathrm{cl}_P(BD)$. Then $Cb(eD/\mathrm{cl}_P(BC)) \subset \mathrm{cl}_P(B) \cap \mathrm{cl}_P(C) = \mathrm{cl}_P(A)$, so $eD \mathop{\smile\!\!\!\!\!\mid}_{\mathrm{cl}_P(A)} \mathrm{cl}_P(BC)$, whence $e \mathop{\smile\!\!\!\!\!\mid}_{\mathrm{cl}_P(A)D} \mathrm{cl}_P(BC)D$. Therefore $e \mathop{\smile\!\!\!\!\!\mid}_{\mathrm{cl}_P(AD)} \mathrm{cl}_P(BCD)$ by Lemma 3.5.5; as $e \in \mathrm{cl}_P(BCD)$ we get $e \in \mathrm{cl}_P(AD)$. ■

COROLLARY 3.5.7 *Suppose* $\mathrm{tp}(a/A)$ *and* $\mathrm{tp}(b/A)$ *are* P*-internal,* $ab \mathop{\smile\!\!\!\!\!\mid}_A B$, *and* a *and* b *are independent over* $\mathrm{cl}_P(aAB) \cap \mathrm{cl}_P(bAB)$. *Then* a *and* b *are independent over* $\mathrm{cl}_P(aA) \cap \mathrm{cl}_P(bA)$.

PROOF: Let $I = \mathrm{cl}_P(aA) \cap \mathrm{cl}_P(bA)$, and note that $I = \mathrm{cl}_P(IA)$. Then $\mathrm{cl}_P(abA) \mathop{\smile\!\!\!\!\!\mid}_{\mathrm{cl}_P(A)} \mathrm{cl}_P(B)$, whence $\mathrm{cl}_P(aA)\mathrm{cl}_P(bA) \mathop{\smile\!\!\!\!\!\mid}_I \mathrm{cl}_P(B)I$, and therefore $\mathrm{cl}_P(aA)\mathrm{cl}_P(bA) \mathop{\smile\!\!\!\!\!\mid}_I \mathrm{cl}_P(IB)$ by P-closedness of I. Hence Lemma 3.5.6 implies that $\mathrm{cl}_P(IB)$ equals $\mathrm{cl}_P(aAB) \cap \mathrm{cl}_P(bAB)$. But $ab \mathop{\smile\!\!\!\!\!\mid}_I \mathrm{cl}_P(IB)$; so $a \mathop{\smile\!\!\!\!\!\mid}_{\mathrm{cl}_P(IB)} b$ yields $a \mathop{\smile\!\!\!\!\!\mid}_I b$. ■

LEMMA 3.5.8 *Suppose* $a \in \mathrm{cl}_P(b)$ *and* $b \mathop{\smile\!\!\!\!\!\mid}_a A$. *Then* $\mathrm{cl}_P(\mathrm{Cb}(a/\mathrm{cl}_P(A))) = \mathrm{cl}_P(\mathrm{Cb}(b/\mathrm{cl}_P(A)))$.

PROOF: Put $B = \mathrm{Cb}(a/\mathrm{cl}_P(A))$ and $B' = \mathrm{Cb}(b/\mathrm{cl}_P(A))$, both subsets of $\mathrm{cl}_P(A)$. Then $b \mathop{\smile\!\!\!\!\!\mid}_{B'} A$, whence $\mathrm{cl}_P(b) \mathop{\smile\!\!\!\!\!\mid}_{\mathrm{cl}_P(B')} \mathrm{cl}_P(A)$ by Lemma 3.5.5. So $a \mathop{\smile\!\!\!\!\!\mid}_{\mathrm{cl}_P(B')} \mathrm{cl}_P(A)$, whence $B \subset \mathrm{cl}_P(B')$ by minimality of the canonical base, and therefore $\mathrm{cl}_P(B) \subseteq \mathrm{cl}_P(B')$. Conversely, $b \mathop{\smile\!\!\!\!\!\mid}_a A$ implies by Lemma 3.5.5 that b and $\mathrm{cl}_P(A)$ are independent over $\mathrm{cl}_P(a)$ and hence over $\mathrm{cl}_P(a)\mathrm{cl}_P(B)$; on the other hand $a \mathop{\smile\!\!\!\!\!\mid}_B \mathrm{cl}_P(A)$ implies $\mathrm{cl}_P(a) \mathop{\smile\!\!\!\!\!\mid}_{\mathrm{cl}_P(B)} \mathrm{cl}_P(A)$, whence $b \mathop{\smile\!\!\!\!\!\mid}_{\mathrm{cl}_P(B)} \mathrm{cl}_P(A)$. By minimality of the canonical base $B' \subset \mathrm{cl}_P(B)$, and hence $\mathrm{cl}_P(B') \subseteq \mathrm{cl}_P(B)$. ■

LEMMA 3.5.9 $\mathrm{Cb}(a/\mathrm{cl}_P(A)) \subseteq \mathrm{cl}_P(\mathrm{Cb}(a/A))$.

PROOF: Put $B = Cb(a/A)$. Then $a \mathop{\smile\!\!\!\!\!\mid}_B A$, whence $a \mathop{\smile\!\!\!\!\!\mid}_{\mathrm{cl}_P(B)} \mathrm{cl}_P(A)$ by Lemma 3.5.5; as $\mathrm{cl}_P(B) \subseteq \mathrm{cl}_P(A)$ we have $\mathrm{Cb}(a/\mathrm{cl}_P(A)) \subseteq \mathrm{cl}_P(B)$. ■

DEFINITION 3.5.10 We call P *locally modular* if for any A and any a and b whose types over A are P-internal, a and b are independent over $\mathrm{cl}_P(aA) \cap \mathrm{cl}_P(bA)$. A theory is *one-based* if the family of all types is locally modular.

PROBLEM 3.5.11 If Q is the family of all P-analysable types and P is locally modular, is necessarily Q locally modular?

REMARK 3.5.12 If P is the family of all types, then $\mathrm{cl}_P(a) = \mathrm{bdd}(a)$. Hence in a one-based theory, $a \mathop{\smile\!\!\!\!\!\!\mid}_{\mathrm{bdd}(a)\cap\mathrm{bdd}(b)} b$ for any a, b. In fact, it is easy to see that $\mathrm{Cb}(a/B) = \mathrm{bdd}(a) \cap \mathrm{bdd}(B)$ for any a, B in a one-based theory.

LEMMA 3.5.13 *If P is locally modular, $\mathrm{tp}(a)$ is P-internal and $\mathrm{tp}(a/A)$ is P-minimal, then $\mathrm{Cb}(a/A) \subseteq \mathrm{cl}_P(a)$ for any set A.*

PROOF: Let I be a Morley sequence in $\mathrm{lstp}(a/A)$ independent of a over A. Then $\mathrm{tp}(I)$ is P-internal, and $\mathrm{Cb}(a/A) = \mathrm{Cb}(a/AI) \in \mathrm{dcl}(I)$ by Corollary 3.3.13, whence $\mathrm{Cb}(a/A) = \mathrm{Cb}(a/I)$. By local modularity $a \mathop{\smile\!\!\!\!\!\!\mid}_{\mathrm{cl}_P(a)\cap\mathrm{cl}_P(I)} I$, and therefore $a \mathop{\smile\!\!\!\!\!\!\mid}_{\mathrm{cl}_P(a)\cap\mathrm{cl}_P(I)} \mathrm{cl}_P(I)$ by Lemma 3.5.5. Since $\mathrm{tp}(a/A)$ is P-minimal, so is $\mathrm{tp}(a/I)$, whence $a \mathop{\smile\!\!\!\!\!\!\mid}_I \mathrm{cl}_P(I)$ and

$$\mathrm{Cb}(a/A) = \mathrm{Cb}(a/I) = \mathrm{Cb}(a/\mathrm{cl}_P(I)) \subseteq \mathrm{cl}_P(a). \quad \blacksquare$$

REMARK 3.5.14 If $\mathrm{tp}(a/A)$ and $\mathrm{tp}(b/A)$ are both P-internal for some locally modular family P and $\mathrm{tp}(a/Ab)$ and $\mathrm{tp}(b/Aa)$ are both P-minimal, Lemma 3.5.13 implies that a and b are independent over $\mathrm{cl}_P(aA) \cap \mathrm{bdd}(bA)$ and over $\mathrm{bdd}(aA) \cap \mathrm{cl}_P(bA)$. Using P-minimality again, we see that they are in fact independent over $\mathrm{bdd}(aA) \cap \mathrm{bdd}(bA)$.

PROPOSITION 3.5.15 *Local modularity is preserved under naming and forgetting parameters.*

PROOF: Preservation under naming parameters is trivial, and that under forgetting parameters follows from Corollary 3.5.7 by conjugating everything to be independent of the parameters. ■

COROLLARY 3.5.16 *Suppose P is locally modular and $\mathrm{tp}(a)$ is P-internal. If $A \subset \mathrm{cl}_P(a)$, then $\mathrm{Cb}(A/\mathrm{cl}_P(B)) \subseteq \mathrm{cl}_P(A)$ for any set B.*

PROOF: By $\emptyset$-invariance of P we may assume $B \mathop{\smile\!\!\!\!\!\!\mid}_A a$. Hence

$$\mathrm{Cb}(A/\mathrm{cl}_P(B)) \subset \mathrm{cl}_P(\mathrm{Cb}(a/\mathrm{cl}_P(B))) \subseteq \mathrm{cl}_P(a)$$

by Lemmas 3.5.8 and 3.5.13. But $\mathrm{cl}_P(a) \mathop{\smile\!\!\!\!\!\!\mid}_{\mathrm{cl}_P(A)} \mathrm{cl}_P(B)$ by Lemma 3.5.5; since $\mathrm{Cb}(A/\mathrm{cl}_P(B)) \subseteq \mathrm{cl}_P(B)$, we get $\mathrm{Cb}(A/\mathrm{cl}_P(B)) \subseteq \mathrm{cl}_P(A)$. ■
The next proposition gives a characterization of local modularity.

PROPOSITION 3.5.17 *P is locally modular if and only if for any two models $\mathfrak{M} \prec \mathfrak{N}$ and any tuple $\bar{a}$ of realizations of types in P over $\mathfrak{M}$ such that $\mathrm{tp}(\bar{a}/\mathfrak{N})$ is P-minimal, we have $\mathrm{Cb}(a/\mathfrak{N}) \subseteq \mathrm{cl}_P(a\mathfrak{M})$.*

PROOF: If P is locally modular, then $\mathrm{Cb}(\bar{a}/\mathfrak{N}) \subseteq \mathrm{cl}_P(\bar{a}\mathfrak{M})$ by Lemma 3.5.13, as $\mathrm{tp}(\bar{a}/\mathfrak{M})$ is obviously P-internal.

Conversely, suppose that the second condition holds and consider a, b, A such that $\mathrm{tp}(a/A)$ and $\mathrm{tp}(b/A)$ are P-internal. Let $\mathfrak{M}'$ be a model containing A with $\mathfrak{M}' \mathop{\smile\!\!\!\!\!\!|}_A a$ such that there are realizations $\bar{a}$ of types in P over $\mathfrak{M}'$ with $a \in \mathrm{dcl}(\mathfrak{M}'\bar{a})$; we choose it such that $\mathfrak{M}'\bar{a} \mathop{\smile\!\!\!\!\!\!|}_{Aa} b$ (whence $\mathfrak{M}' \mathop{\smile\!\!\!\!\!\!|}_A ab$). Let $\mathfrak{M}$ be a model containing $\mathfrak{M}'$ with $\mathfrak{M} \mathop{\smile\!\!\!\!\!\!|}_{\mathfrak{M}'} b$ such that there are realizations $\bar{b}$ of types in P over $\mathfrak{M}$ with $b \in \mathrm{dcl}(\mathfrak{M}\bar{b})$; we choose it such that $\mathfrak{M}\bar{b} \mathop{\smile\!\!\!\!\!\!|}_{\mathfrak{M}'b} \bar{a}$. Then $\mathfrak{M} \mathop{\smile\!\!\!\!\!\!|}_{\mathfrak{M}'} ab$ and hence $\mathfrak{M} \mathop{\smile\!\!\!\!\!\!|}_A ab$; moreover $\bar{a} \mathop{\smile\!\!\!\!\!\!|}_{\mathfrak{M}a} b$ and $\bar{b} \mathop{\smile\!\!\!\!\!\!|}_{\mathfrak{M}b} \bar{a}$. Put $C = \mathrm{cl}_P(\mathfrak{M}\bar{b}) \cap \mathrm{bdd}(\mathfrak{M}\bar{b}\bar{a})$, and let $\mathfrak{N} \mathop{\smile\!\!\!\!\!\!|}_C \bar{a}$ be a model containing C. Then $\mathrm{tp}(\bar{a}/\mathfrak{N})$ is P-minimal, and $\mathrm{Cb}(\bar{a}/\mathfrak{N}) = \mathrm{Cb}(\bar{a}/\mathrm{cl}_P(\mathfrak{N}))$.

By assumption $\mathrm{Cb}(\bar{a}/\mathfrak{N}) \subseteq \mathrm{cl}_P(\mathfrak{M}\bar{a})$, so $\mathrm{Cb}(a/\mathfrak{N}) \subset \mathrm{cl}_P(\mathfrak{N}) \cap \mathrm{cl}_P(\mathfrak{M}\bar{a})$. As $\mathrm{cl}_P(\mathfrak{N}) \mathop{\smile\!\!\!\!\!\!|}_{\mathrm{cl}_P(C)} \mathrm{cl}_P(\mathfrak{M}\bar{a})$ and $\mathrm{cl}_P(C) = \mathrm{cl}_P(\mathfrak{M}\bar{b})$, we have

$$\mathrm{cl}_P(\mathfrak{N}) \cap \mathrm{cl}_P(\mathfrak{M}\bar{a}) = \mathrm{cl}_P(C) \cap \mathrm{cl}_P(\mathfrak{M}\bar{a}) = \mathrm{cl}_P(\mathfrak{M}\bar{a}) \cap \mathrm{cl}_P(\mathfrak{M}\bar{b}).$$

As $\mathrm{cl}_P(\mathfrak{M}\bar{b}) \mathop{\smile\!\!\!\!\!\!|}_{\mathrm{cl}_P(\mathfrak{M}b)} \mathrm{cl}_P(\mathfrak{M}\bar{a})$ and $\mathrm{cl}_P(\mathfrak{M}\bar{a}) \mathop{\smile\!\!\!\!\!\!|}_{\mathrm{cl}_P(\mathfrak{M}a)} \mathrm{cl}_P(\mathfrak{M}b)$, we get

$$\mathrm{cl}_P(\mathfrak{M}\bar{b}) \cap \mathrm{cl}_P(\mathfrak{M}\bar{a}) = \mathrm{cl}_P(\mathfrak{M}b) \cap \mathrm{cl}_P(\mathfrak{M}\bar{a}) = \mathrm{cl}_P(\mathfrak{M}b) \cap \mathrm{cl}_P(\mathfrak{M}a).$$

Therefore $\bar{a}$ and $\mathfrak{N}$ are independent over $\mathrm{cl}_P(\mathfrak{M}a) \cap \mathrm{cl}_P(\mathfrak{M}b)$, as are a and b. By Lemma 3.5.7 we get $a \mathop{\smile\!\!\!\!\!\!|}_{\mathrm{cl}_P(Aa)\cap\mathrm{cl}_P(Ab)} b$, and P is locally modular. ■

If the theory is one-based, this simplifies nicely:

COROLLARY 3.5.18 *A theory is one-based if and only if every real type* $\mathrm{tp}(\bar{a}/A)$ *is based on* $\mathrm{bdd}(\bar{a})$. *In particular, in a one-based theory every finitary type is based on a finite set.* ■

EXERCISE 3.5.19 Let Q be an $\emptyset$-invariant subfamily of P, and suppose P is locally modular. Show that Q is locally modular.

EXERCISE 3.5.20 Call two $\emptyset$-invariant families P and Q of types *perpendicular* if every type in P is foreign and co-foreign to every type in Q. Suppose P and Q are perpendicular and locally modular. Show that $P \cup Q$ is locally modular.

3.6 ELIMINATION OF HYPERIMAGINARIES

One might ask to what extent hyperimaginaries are a necessary feature of simple theories, just as imaginaries are a necessary feature of stable theories. In this section, we shall collect various conditions and conjectures implying elimination of hyperimaginaries, and different levels up to which this might be achievable. All theories will be supposed to be simple (although we may repeat this global condition from time to time).

The following example shows that even in a stable theory, we cannot hope to eliminate type-definable bounded equivalence relations in favour of infinite intersections of definable equivalence relations:

EXAMPLE 3.6.1 Let $\mathfrak{M}$ be a structure with domain $\mathbb{R}$ and unary predicates $U_a = \{x \in \mathbb{Q} : x \leq a\}$, for all $a \in \mathbb{Q}$. Let E be the relation type-defined by the conjunction of the formulas

$$[U_a(x) \to U_b(y)] \wedge [U_a(y) \to U_b(x)],$$

for all rational $a < b$. It is easy to check, using density of $\mathbb{Q}$, that E is an equivalence relation, whose classes consist of all elements infinitely close to some real number $r \in \mathbb{R}$ (in any elementary extension of $\mathfrak{M}$). Thus E is bounded. However, $\mathfrak{M}$ admits elimination of quantifiers and is stable. Furthermore, E is not given by an infinite intersection of definable equivalence relations, since the only definable equivalence relations on $\mathfrak{M}$ are equality, finite conjunctions of $U_a(x) \leftrightarrow U_a(y)$ for various a, and the trivial equivalence relation.

However, one might at least hope to eliminate type-definable equivalence relations restricted to a complete type.

DEFINITION 3.6.2 A theory T has *elimination of hyperimaginaries* if every hyperimaginary a is interdefinable with is a set A of imaginary elements.

The following Lemma shows that we can eliminate hyperimaginaries which are sandwiched close to a set of imaginaries.

LEMMA 3.6.3 *Let a be a hyperimaginary element and suppose there is a set A of imaginary elements with* $a \in \mathrm{dcl}(A) \subseteq \mathrm{acl}(a)$. *Then there is a set B of imaginary elements interdefinable with a.*

PROOF: Suppose $A = \{a_i : i \in I\}$. For every finite $J \subseteq I$ let A_J be the set of a-conjugates of $(a_j : j \in J)$, an imaginary element, and put $B = (A_J : J \subseteq I \text{ finite})$. Then $B \subset \mathrm{dcl}(a)$; we claim that $a \in \mathrm{dcl}(B)$. So consider an automorphism σ fixing B pointwise. Then $\sigma(a_j : j \in J) \in \sigma(A_J) = A_J$, and therefore $\mathrm{tp}(\sigma(a_j : j \in J)/a) = \mathrm{tp}(a_j : j \in J/a)$ for all finite $J \subseteq I$. Hence $\mathrm{tp}(A/a) = \mathrm{tp}(\sigma(A)/a)$; so $a \in \mathrm{dcl}(A)$ implies $\sigma(a) = a$. Thus $a \in \mathrm{dcl}(B)$. ■

Let us note that in order to eliminate hyperimaginaries, it is sufficient to eliminate hyperimaginary canonical bases for types of real tuples.

LEMMA 3.6.4 *Suppose for every finitary real Lascar strong type p in a simple theory T there is a set A of imaginary elements such that* $\mathrm{Cb}(p) = \mathrm{dcl}(A)$. *Then T eliminates hyperimaginaries.*

PROOF: Let $\bar{a}_E$ be a hyperimaginary, where $\bar{a}$ is a possibly infinite tuple of real elements. By Lemma 3.1.3 we may assume that $\bar{a} = (a_i : i < \omega)$. For every $n < \omega$ put $a^n = (a_0, \ldots, a_n)$. By assumption, $\mathrm{Cb}(a^n/\bar{a}_E)$ is interdefinable with a set C_n of imaginary elements; clearly $C_i \subseteq \mathrm{dcl}(C_j)$ for $i \leq j < \omega$. Note that $\mathrm{tp}(a^j/C_i)$ is a Lascar strong type for all $i \geq j$, and put $C = \bigcup_{n<\omega} C_n$.

If $\mathrm{tp}(a^j/C) = \mathrm{tp}(a'/C)$ for some a', then $\mathrm{tp}(a^j/C_i) = \mathrm{tp}(a'/C_i)$ for all $i < \omega$, whence $\mathrm{lstp}(a^j/C_i) = \mathrm{lstp}(a'/C_i)$ for all $i \geq j$. Hence $\mathrm{lstp}(a^j/C) = \mathrm{lstp}(a'/C)$ by Corollary 2.7.9, and $\mathrm{tp}(a^j/C)$ is a Lascar strong type for all $j < \omega$, as is $\mathrm{tp}(\bar{a}/C)$. Furthermore, $\bar{a} \perp_C \bar{a}_E$, so $\mathrm{Cb}(\bar{a}/\bar{a}_E) \subseteq \mathrm{dcl}(C)$. But clearly $C_n \subseteq \mathrm{Cb}(\bar{a}/\bar{a}_E)$, whence $\mathrm{dcl}(C) = \mathrm{Cb}(\bar{a}/\bar{a}_E)$; as $\bar{a}_E \in \mathrm{Cb}(\bar{a}/\bar{a}_E) \subseteq \mathrm{bdd}(\bar{a}_E)$ by Remark 3.3.14, we are done by Lemma 3.6.3. ■

In particular, a *stable* theory eliminates hyperimaginaries.

A consequence of elimination of hyperimaginaries is the equality of Lascar strong type and strong type:

PROPOSITION 3.6.5 *Suppose T is a simple theory with elimination of hyperimaginaries. Then for any A and a, a', $\mathrm{stp}(a/A) = \mathrm{stp}(a'/A)$ if and only if $\mathrm{lstp}(a/A) = \mathrm{lstp}(a'/A)$.*

PROOF: As equality of Lascar strong type is a bounded equivalence relation E, the hyperimaginary $a_E \in \mathrm{bdd}(A)$. By elimination of hyperimaginaries, there is a set B of imaginaries with $\mathrm{dcl}(a_E) = \mathrm{dcl}(B)$. Thus $B \subseteq \mathrm{bdd}(A)$, whence $B \subseteq \mathrm{acl}(A)$. Therefore $\mathrm{stp}(a/A) = \mathrm{stp}(a'/A)$ implies $tp(a/B) = \mathrm{tp}(a'/B)$, whence $\mathrm{tp}(a/a_E) = \mathrm{tp}(a'/a_E)$, and $E(a, a')$ holds. But this means $\mathrm{lstp}(a/A) = \mathrm{lstp}(a'/A)$. ■

REMARK 3.6.6 Lascar strong type is the same as strong type in a simple theory if and only if the Independence Theorem holds over algebraically closed sets.

PROOF: If Lascar strong type is the same as strong type, then the Independence Theorem holds for strong types, i.e. for types over algebraically closed sets. Conversely, suppose that $A = \mathrm{acl}(A)$, $p \in S(A)$ and p_1 and p_2 are two extensions of p to $\mathrm{bdd}(A)$. As $\mathrm{bdd}(A) \perp_A \mathrm{bdd}(A)$, the Independence Theorem yields that $p_1 \cup p_2$ is consistent and does not fork over A. Therefore $p_1 = p_2$. ■

EXAMPLE 3.6.7 Let $\mathfrak{M}$ be the disjoint union of the real line $\mathbb{R}$ and the unit circle C. The language consists of addition on $\mathbb{R}$, the additive group action of $\mathbb{R}$ on C identifying C with $\mathbb{R}/2\pi$, and a ternary relation $U(x, y, z)$ which holds if $x, y \in C$, $z \in \mathbb{R}$ and the length of the shorter arc from x to y is less than z. Then for any $r \in \mathbb{R}$ there is n_r such that among any n_r points $\{x_1, \ldots, x_{n_r}\}$ on C, two must satisfy $U(x_i, x_j, r)$. Therefore $\bigwedge_{r\in\mathbb{R}} U(x, y, r)$ defines a bounded equivalence relation on C. But $\mathrm{stp}(c/\mathbb{R}) = \mathrm{stp}(c'/\mathbb{R})$ for any $c, c' \in C$. So Lascar strong type is different from strong type over $\mathbb{R}$. (A similar argument works over the line — which must be a divisible ordered abelian group — in any elementarily equivalent structure.)

The most important occurence of hyperimaginaries is as the canonical basis of a type. In a stable theory, we know what the canonical basis of a strong type p consists of: it is the definable closure of the set of names of φ-definitions for p. The more audacious will therefore conjecture:

CONJECTURE 3.6.8 STABLE FORKING CONJECTURE *Let T be a simple theory, $\bar{a}, \bar{b}$ imaginary tuples, and C a set of imaginary elements. If $\bar{a} \not\perp_C \bar{b}$, then there is a stable formula $\varphi(\bar{x}, \bar{y}\bar{z})$ and $\bar{c} \in C$, such that $\varphi(\bar{x}, \bar{b}\bar{c})$ is satisfied by $\bar{a}$ and forks over C.*

Here, we call a formula *stable* if it defines a stable relation. We shall see below in Lemma 3.6.11 that for a stable formula φ every type has a φ-definition. The stable forking conjecture obviously implies the existence of a lot of stable formulas, a property true for all the known examples, but highly unclear in general. However, the following (*a priori* weaker) condition is sufficient for φ-definability.

DEFINITION 3.6.9 Let p be Lascar strong type. A formula $\varphi(x, y)$ is *p-stable* if for any B and any two Lascar strong types $q_1, q_2 \in S(B)$ with $q_1 \sim p \sim q_2$, we have $q_1 \lceil \varphi = q_2 \lceil \varphi$ (i.e. q_1 and q_2 agree on all instances of φ over B).
A formula $\varphi(x, a)$ is *p-normal* if whenever $\varphi(x, a')$ is a conjugate of $\varphi(x, a)$ which is in some $q \sim p$, then $\varphi(x, a')$ is equivalent to $\varphi(x, a)$.
Let $\psi(x, y)$ be a formula and $k < \omega$. A formula $\varphi(x, a)$ is *(ψ, k)-normal* if for any conjugate $\varphi(x, a')$ of $\varphi(x, a)$ either $D(\varphi(x, a) \wedge \varphi(x, a'), \psi, k) < D(\varphi(x, a), \psi, k)$, or $\varphi(x, a')$ is equivalent to $\varphi(x, a)$.

LEMMA 3.6.10 *Let T be a simple theory. If φ is p-stable, then p has a φ-definition.*

PROOF: We may assume that p is a Lascar strong type over its canonical base. By Lemma 2.3.15 there is a partial type $\Phi(\bar{y})$ such that $\Phi(\bar{b})$ holds if and only if there is a non-forking extension of p containing $\varphi(\bar{x}, \bar{b})$. By p-stability, this means that all non-forking extensions of p to $\bar{b}$ must contain $\varphi(\bar{x}, \bar{b})$. Similarly, there is a partial type $\Phi'(\bar{y})$ such that $\Phi'(\bar{b})$ holds if and only if $\neg\varphi(\bar{x}, \bar{b})$ is in some non-forking extension of p. So $\Phi \wedge \Phi'$ is contradictory and $\Phi \vee \Phi'$ is universally valid. By compactness, there are finite subtypes $\Phi_0 \subseteq \Phi$ and $\Phi'_0 \subseteq \Phi'$ with $\Phi_0 \vdash \Phi$ and $\Phi'_0 \vdash \Phi'$; clearly $\bigwedge \Phi_0(\bar{y})$ is the required φ-definition for p. ■

LEMMA 3.6.11 *Let T be a simple theory. A formula $\varphi(\bar{x}, \bar{y})$ is stable if and only if it is p-stable for every Lascar strong type p.*

PROOF: Suppose φ is not p-stable for some Lascar strong type p over A. Then there is $\bar{b}$ and two non-forking extensions of p to $A\bar{b}$, one containing $\varphi(\bar{x}, \bar{b})$ and the other containing $\neg\varphi(\bar{x}, \bar{b})$. Let $(b_i : i < \omega)$ be a Morley sequence in $\text{lstp}(\bar{b}/A)$. By the Independence Theorem 2.5.11 for any $j < \omega$ there is some non-forking extension q_j of p to $A \cup \{\bar{b}_i : i < \omega\}$ with $\varphi(\bar{x}, \bar{b}_i) \in q_j$ if and only if $i < j$. Taking realizations $\bar{a}_j \models q_j$ for all $j < \omega$, Ramsey's Theorem and compactness yield the existence of an indiscernible sequence $(\bar{a}_i\bar{b}_i : i < \omega)$ such that $\models \varphi(\bar{a}_j, \bar{b}_i)$ if and only if $i < j$, so φ does not define a stable relation.

Conversely, suppose φ is not stable. So there is an indiscernible sequence $(\bar{a}_i\bar{b}_i : i < \omega 2)$ such that $\models \varphi(\bar{a}_i, \bar{b}_j)$ if and only if $i < j$. Put $A = \{\bar{a}_i\bar{b}_i : i < \omega\}$, and $p = \text{tp}(\bar{a}_{\omega+1}/A)$; note that p is in fact a Lascar strong type by finite satisfiability in A and Lemma 3.3.11. Then $(\bar{a}_i\bar{b}_i : \omega \leq i < \omega 2)$ is a Morley sequence over A by Lemma 2.5.2; in particular $\bar{b}_{\omega+1}$ is independent of $\bar{a}_\omega\bar{a}_{\omega+2}$ over A. However, $\models \neg\varphi(\bar{a}_\omega, \bar{b}_{\omega+1}) \wedge \varphi(\bar{a}_{\omega+2}, \bar{b}_{\omega+1})$. Hence p has two non-forking extensions to $A\bar{b}_{\omega+1}$ which differ on a φ-formula, and φ is not p-stable. ■

LEMMA 3.6.12 *Let $p \in S(A)$ be a Lascar strong type.*

1. *If $\varphi(x, a) \in p$ is (ψ, k)-normal and $D(p, \psi, k) = D(\varphi, \psi, k)$, then $\varphi(x, a)$ is p-normal.*

2. *If $\varphi(x, a)$ is p-normal, then $\varphi(x, a)$ is $\text{Cb}(p)$-invariant.*

3. *If $\varphi(x, a)$ is p-normal and a is a canonical parameter for $\varphi(x, a)$, then there is a formula $\vartheta(y) \in \text{tp}(b)$ such that $\varphi(x, a') \wedge \vartheta(a') \in q \sim p$ implies $a' = a$. In particular, $\varphi(x, y) \wedge \vartheta(y)$ is p-stable, with canonical parameter a.*

PROOF:

1. If $p' \sim_0 p$ and $\varphi(x, a') \in p'$ is a conjugate of $\varphi(x, a)$, then $p \cup p'$ extends to a non-forking extension of p, and

$$D(\varphi(x, a) \wedge \varphi(x, a'), \psi, k) \geq D(p \cup p', \psi, k) = D(p, \psi, k)$$
$$= D(\varphi(x, a), \psi, k).$$

So $\varphi(x, a')$ is equivalent to $\varphi(x, a)$. The assertion follows.

2. Suppose $p' \sim_0 p$ is a conjugate of p, and $\varphi(x, a')$ is the corresponding conjugate of $\varphi(x, a)$. By p-normality, $\varphi(x, a')$ is equivalent to $\varphi(x, a)$. Therefore any automorphism stabilizing the $\sim$-class of p, in other words, which fixes $\text{Cb}(p)$, maps $\varphi(x, a)$ to an equivalent formula.

3. By the previous part, $\text{tp}(a') = \text{tp}(a)$ and "$p(x) \cup \{\varphi(x, a')\}$ does not fork over p" together imply $a = a'$. By compactness, there is a finite bit $\vartheta(y) \in \text{tp}(a)$ which suffices for the implication; the rest follows. ■

DEFINITION 3.6.13 A simple theory T has *stable forking* if it satisfies the stable forking conjecture.

PROPOSITION 3.6.14 *If T is a simple theory with stable forking, then for any imaginary Lascar strong type $p \in S(A)$ (over an imaginary parameter set A) the canonical base* $\text{Cb}(p)$ *is in the bounded closure of the canonical parameters for the φ-definitions of p, where φ runs over all stable $\mathfrak{L}$-formulas.*

PROOF: Put $C = \text{dcl}(\text{Cb}(d_p\varphi) : \varphi$ a stable formula$)$, a set of imaginary elements. Clearly any automorphism fixing the $\sim$-class of p must fix $\text{Cb}(d_p\varphi)$, so $C \subseteq \text{Cb}(p)$. Now suppose $\bar{b} \models p$ and $\bar{b} \not\perp_C A$. By stable forking, there is a stable formula $\varphi(\bar{x}, \bar{y})$ and a tuple $\bar{a} \in A$ such that $\varphi(\bar{x}, \bar{a})$ is satisfied by $\bar{b}$ and forks over C. But then $\text{Cb}(d_p\varphi) \in C$, so any C-indiscernible sequence $(\bar{a}_i : i < \omega)$ of type $\text{tp}(\bar{a}/C)$ satisfies $\models d_p\varphi(\bar{a}_i)$, and $\bigwedge_{i<\omega} \varphi(\bar{x}, \bar{a}_i)$ is satisfied by any realization of a non-forking extension of p to $A \cup \{\bar{a}_i : i < \omega\}$. Therefore $\varphi(\bar{x}, \bar{a})$ cannot fork over C, a contradiction. Thus $C \subseteq \text{Cb}(p) \subseteq \text{bdd}(C)$. ■

COROLLARY 3.6.15 *If T is a simple theory with stable forking, $\bar{b}$ is an imaginary tuple and* $\text{tp}(a/A)$ *is not foreign to* $\text{tp}(\bar{b}/A)$, *then there is an imaginary element $a_0 \in \text{dcl}(Aa) - \text{acl}(A)$.*

PROOF: By Proposition 3.6.14 there is an imaginary set C with $C \subseteq \text{Cb}(\bar{b}/Aa) \subseteq \text{bdd}(C)$. So we take a_1 to be any tuple in $C - \text{acl}(A)$ and a_0 the set of Aa-conjugates of a_1. ■

COROLLARY 3.6.16 *If T is a simple theory with stable forking and p is Σ-analysable, where Σ is a collection of imaginary types, then every realization of p has a Σ-analysis consisting of imaginary elements.* ■

Of course Corollaries 3.6.15 and 3.6.16 are trivial for theories with elimination of hyperimaginaries.

REMARK 3.6.17 In the notation of the proof of Proposition 3.6.14, if q_1 and q_2 are two non-forking extensions of $p\lceil C$ to the same set B, then they agree on all stable formulas (even though they need not be $\sim$-related).

PROOF: There are C-conjugates p_1 and p_2 of p with $p_1 \sim q_1$ and $p_2 \sim q_2$. By C-conjugacy, p_1, p and p_2, and hence also q_1 and q_2, have the same φ-definitions $d_p\varphi$ for all stable formulas φ. ■

REMARK 3.6.18 Let T be a one-based simple theory with elimination of hyperimaginaries. Then T has stable forking.

PROOF: By elimination of hyperimaginaries, $\mathrm{bdd}(A) = \mathrm{acl}(A)$ for all sets A, and for the algebraic closure we need only consider imaginary elements. As the theory is one-based, $a \mathop{\smile\hskip-0.9em\raise0.6ex\hbox{$\mid$}}_{\mathrm{acl}(a)\cap\mathrm{acl}(b)} b$ for any tuples a and b. Suppose $a \not\mathop{\smile\hskip-0.9em\raise0.6ex\hbox{$\mid$}}_C b$. So

$$\mathrm{acl}(a) \cap \mathrm{acl}(bC) = \mathrm{Cb}(a/bC) \not\subseteq \mathrm{acl}(C)$$

and there is $a_0 \in [\mathrm{acl}(a) \cap \mathrm{acl}(bC)] - \mathrm{acl}(C)$. Hence there are some tuple $c \in C$ and algebraic formulas $\varphi(u, x)$ and $\psi(u, yz)$ (i.e. formulas which have only finitely many solutions in u for any choice of x, y, z), such that $a_0 \models \varphi(u, a) \wedge \psi(u, bc)$ and $\psi(u, bc)$ has no solution in $\mathrm{acl}(C)$.

CLAIM. $\theta(x, yz) = \exists u\, [\varphi(u, x) \wedge \psi(u, yz)]$ is a stable formula.

PROOF OF CLAIM: Let $(a_i, b_i c_i : i < \omega)$ be a sequence such that $\models \theta(a_i, b_j c_j)$ if and only if $i < j$. By algebraicity of φ and ψ we may thin out the sequence and suppose that the element witnessing the existential quantifier for $i < j$ is always the same. But then clearly $\models \theta(a_i, b_j c_j)$ for all i, j, a contradiction. ■

As $\psi(u, bc)$ has no solution in $\mathrm{acl}(C)$, for any realization a' of $\theta(x, bc)$ there is $a_0' \in [\mathrm{acl}(a') \cap \mathrm{acl}(bc)] - \mathrm{acl}(C)$. Hence $a' \not\mathop{\smile\hskip-0.9em\raise0.6ex\hbox{$\mid$}}_C b$, and $\theta(x, bc)$ forks over C. ■

3.7 THE LASCAR GROUP

In this section T is not necessarily simple.

DEFINITION 3.7.1 The *Galois group* $\mathrm{Gal}(\mathfrak{C})$ of the structure $\mathfrak{C}$ is the quotient group $\mathrm{Aut}(\mathfrak{C})/\mathrm{Autf}(\mathfrak{C})$. We denote the canonical surjective homomorphism $\mathrm{Aut}(\mathfrak{C}) \to \mathrm{Gal}(\mathfrak{C})$ by μ.

THEOREM 3.7.2 *Let $\mathfrak{C}'$ be an elementary extension of $\mathfrak{C}$, and $\sigma \in \mathrm{Aut}(\mathfrak{C}')$. Choose elementary substructures $\mathfrak{N}'$ of $\mathfrak{C}'$ and $\mathfrak{M}, \mathfrak{N}$ of $\mathfrak{C}$ with $\mathrm{tp}(\mathfrak{N}/\mathfrak{M}) = \mathrm{tp}(\mathfrak{N}'/\mathfrak{M})$, and an automorphism $\tau \in \mathrm{Aut}(\mathfrak{C})$ with $\mathrm{tp}(\tau(\mathfrak{N})/\mathfrak{M}) = \mathrm{tp}(\sigma(\mathfrak{N}')/\mathfrak{M})$. Then $\tau/\mathrm{Autf}(\mathfrak{C}) \in \mathrm{Gal}(\mathfrak{C})$ depends only on σ; we denote it as $\alpha_{\mathfrak{C}',\mathfrak{C}}(\sigma)$. The map $\alpha_{\mathfrak{C}',\mathfrak{C}}$ is a homomorphism from $\mathrm{Aut}(\mathfrak{C}')$ to $\mathrm{Gal}(\mathfrak{C})$. If $\mathfrak{C}'$ is $|\mathfrak{C}|^+$-saturated and -homogeneous, then $\alpha_{\mathfrak{C}',\mathfrak{C}}$ induces an isomorphism $\gamma_{\mathfrak{C}',\mathfrak{C}}$ between $\mathrm{Gal}(\mathfrak{C}')$ and $\mathrm{Gal}(\mathfrak{C})$.*

PROOF: Suppose first that τ' is another choice of automorphism in $\mathrm{Aut}(\mathfrak{C})$ with $\mathrm{tp}(\tau'(\mathfrak{N})/\mathfrak{M}) = \mathrm{tp}(\sigma(\mathfrak{N}')/\mathfrak{M})$. Then $\mathrm{tp}(\tau'(\mathfrak{N})/\mathfrak{M}) = \mathrm{tp}(\tau(\mathfrak{N})/\mathfrak{M})$, so there is $\tau'' \in \mathrm{Autf}_{\mathfrak{M}}(\mathfrak{C})$ with $\tau''(\tau'(\mathfrak{N})) = \tau(\mathfrak{N})$. Thus $\tau''\tau'(\mathfrak{N}) = \tau(\mathfrak{N})$ and $\tau^{-1}\tau''\tau'$ fixes $\mathfrak{N}$. So it is in $\mathrm{Autf}(\mathfrak{C})$; as $\tau'' \in \mathrm{Autf}(\mathfrak{C})$, we get $\tau/\mathrm{Autf}(\mathfrak{C}) = \tau'/\mathrm{Autf}(\mathfrak{C})$.

Secondly, suppose $\mathfrak{N}_0 \prec \mathfrak{C}$ and $\mathfrak{N}'_0 \prec \mathfrak{C}'$ are two other models with $\mathrm{tp}(\mathfrak{N}_0/\mathfrak{M}) = \mathrm{tp}(\mathfrak{N}'_0/\mathfrak{M})$, and $\tau_0 \in \mathrm{Aut}(\mathfrak{C})$ satisfies $\mathrm{tp}(\tau_0(\mathfrak{N}_0)/\mathfrak{M}) = \mathrm{tp}(\sigma(\mathfrak{N}'_0)/\mathfrak{M})$. If $\mathfrak{N}_1 \prec \mathfrak{C}$ and $\mathfrak{N}'_1 \prec \mathfrak{C}'$ have the same type over $\mathfrak{M}$, with $\mathfrak{N}\mathfrak{N}_0 \subset \mathfrak{N}_1$ and $\mathfrak{N}'\mathfrak{N}'_0 \subset \mathfrak{N}'_1$, choose $\tau_1 \in \mathrm{Aut}(\mathfrak{C})$ with $\mathrm{tp}(\tau(\mathfrak{N}_1)/\mathfrak{M}) = \mathrm{tp}(\sigma(\mathfrak{N}'_1)/\mathfrak{M})$. Then $\mathrm{tp}(\tau_1(\mathfrak{N})/\mathfrak{M}) = \mathrm{tp}(\sigma(\mathfrak{N}')/\mathfrak{M}) = \mathrm{tp}(\tau(\mathfrak{N})/\mathfrak{M})$ and $\mathrm{tp}(\tau_1(\mathfrak{N}_0)/\mathfrak{M}) = \mathrm{tp}(\sigma(\mathfrak{N}'_0)/\mathfrak{M}) = \mathrm{tp}(\tau_0(\mathfrak{N}_0)/\mathfrak{M})$, whence by the first part $\tau/\mathrm{Autf}(\mathfrak{C}) = \tau_1/\mathrm{Autf}(\mathfrak{C}) = \tau_0/\mathrm{Autf}(\mathfrak{C})$.

Next, suppose $\mathfrak{M}_0$ is another small elementary substructure of $\mathfrak{C}$, and $\mathfrak{N}_0 \prec \mathfrak{C}$ and $\mathfrak{N}'_0 \prec \mathfrak{C}'$ have the same type over $\mathfrak{M}_0$. By the second part we may assume that they have the same type over a model $\mathfrak{M}'$ containing $\mathfrak{M} \cup \mathfrak{M}_0$, and $\mathfrak{N}_0 = \mathfrak{N}$, $\mathfrak{N}'_0 = \mathfrak{N}'$. So if $\mathrm{tp}(\tau_0(\mathfrak{N})/\mathfrak{M}') = \mathrm{tp}(\sigma(\mathfrak{N})/\mathfrak{M}')$, then $\mathrm{tp}(\tau_0(\mathfrak{N})/\mathfrak{M}) = \mathrm{tp}(\tau(\mathfrak{N})/\mathfrak{M})$ and $\tau/\mathrm{Autf}(\mathfrak{C}) = \tau_0/\mathrm{Autf}(\mathfrak{C})$. This shows that $\alpha_{\mathfrak{C}',\mathfrak{C}}$ is well-defined.

Now consider $\sigma' \in \mathrm{Aut}(\mathfrak{C}')$, and choose $\tau' \in \mathrm{Aut}(\mathfrak{C})$ such that $\mathrm{tp}(\tau'(\tau(N))/\mathfrak{M}) = \mathrm{tp}(\sigma'(\sigma(\mathfrak{N}'))/\mathfrak{M})$. Then $\alpha_{\mathfrak{C}',\mathfrak{C}}(\sigma') = \tau'$, and clearly $\alpha_{\mathfrak{C}',\mathfrak{C}}(\sigma'\sigma) = \tau'\tau$. As $\alpha_{\mathfrak{C}',\mathfrak{C}}(1) = 1$, we see that $\alpha_{\mathfrak{C}',\mathfrak{C}}$ is a homomorphism.

Finally, suppose that $\mathfrak{C}'$ is $|\mathfrak{C}|^+$-saturated and -homogeneous. So if $\tau \in \mathrm{Aut}(\mathfrak{C})$, then there is $\sigma \in \mathrm{Aut}(\mathfrak{C}')$ extending τ, whence $\alpha_{\mathfrak{C}',\mathfrak{C}}(\sigma) = \tau$ and $\alpha_{\mathfrak{C}',\mathfrak{C}}$ is surjective. Clearly, if $\tau \in \mathrm{Autf}(\mathfrak{C})$, then the continuation σ is in $\mathrm{Autf}(\mathfrak{C}')$, so $\ker(\alpha_{\mathfrak{C}',\mathfrak{C}}) \leq \mathrm{Autf}(\mathfrak{C}')$. Conversely, if $\sigma \in \mathrm{Aut}(\mathfrak{C}')$ fixes some model $\mathfrak{N}'$, we can take any $\mathfrak{M} \prec \mathfrak{C}$ and $\mathfrak{N} \models \mathrm{tp}(\mathfrak{N}'/\mathfrak{M})$ in $\mathfrak{C}$, to see that $\alpha_{\mathfrak{C}',\mathfrak{C}}(\sigma) = 1$. ∎

LEMMA 3.7.3 *Let $\sigma \in Aut(\mathfrak{C})$. Then the following are equivalent:*

1. *$\sigma \in Autf(\mathfrak{C})$.*
2. *For any $\emptyset$-invariant set X and any $\emptyset$-invariant bounded equivalence relation E, we have $E(x, \sigma(x))$ for all $x \in X$.*
3. *For some (any) elementary submodel $\mathfrak{M} \prec \mathfrak{C}$ we have* $\mathrm{lstp}(\mathfrak{M}) = \mathrm{lstp}(\sigma(\mathfrak{M}))$.

PROOF: 1. ⇒ 2. is Proposition 2.7.5. 2. ⇒ 3. is obvious, as equality of Lascar strong type is the finest $\emptyset$-invariant bounded equivalence relation over $\emptyset$. Now assume $\mathrm{lstp}(\mathfrak{M}) = \mathrm{lstp}(\sigma(\mathfrak{M}))$. So there is $\tau \in \mathrm{Autf}(\mathfrak{C})$ with $\tau(\mathfrak{M}) = \sigma(\mathfrak{M})$. Hence $\tau^{-1}\sigma(\mathfrak{M}) = \mathfrak{M}$ and $\tau^{-1}\sigma \in \mathrm{Autf}(\mathfrak{C})$, whence $\sigma \in \mathrm{Autf}(\mathfrak{C})$. ∎

REMARK 3.7.4 It follows that $\mathrm{Gal}(\mathfrak{C})$ acts on the collection of equivalence classes modulo $\emptyset$-invariant bounded equivalence relations.

In a non-simple theory, a bounded $\emptyset$-invariant equivalence relation need not be type-definable.

We can now define a topology on $\mathrm{Gal}(\mathfrak{C})$.

DEFINITION 3.7.5 A subset X of Gal($\mathfrak{C}$) is *closed* if whenever $\mathcal{U}$ is an ultrafilter on a set I and $\sigma_i \in$ Aut($\mathfrak{C}$) for $i \in I$ with $\mu(\sigma_i) \in X$, then, for $\mathfrak{C}' = \mathfrak{C}^I/\mathcal{U}$, we have $\alpha_{\mathfrak{C}',\mathfrak{C}}((\sigma_i : i \in I)/\mathcal{U}) \in X$.

LEMMA 3.7.6 *Definition 3.7.5 defines a topology on Gal($\mathfrak{C}$). Furthermore, the following are equivalent for any subset $X \subseteq$ Gal($\mathfrak{C}$):*

1. *X is closed.*

2. *For any (possibly infinite) tuple $\bar{a} \in \mathfrak{C}$, the set*

$$\{\sigma(\bar{a}) : \sigma \in Aut(\mathfrak{C}) \text{ and } \mu(\sigma) \in X\}$$

is type-definable over some (any) small submodel of $\mathfrak{C}$.

3. *There are a tuple $\bar{a}$ and a partial type $\pi(\bar{x})$ (possibly with parameters) such that $\mu^{-1}(X) = \{\sigma \in Aut(\mathfrak{C}) : \sigma(\bar{a}) \models \pi\}$. Moreover, we may choose any small model $\mathfrak{M} \prec \mathfrak{C}$ for $\bar{a}$, and π to be a partial type over $\mathfrak{M}$.*

PROOF: Closed sets are clearly closed under intersections and finite unions, so Definition 3.7.5 defines a topology.

1. $\Rightarrow$ 2. Let X be closed, and $\bar{a}$ a tuple in $\mathfrak{C}$. Then $\bar{a}$ is contained in a small model $\mathfrak{M}$; clearly it is sufficient to prove the assertion for $\bar{a} = \mathfrak{M}$. Put $Y = \mu^{-1}(X)$ and $S = \{\sigma(\mathfrak{M}) : \sigma \in Y\}$.

CLAIM. S is type-definable over $\mathfrak{M}$.

PROOF OF CLAIM: Suppose $q \in S(\mathfrak{M})$ is finitely satisfiable in S, choose a realization $\mathfrak{N} \prec \mathfrak{C}$ of q and $\sigma \in$ Aut($\mathfrak{C}$) with $\sigma(\mathfrak{M}) = \mathfrak{N}$. If I is the collection of finite subsets of q, then for every $i \in I$ there are a realization $\mathfrak{N}_i \in S$ of i and an ultrafilter $\mathcal{U}$ on I with $(\mathfrak{N}_i : i \in I)/\mathcal{U} \models q$ in the ultraproduct $\mathfrak{C}' = \mathfrak{C}^I/\mathcal{U}$. Put $\mathfrak{N}' = (\mathfrak{N}_i : i \in I)/\mathcal{U}$. If $\sigma_i \in Y$ is such that $\sigma_i(\mathfrak{M}) = \mathfrak{N}_i$ and $\sigma' = \prod_{i\in I} \sigma_i/\mathcal{U}$, then $\sigma' \in$ Aut($\mathfrak{C}'$) and $\sigma'(\mathfrak{M}) = \mathfrak{N}'$. As tp($\mathfrak{N}'/\mathfrak{M}$) $= q =$ tp($\mathfrak{N}/\mathfrak{M}$), we have $\alpha_{\mathfrak{C}',\mathfrak{C}}(\sigma') = \sigma/$Autf($\mathfrak{C}$). As X is closed, $\sigma/$Autf($\mathfrak{C}$) $\in X$, whence $\mathfrak{N} \in S$. ■

Hence 2. must hold.

2. $\Rightarrow$ 3. Assume 2. and choose a model $\mathfrak{M} \prec \mathfrak{C}$. By 2. the set $\{\sigma(\mathfrak{M}) : \sigma \in \mu^{-1}(X)\}$ is given by a partial type π over $\mathfrak{M}$. Clearly $\sigma \in \mu^{-1}(X)$ implies $\sigma(\mathfrak{M}) \models \pi$. For the converse, consider $\sigma \in$ Aut($\mathfrak{C}$) with $\sigma(\mathfrak{M}) \models \pi$. So there is $\sigma' \in \mu^{-1}(X)$ with $\sigma'(\mathfrak{M}) = \sigma(\mathfrak{M})$, whence $\sigma^{-1}\sigma'(\mathfrak{M}) = \mathfrak{M}$ and $\sigma^{-1}\sigma \in$ Autf($\mathfrak{C}$). Therefore $\sigma \in \mu^{-1}(X)$.

3. $\Rightarrow$ 1. Suppose $\bar{a}$ and π satisfy 3.; by adding dummy variables, we may assume that $\bar{a} = \mathfrak{M}$ is a model and π is over $\mathfrak{M}$. Let $\mathcal{U}$ be an ultrafilter in a set I, and suppose $\sigma_i \in \mu^{-1}(X)$ for all $i \in I$. Put $\mathfrak{C}' = \mathfrak{C}^I/\mathcal{U}$ and $\sigma' = \prod_{i\in I} \sigma_i/\mathcal{U}$. Then $\mathfrak{C}' \models \pi(\sigma'(\mathfrak{M}))$. If $\mathfrak{N} \prec \mathfrak{C}$ realizes tp($\sigma'(\mathfrak{M})/\mathfrak{M}$), then $\mathfrak{N} \models \pi$, so there

is $\sigma \in \mu^{-1}(X)$ with $\sigma(\mathfrak{M}) = \mathfrak{N}$. Hence $\alpha_{\mathfrak{C}',\mathfrak{C}}(\sigma') = \sigma/\text{Autf}(\mathfrak{C})$, which is in X. As I and $\mathcal{U}$ were arbitrary, X is closed. ∎

For a set $X \subseteq \text{Gal}(\mathfrak{C})$ we shall write $\text{cl}(X)$ for the closure of X in this topology.

LEMMA 3.7.7 *Let G be a closed subgroup of Gal($\mathfrak{C}$). Then there is a type-definable equivalence relation $R(x, y)$ and a complete type $p(x)$ over $\emptyset$, such that $\mu(\sigma) \in G$ if and only if $R(\sigma(a), a)$ holds for some $a \models p$. In fact, we can choose for a any elementary substructure of $\mathfrak{C}$.*

PROOF: By Lemma 3.7.6 for any elementary substructure $\mathfrak{M} \prec \mathfrak{C}$ there is a partial type $\pi(X, \mathfrak{M})$ over $\mathfrak{M}$ with $\mu^{-1}(G) = \{\sigma \in \text{Aut}(\mathfrak{C}) : \sigma(\mathfrak{M}) \models \pi(X, \mathfrak{M})\}$.

CLAIM. $\pi(X, Y)$ is a type-definable equivalence relation on $\text{tp}(\mathfrak{M})$.

PROOF OF CLAIM: $1 \in G$ implies $\models \pi(\mathfrak{M}, \mathfrak{M})$. If $\models \pi(\mathfrak{N}, \mathfrak{M})$ for some $\mathfrak{N} \models \text{tp}(\mathfrak{M})$, then there are $\sigma \in \text{Aut}(\mathfrak{C})$ with $\sigma(\mathfrak{M}) = \mathfrak{N}$; by definition of π we get $\mu(\sigma) \in G$ and $\sigma^{-1} \in \mu^{-1}(G)$. So $\models \pi(\sigma^{-1}(\mathfrak{M}), \mathfrak{M})$, whence $\models \pi(\mathfrak{M}, \sigma(\mathfrak{M}))$, that is $\models \pi(\mathfrak{M}, \mathfrak{N})$. Finally, if $\pi(\mathfrak{M}'', \mathfrak{M}')$ and $\pi(\mathfrak{M}', \mathfrak{M})$ hold for some $\mathfrak{M}, \mathfrak{M}', \mathfrak{M}'' \models \text{tp}(\mathfrak{M})$, then there are $\sigma, \sigma' \in \text{Aut}(\mathfrak{C})$ with $\mathfrak{M}' = \sigma(\mathfrak{M})$ and $\mathfrak{M}'' = \sigma'(\mathfrak{M})$. Hence $\models \pi(\sigma'(\mathfrak{M}), \sigma(\mathfrak{M}))$ and $\models \pi(\sigma(\mathfrak{M}), \mathfrak{M})$; we also get (by applying σ^{-1}) that $\pi(\sigma^{-1}\sigma'(\mathfrak{M}), \mathfrak{M})$. By definition, $\mu(\sigma), \mu(\sigma^{-1}\sigma') \in G$, whence $\mu(\sigma') \in G$ and $\models \pi(\sigma'(\mathfrak{M}), \mathfrak{M})$, that is $\models \pi(\mathfrak{M}'', \mathfrak{M})$. ∎

So we take $p = \text{tp}(\mathfrak{M})$, $a = \mathfrak{M}$ and $R = \pi$. Note that R must be bounded, since $\neg R(\sigma'(\mathfrak{M}), \sigma(\mathfrak{M}))$ implies $\sigma^{-1}\sigma' \notin \mu^{-1}(G)$, and G has bounded index in Gal($\mathfrak{C}$) (the latter being bounded). ∎

LEMMA 3.7.8 *1. Gal($\mathfrak{C}$) is quasicompact.*

2. Translation and inversion are continuous.

3. If $\sigma, \tau \in$ Gal($\mathfrak{C}$) and $\sigma \in \text{cl}(\tau)$, then $\tau \in \text{cl}(\sigma)$.

4. $\text{cl}(1)$ is a subgroup of Gal($\mathfrak{C}$).

PROOF:

1. Suppose $(U_i : i \in I)$ is an open cover of Gal($\mathfrak{C}$) without a finite subcover. For any finite subset $j \subset I$ choose $\sigma_j \in \text{Gal}(\mathfrak{C}) - \bigcup_{i \in j} U_i$. Let J be the collection of finite subsets of I, and $\mathcal{I}$ the family of those $J_0 \subset J$ such that there is a finite $j_0 \subset I$ with $j_0 \not\subseteq j$ for all $j \in J_0$. Then $\emptyset \in \mathcal{I}$, $J \notin \mathcal{I}$, and $J_0 \cup J_1 \in \mathcal{I}$ for all $J_0, J_1 \in \mathcal{I}$, so $\mathcal{I}$ is an ideal, and there is an ultrafilter $\mathcal{U}$ on J intersecting $\mathcal{I}$ trivially. Put $\sigma = \alpha_{\mathfrak{C}^J/\mathcal{U},\mathfrak{C}}(\prod_{j \in J} \sigma_j/\mathcal{U})$. As $\sigma \in \text{Gal}(\mathfrak{C})$, there is some $i \in I$ such that $\sigma \in U_i$. Since the complement of U_i is closed,

$J' = \{j \in J : \sigma_j \in U_i\} \in \mathcal{U}$. But $\sigma_j \notin U_i$ for all $j \ni i$, so $i \notin j$ for all $j \in J'$. Thus $J' \in \mathcal{I}$, a contradiction.

2. This follows from the definition of closedness, together with the fact that $(\prod_{i \in I} \sigma_i/\mathcal{U})(\prod_{i \in I} \tau_i/\mathcal{U}) = \prod_{i \in I} \sigma_i\tau_i/\mathcal{U}$ for all index sets I, sequences $(\sigma_i : i \in I)$ and $(\tau_i : i \in I)$ in Gal($\mathfrak{C}$), and ultrafilters $\mathcal{U}$ on I. Note that in particular, for any $X \subseteq \text{Gal}(\mathfrak{C})$ and $\sigma \in \text{Gal}(\mathfrak{C})$ we have $\sigma\text{cl}(X) = \text{cl}(\sigma X)$.

3. For $\sigma, \tau \in \text{Gal}(\mathfrak{C})$ write $\sigma \geq \tau$ if $\tau \in \text{cl}(\sigma)$. By quasicompactness of Gal($\mathfrak{C}$), there is some $\sigma_0 \in \text{Gal}(\mathfrak{C})$ which is minimal for the order $\geq$. But every element of Gal($\mathfrak{C}$) is a translate of σ_0 and the topology is translation invariant, so all elements are minimal for $\geq$. Hence $\sigma \geq \tau$ implies $\tau \geq \sigma$, so σ and τ have the same closure.

4. Suppose σ and τ are in $\text{cl}(1)$. By translation invariance, $\sigma\tau \in \text{cl}(\sigma 1) = \text{cl}(\sigma) = \text{cl}(1)$. Similarly, $1 = \sigma^{-1}\sigma \in \text{cl}(\sigma^{-1}1) = \text{cl}(\sigma^{-1})$, so $\sigma^{-1} \in \text{cl}(1)$ by part 3. ■

LEMMA 3.7.9 *The following are equivalent for any $\sigma \in Aut(\mathfrak{C})$:*

1. $\mu(\sigma) \in \text{cl}(1)$.

2. For any $\emptyset$-invariant set X and any type-definable bounded equivalence relation R over $\emptyset$, we have $R(x, \sigma(x))$ for all $x \in X$.

3. For some (any) elementary submodel $\mathfrak{M} \prec \mathfrak{C}$ and the finest type-definable bounded equivalence relation $R_{\mathfrak{M}}$ over $\emptyset$ on $\text{tp}(\mathfrak{M})$ we have $R_{\mathfrak{M}}(\mathfrak{M}, \sigma(\mathfrak{M}))$.

Furthermore, if $\sigma \notin \text{cl}(\sigma')$, then there are disjoint open sets $U \ni \sigma$ and $U' \ni \sigma'$.

PROOF: 1. ⇒ 2. Fix $x \in X$ and consider the set $Y = \{\tau \in \text{Aut}(\mathfrak{C}) : R(x, \tau(x))\}$. As $R(\tau(x), \tau'(\tau(x))$ for any $\tau' \in \text{Autf}(\mathfrak{C})$ by Lemma 3.7.3, we get $R(x, \tau'\tau(x))$. Therefore $Y = \mu^{-1}(\mu(Y))$, and $\mu(Y)$ is a closed subset of Gal($\mathfrak{C}$) containing 1 by Lemma 3.7.6. So $\mu(\sigma) \in \text{cl}(1)$ implies $\mu(\sigma) \in \mu(Y)$, whence $\sigma \in Y$ and $R(x, \sigma(x))$.

2. ⇒ 3. is obvious. For 3. ⇒ 1. by Lemma 3.7.7 given any elementary substructure $\mathfrak{M} \prec \mathfrak{C}$ there is a type-definable bounded equivalence relation R on $\text{tp}(\mathfrak{M})$ such that $\mu^{-1}(\text{cl}(1)) = \{\sigma \in \text{Aut}(\mathfrak{C}) :\models R(\sigma(\mathfrak{M}), \mathfrak{M})\}$. In fact, by 1. ⇒ 2. we have $R = R_{\mathfrak{M}}$; in particular $R_{\mathfrak{M}}(\mathfrak{M}, \sigma(\mathfrak{M}))$ if and only if $\mu(\sigma) \in \text{cl}(1)$.

To show the "furthermore", suppose $\sigma \notin \text{cl}(\sigma')$, so $\sigma'^{-1}\sigma \notin \text{cl}(1)$. Choose τ and τ' in Aut($\mathfrak{C}$) with $\mu(\tau) = \sigma$ and $\mu(\tau') = \sigma'$. Let $\mathfrak{M}$ and $R_{\mathfrak{M}}$ be as in 3., and choose a formula $\varphi(X, Y) \in R_{\mathfrak{M}}$ with $\models \neg\varphi(\tau(\mathfrak{M}), \tau'(\mathfrak{M}))$. As

$R_{\mathfrak{M}}$ is a type-definable equivalence relation, there is a formula $\psi \in R_{\mathfrak{M}}$ with $\psi(X,Y) \wedge \psi(Y,Z) \vdash \varphi(X,Z)$. Put

$$C_1 = \{\tau'' \in \mathrm{Aut}(\mathfrak{C}) : \tau''(\mathfrak{M}) \models \exists Y\,[\neg\psi(\tau(\mathfrak{M}),Y) \wedge R_{\mathfrak{M}}(\tau''(\mathfrak{M}),Y)]\}$$
$$C_2 = \{\tau'' \in \mathrm{Aut}(\mathfrak{C}) : \tau''(\mathfrak{M}) \models \exists Y\,[\neg\psi(Y,\tau'(\mathfrak{M})) \wedge R_{\mathfrak{M}}(Y,\tau''(\mathfrak{M}))]\}.$$

It is easy to see that C_1 and C_2 consist of cosets of $\mathrm{Autf}(\mathfrak{C})$, so $\mu(C_1)$ and $\mu(C_2)$ are closed in $\mathrm{Gal}(\mathfrak{C})$ by Lemma 3.7.6. Furthermore, since $\models \neg\varphi(\tau(\mathfrak{M}),\tau'(\mathfrak{M}))$ we must have $\models \neg\psi(\tau(\mathfrak{M}),\mathfrak{M}') \vee \neg\psi(\mathfrak{M}',\tau'(\mathfrak{M}))$ for all $\mathfrak{M}' \models \mathrm{tp}(\mathfrak{M})$, so $C_1 \cup C_2 = \mathrm{Aut}(\mathfrak{C})$ and $\mu(C_1) \cup \mu(C_2) = \mathrm{Gal}(\mathfrak{C})$. Finally, as $R_{\mathfrak{M}} \vdash \psi$, we have $\tau \notin C_1$ and $\tau' \notin C_2$. Putting $U = \mathrm{Gal}(\mathfrak{C}) - \mu(C_1)$ and $U' = \mathrm{Gal}(\mathfrak{C}) - \mu(C_2)$, we are done. ■

PROPOSITION 3.7.10 *1. $Gal(\mathfrak{C})$ is a quasicompact topological group.*

2. *If $X \subset Gal(\mathfrak{C})$ and $\sigma \in Gal(\mathfrak{C})$, then σ is in the topological closure of X if and only if there are a set I, an ultrafilter $\mathcal{U}$ on I, and a sequence $(\tau_i : i \in I)$ of elements in $Aut(\mathfrak{C})$ with $\mu(\tau_i) \in X$ for all $i \in I$, such that $\alpha_{\mathfrak{C}^I/\mathcal{U},\mathfrak{C}}(\prod_{i\in I} \tau_i/\mathcal{U}) = \sigma$.*

3. *If $\mathfrak{C}'$ is a $|\mathfrak{C}|^+$-saturated and -homogeneous elementary extension of $\mathfrak{C}$, then $\gamma_{\mathfrak{C}',\mathfrak{C}}$ is a homeomorphism.*

4. *If $Aut(\mathfrak{C})$ is endowed with the topology of pointwise convergence, then $\mu : Aut(\mathfrak{C}) \to Gal(\mathfrak{C})$ is continuous.*

PROOF:

1. We have to show that multiplication is continuous. So consider $\sigma, \tau \in \mathrm{Gal}(\mathfrak{C})$ and an open neighbourhood U of $\sigma\tau$. By translation invariance of the topology, it is sufficient to show that for every open neighbourhood U of 1 there is an open neighbourhood U' of 1 such that $\sigma\tau \in U$ for all $\sigma, \tau \in U'$. So let $\mathfrak{O}$ be the collection of all open neighbourhoods of 1, and suppose there is $U \in \mathfrak{O}$ such that for every $O \in \mathfrak{O}$ there are $\sigma_O, \tau_O \in O$ with $\sigma_O\tau_O \notin U$. Let $\mathcal{U}$ be an ultrafilter on $\mathfrak{O}$ extending the filter $\{\{O \in \mathfrak{O} : O \subseteq O_0\} : O_0 \in \mathfrak{O}\}$.

   CLAIM. If $\sigma = \alpha_{\mathfrak{C}^I/\mathcal{U},\mathfrak{C}}(\prod_{O\in\mathfrak{O}} \sigma_O/\mathcal{U})$, then $\mathrm{cl}(1) = \mathrm{cl}(\sigma)$.

   PROOF OF CLAIM: First we show that if U' is an open neighbourhood of σ, then $\{O \in \mathfrak{O} : \sigma_O \in U'\} \in \mathcal{U}$. So suppose otherwise, fix some $\tau \notin U'$, and put $\sigma'_O = \sigma_O$ if $\sigma_O \notin U'$, and $\sigma'_O = \tau$ if $\sigma_O \in U'$. Then

$$\sigma = \alpha_{\mathfrak{C}^I/\mathcal{U},\mathfrak{C}}(\prod_{O\in\mathfrak{O}} \sigma_O) = \alpha_{\mathfrak{C}^I/\mathcal{U},\mathfrak{C}}(\prod_{O\in\mathfrak{O}} \sigma'_O) \notin U'$$

by closedness of the complement of U', a contradiction.

It follows that for any open neighbourhood U' of σ and O of 1 the intersection $U' \cap O$ is non-empty. Hence $\mathrm{cl}(1) = \mathrm{cl}(\sigma)$ by Lemma 3.7.9. ■

Similarly, we see that $\mathrm{cl}(1) = \mathrm{cl}(\tau)$, where $\tau = \alpha_{\mathfrak{C}^I/\mathcal{U},\mathfrak{C}}(\prod_{O \in \mathfrak{O}} \tau_O/\mathcal{U})$. As $\mathrm{cl}(1)$ is closed under multiplication, $\mathrm{cl}(\sigma\tau) = \mathrm{cl}(1)$. On the other hand, by closedness of the complement, $\tau\sigma \notin U$, contradicting $1 \in U$.

2. "If" is clear by definition of closure, so we show "only if". Suppose for $i \in I$ we are given an ultrafilter $\mathcal{U}_i$ on a set J_i and a sequence $(\sigma_i^j : j \in J_i)$ in $\mu^{-1}(X)$. Put $\sigma_i = \alpha_{\mathfrak{C}^{J_i}/\mathcal{U}_i,\mathfrak{C}}(\prod_{j \in J_i} \sigma_i^j/\mathcal{U}_i)$. Now consider an ultrafilter $\mathcal{U}$ on I. Put $I_0 = \bigcup J_i \times \{i\}$ and
$$\mathcal{U}_0 = \{S \subseteq I_0 : \{i \in I : \exists T \in \mathcal{U}_i\ T \times \{i\} \subseteq S\} \in \mathcal{U}\}.$$
Then $\mathcal{U}_0$ is an ultrafilter on I_0, the ultraproduct $\prod_{i \in I}(\mathfrak{C}^{J_i}/\mathcal{U}_i)/\mathcal{U}$ is canonically isomorphic to $\prod_{(j,i) \in I_0} \mathfrak{C}^{I_0}/\mathcal{U}_0$, and this isomorphism maps $\prod_{i \in I}(\prod_{j \in J_i} \sigma_i^j/\mathcal{U}_i)/\mathcal{U}$ to $\prod_{(j,i) \in I_0} \sigma_i^j/\mathcal{U}_0$. If $\tau_i \in \mu^{-1}(\sigma_i)$ for all $i \in I$, it can now be checked that $\alpha_{\mathfrak{C}^I/\mathcal{U},\mathfrak{C}}(\prod_{i \in I} \tau_i/\mathcal{U}) = \alpha_{\mathfrak{C}^{I_0}/\mathcal{U}_0}(\prod_{(i,j) \in I_0} \sigma_i^j/\mathcal{U}_0)$. Thus the set of points in $\mathrm{Gal}(\mathfrak{C})$ of the form $\alpha_{\mathfrak{C}^I/\mathcal{U},\mathfrak{C}}(\prod_{i \in I} \sigma_i/\mathcal{U})$ for some set I, some ultrafilter $\mathcal{U}$ on I, and some sequence $(\sigma_i : i \in I)$ in $\mu^{-1}(X)$ is closed under the ultraproduct construction, and hence closed in the topology.

3. If X is a closed subset of $\mathrm{Gal}(\mathfrak{C})$, then by Lemma 3.7.6 there is a elementary substructure $\mathfrak{M} \prec \mathfrak{C}$ and a partial type π over $\mathfrak{M}$ such that $\mu_{\mathfrak{C}}^{-1}(X) = \{\sigma \in \mathrm{Aut}(\mathfrak{C}) : \sigma(\mathfrak{M}) \models \pi\}$. But then $\alpha_{\mathfrak{C}',\mathfrak{C}}^{-1}(X) = \{\sigma' \in \mathrm{Aut}(\mathfrak{C}') : \sigma'(\mathfrak{M}) \models \pi\}$, so $\gamma_{\mathfrak{C}',\mathfrak{C}}^{-1}(X)$ is closed by Lemma 3.7.6. Continuity of $\gamma_{\mathfrak{C}',\mathfrak{C}}^{-1}$ is shown similarly.

4. Suppose $X \subseteq \mathrm{Gal}(\mathfrak{C})$ is closed, and suppose $\sigma \in \mathrm{Aut}(\mathfrak{C})$ is in the closure of $\mu^{-1}(X)$. So for every finite tuple $\bar{a} \in \mathfrak{C}$ there is $\sigma_{\bar{a}} \in \mu^{-1}(X)$ with $\sigma_{\bar{a}}(\bar{a}) = \sigma(\bar{a})$. Let I be the set of finite tuples in $\mathfrak{C}$ and $\mathcal{U}$ an ultrafilter on I extending the filter $\{\{i \in I : i \supseteq i_0\} : i_0 \in I\}$. Then $\prod_{i \in I} \sigma_i/\mathcal{U}$ extends σ, and $\alpha_{\mathfrak{C}^I/\mathcal{U},\mathfrak{C}}(\prod_{i \in I} \sigma_i/\mathcal{U}) = \sigma$. ■

DEFINITION 3.7.11 If G is a group acting on a set X and $Y \subseteq X$, the *stabilizer* $\mathrm{stab}(Y)$ is the subgroup $\{g \in G : \forall x \in Y\ gx = x\}$.

PROPOSITION 3.7.12 *Let G be a subgroup of $\mathrm{Gal}(\mathfrak{C})$. Then G is closed if and only if G is the stabilizer of some bounded hyperimaginary.*

PROOF: Suppose $G = \mathrm{stab}(e)$, where $e \in \mathrm{bdd}(\emptyset)$. If $\bar{a}$ is any tuple from $\mathfrak{C}$ and $\mu(\sigma) \in G$, then
$$\mathrm{tp}(\bar{a}/e) = \mathrm{tp}(\sigma(\bar{a})/\sigma(e)) = \mathrm{tp}(\sigma(\bar{a})/e),$$

so $\{\sigma(\bar{a}) : \sigma \in \mu^{-1}(G)\}$ is type-definable. Hence G is closed by Lemma 3.7.6.

Conversely, suppose G is closed. By Lemma 3.7.7 there is a type p, some $a \models p$ and a type-definable bounded equivalence relation R on p, all over $\emptyset$, with $\mu^{-1}(G) = \{\sigma \in \mathrm{Aut}(\mathfrak{C}) : R(\sigma(a), a)\}$. By Remark 3.1.1 we can extend R to a type-definable equivalence relation defined on the whole of $\mathfrak{C}$. Hence $G = \mathrm{stab}(a_R)$. ■

LEMMA 3.7.13 *If G is a closed subgroup of Gal($\mathfrak{C}$), then the equivalence relation E_n on $\mathfrak{C}^n$ stating that $\bar{a}$ and $\bar{b}$ lie in the same $\mu^{-1}(G)$-orbit is type-definable. If $G = \mathrm{cl}(1)$, or more generally if G is normal in Gal($\mathfrak{C}$), then E_n is type-definable over $\emptyset$.*

PROOF: Since G is closed, there is a hyperimaginary element e with $G = \mathrm{stab}(e)$, and E_n is e-invariant. If moreover G is normal in Gal($\mathfrak{C}$), then E_n is actually $\emptyset$-invariant. We shall show that E_n is closed in $S_{2n}(e)$ (or $S_{2n}(\emptyset)$, respectively). So suppose $E_n(\bar{a}_i, \bar{b}_i)$ for all $i \in I$, and consider $\sigma_i \in \mu^{-1}(G)$ with $\sigma_i(\bar{a}_i) = \bar{b}_i$. Then for any ultrafilter $\mathcal{U}$ on I we can find $\sigma \in \mathrm{Aut}(\mathfrak{C})$, an elementary substructure $\mathfrak{M} \prec \mathfrak{C}$, and $\bar{a}' \in \mathfrak{C}$ with

$$\mathrm{tp}(\mathfrak{M}, (\prod_{i\in I} \sigma_i/\mathcal{U})(\mathfrak{M}), \prod_{i\in I} \bar{a}_i/\mathcal{U}, \prod_{i\in I} \bar{b}_i/\mathcal{U}) = \mathrm{tp}(\mathfrak{M}, \sigma(\mathfrak{M}), \bar{a}', \sigma(\bar{a}')).$$

It follows that $\mu(\sigma) = \alpha_{\mathfrak{C}^I/\mathcal{U},\mathfrak{C}}(\prod_{i\in I} \sigma_i/\mathcal{U})$, and this is in G by closedness of G. Hence $E_n(\bar{a}', \sigma(\bar{a}'))$, whence $E_n(\prod_{i\in I} \bar{a}_i/\mathcal{U}, \prod_{i\in I} \bar{b}_i/\mathcal{U})$. Thus E_n is closed under ultraproducts, and therefore a partial type (over e or over $\emptyset$). ■

REMARK 3.7.14 For $G = \mathrm{cl}(1)$ this is a generalization of Corollary 2.7.9 to arbitrary theories, replacing the notion of Lascar strong type (orbits of tuples under Autf($\mathfrak{C}$)) by the orbits under $\mu^{-1}(\mathrm{cl}(1))$.

DEFINITION 3.7.15 The *closed Galois group* $\mathrm{Gal}_c(\mathfrak{C})$ of $\mathfrak{C}$ is the quotient group $\mathrm{Gal}(\mathfrak{C})/\mathrm{cl}(1)$.

REMARK 3.7.16 By Lemmas 3.7.8 and 3.7.9 and Proposition 3.7.10 the closed Galois group of $\mathfrak{C}$ is a compact Hausdorff topological group. Moreover, $\mathrm{Gal}_c(\mathfrak{C}) = \mathrm{Gal}(\mathfrak{C})/\mathrm{stab}(\mathrm{bdd}(\emptyset))$.

DEFINITION 3.7.17 The theory T is G-compact if for a sufficiently saturated and homogeneous model $\mathfrak{C}$ of T the Galois group Gal($\mathfrak{C}$) is Hausdorff.

Equivalently, we may require $\{1\}$ to be closed in Gal($\mathfrak{C}$).

PROPOSITION 3.7.18 *A simple theory is G-compact.*

PROOF: By Proposition 2.7.5 equality of Lascar strong type over $\emptyset$ is a (the finest) type-definable bounded equivalence relation over $\emptyset$ in a simple theory. The assertion now follows from Lemmas 3.7.3 and 3.7.9. ■

PROBLEM 3.7.19 Are there theories which are not G-compact?

PROPOSITION 3.7.20 *Let T be countable and G-compact. Let E be an $\emptyset$-invariant equivalence relation on a complete real type $p(x)$ over $\emptyset$ with less than continuum many classes, and suppose E is obtained from definable relations by taking countable unions and intersections. Then E is definable on $p(x)$ and has only finitely many classes.*

REMARK 3.7.21 For instance, E might be the transitive closure of a type-definable relation.

PROOF: Fix an E-class a_E for some (possibly infinite) real tuple $a \models p$, and put $G = \mathrm{stab}(a_E)$. Then G is a Borel subset of the compact separable group $\mathrm{Gal}(\mathfrak{C})$. As the index of G in $\mathrm{Gal}(\mathfrak{C})$ is less than 2^ω, we see that G is not meagre. Hence G contains a non-empty open subset U of $\mathrm{Gal}(\mathfrak{C})$, up to a meagre set M. Then $U^{-1}U$ is an open neighbourhood of 1 contained in G, so G is open. By (topological) compactness G has finite index in $\mathrm{Gal}(\mathfrak{C})$, and is closed.

By Lemma 3.7.13 the equivalence relation given by $\mathrm{tp}(a/a_e) = \mathrm{tp}(a'/a_E)$ is type-definable over a_E, and in particular over a. Hence E is type-definable, with finitely many classes on p. It follows by compactness that E is definable on p (and only uses a finite subtuple of a, if a is infinite. ■

The final theorem of this section eliminates infinitary bounded hyperimaginaries.

THEOREM 3.7.22 *Let e be a bounded hyperimaginary. Then e is interdefinable with a sequence of finitary bounded hyperimaginaries.*

PROOF: Suppose $e = a_E$, where a is a possibly infinite tuple of real elements and E is a type-definable bounded equivalence relation over $\emptyset$. (We may suppose that E is bounded on the whole of $\mathfrak{C}$, not just on $\mathrm{tp}(a)$, by modifying E outside $\mathrm{tp}(a)$ if necessary.) Let G be the stabilizer in $\mathrm{Gal}(\mathfrak{C})$ of all E-classes. Then $\mathrm{Gal}(\mathfrak{C})/G$ is a compact Hausdorff topological group.

FACT 3.7.23 *[164] A compact Hausdorff topological group G is a projective limit of compact Lie groups (i.e. there is a family $\{G_i : i \in I\}$ of closed normal subgroups of G such that $\bigcap_{i \in I} G_i = \{1\}$ and G/G_i is a compact (not necessarily connected) Lie group, for all $i \in I$.*

It follows that there is a family $\{G_i : i \in I\}$ of closed normal subgroups of $\mathrm{Gal}(C)$ with $\bigcap_{i \in I} G_i = G$, such that every quotient $\mathrm{Gal}(\mathfrak{C})/G_i$ is a compact Lie group.

CLAIM. A compact Lie group has the descending chain condition on closed subgroups.

PROOF OF CLAIM: A closed subgroup H of a Lie group G is also a Lie group, so if H has infinite index in G, then $\dim(H) < \dim(G) < \omega$ (where *dim* means dimension as a real manifold). And if H has finite index in G, then H contains the connected component G^0 of G, which has finite index in G by compactness of G. The assertion follows. ■

By Lemma 3.7.13 the relation on $\mathrm{tp}(a)^2$ to lie in the same $\mu^{-1}(G_i)$-orbit is a type-definable equivalence relation E_i over $\emptyset$; as $\bigcap_{i\in I} G_i = G$, we have $\bigwedge_{i\in I} E_i = E$. Hence a_E is interdefinable with $(a_{E_i} : i \in I)$.

CLAIM. For every real tuple $\bar{a} \in \mathfrak{C}$ and every closed $G \subseteq \mathrm{Gal}(\mathfrak{C})$ the group $\mu(\mathrm{Aut}_{\bar{a}}(\mathfrak{C}))G$ is closed.

PROOF OF CLAIM: Let $(\sigma_i : i \in I)$ be a sequence of elements in $\mu(\mathrm{Aut}_{\bar{a}}(\mathfrak{C}))G$. Then for every $i \in I$ there are automorphisms $\tau_i \in \mathrm{Aut}_{\bar{a}}(\mathfrak{C})$ and $\tau_i' \in \mu^{-1}(G)$ with $\mu(\tau_i\tau_i') = \sigma_i$. If $\mathcal{U}$ is an ultrafilter on I, then $\prod_{i\in I}\tau_i \in \mathrm{Aut}_{\bar{a}}(\mathfrak{C}^I/\mathcal{U})$, whence $\alpha_{\mathfrak{C}^I/\mathcal{U},\mathfrak{C}}(\prod_{i\in I}\tau_i) \in \mu(\mathrm{Aut}_{\bar{a}}(\mathfrak{C}))$. Furthermore $\alpha_{\mathfrak{C}^I/\mathcal{U},\mathfrak{C}}(\prod_{i\in I}\tau_i') \in G$ by closedness of G. Hence $\alpha_{\mathfrak{C}^I/\mathcal{U},\mathfrak{C}}(\prod_{i\in I}\tau_i\tau_i') \in \mu(\mathrm{Aut}_{\bar{a}}(\mathfrak{C}))G$. ■

CLAIM. For each $i \in I$ there is a finitary hyperimaginary interdefinable with a_{E_i}.

PROOF OF CLAIM: Since $\mu(\mathrm{Aut}_a(\mathfrak{C}))G_i/G_i = \mathrm{stab}(a_{E_i})$, for any subtuple $\bar{a}$ of a we have $\mu(\mathrm{Aut}_{\bar{a}}(\mathfrak{C}))G_i/G_i \geq \mathrm{stab}(a_{E_i})$. It follows from the descending chain condition on $\mathrm{Gal}(\mathfrak{C})/G_i$ that there must be some finite subtuple $\bar{a}_i$ of a with $\mu(\mathrm{Aut}_{\bar{a}_i}(\mathfrak{C}))G_i/G_i = \mathrm{stab}(a_{E_i})$. Hence $a_{E_i} \in \mathrm{dcl}(\bar{a}_i)$. By Corollary 3.1.17 there is a type-definable equivalence relation F_i such that a_{E_i} is interdefinable with $(\bar{a}_i)_{F_i}$. ■

Hence a_E is interdefinable with the sequence $((\bar{a}_i)_{F_i} : i \in I)$ of finitary hyperimaginaries. ■

3.8 BIBLIOGRAPHICAL REMARKS

The model theory of hyperimaginaries (Sections 3.1 to 3.3) was developed by Hart, Kim and Pillay [40]. Section 3.4 on internality and analysability and Section 3.5 generalize the stable results of Hrushovski [46, 47]; our approach via P-closure comes from [161], and the simple version from [163]. Example 3.4.4 was essentially given by Pillay.

Example 3.6.1 is due to Lascar and Poizat [112]; they also prove elimination of hyperimaginaries for stable theories. Example 3.6.7 was given by Poizat (see [87]). Definition 3.6.9 to Lemma 3.6.12 are due to Kim [68] and Pillay [111]. I don't know whether anyone wants to be connected to the Stable Forking Conjecture, which in a very strong form epithomizes the fact that all known examples of a simple structure are closely related to a stable one.

Lascar defined the Lascar group in [82, 87]; he and Pillay studied it for a simple theory in [90].

Chapter 4

GROUPS

Groups form an important class of examples of simple theories; both abstractly and in the applications. They also sometimes appear unexpectedly out of general structural considerations in a context where *a priori* no group was given. Moreover, they are amenable to a more detailed model-theoretic study: due to the homogeneity imposed by the group law, a group in a simple theory often has a more friendly behaviour than a general simple structure.

We shall first treat the case of a definable or type-definable group whose domain consists of real (or imaginary) tuples. This will be generalized to groups of hyperimaginary elements in the following sections. Throughout, the ambient structure $\mathfrak{C}$ will be a monster model of a simple theory T, except where otherwise stated.

4.1 TYPE-DEFINABLE GROUPS

In this section G will be a type-definable set, $*$ a definable binary function, both over $\emptyset$, and 1 an element in G, such that $\langle G^{\mathfrak{C}}, *, 1\rangle$ forms a group. As $\mathfrak{C}$ is κ-saturated, the set of realizations of G will form a group in any model of T. Elements of G will be denoted by $g, h, g', h', \ldots$. We shall usually write gh rather than $g * h$.

DEFINITION 4.1.1 $S_G(A)$ is the set of types over A extending the partial type "$x \in G$". An element g is *left generic* for G over A if for any $h \mathop{\smile\!\!\!\!\!\!\,|}_A g$ in G we have $hg \mathop{\smile\!\!\!\!\!\!\,|} A, h$. A type $p \in S_G(A)$ is *left generic* if its realizations are.

Similarly, we can define *right generic* types and elements for G.

LEMMA 4.1.2 *Let g be generic for G over A.*

1. *If $h \in \mathrm{acl}(A) \cap G$, then $\mathrm{tp}(hg/A)$ is left generic for G.*

2. *If $B \supset A$ with $g \mathop{\smile\!\!\!\!\!\!\,|}_A B$, then $\mathrm{tp}(g/B)$ is left generic for G.*

3. *If $B \subset A$ with $g \mathop{\smile\!\!\!\!\!\!\mid}_B A$, then* $\mathrm{tp}(g/B)$ *is left generic for* G.

4. $\mathrm{tp}(g^{-1}/A)$ *is left generic for* G.

5. *A type is left generic for G if and only if it is right generic for G.*

In view of Lemma 4.1.2.5 we shall just speak of *generic* types and elements.

PROOF:

1. Suppose $h' \in G$ with $h' \mathop{\smile\!\!\!\!\!\!\mid}_A hg$. Then $h'h \in G$ and $g \mathop{\smile\!\!\!\!\!\!\mid}_A h'h$, as $g \in \mathrm{acl}(A, hg)$ and $h'h \in \mathrm{acl}(A, h')$. Therefore $h'hg \mathop{\smile\!\!\!\!\!\!\mid} A, h'h$ by left genericity, whence $h'hg \mathop{\smile\!\!\!\!\!\!\mid} A, h'$, as $h' \in \mathrm{acl}(A, h'h)$.

2. Consider $h \in G$ with $h \mathop{\smile\!\!\!\!\!\!\mid}_B g$. As $g \mathop{\smile\!\!\!\!\!\!\mid}_A B$ we get $g \mathop{\smile\!\!\!\!\!\!\mid}_A B, h$, whence $hg \mathop{\smile\!\!\!\!\!\!\mid} A, h$. Furthermore $g \mathop{\smile\!\!\!\!\!\!\mid}_{A,h} B$ and so $hg \mathop{\smile\!\!\!\!\!\!\mid}_{A,h} B$, whence $hg \mathop{\smile\!\!\!\!\!\!\mid} B, h$.

3. Consider $h \in G$ with $h \mathop{\smile\!\!\!\!\!\!\mid}_B g$. We may assume that $h \mathop{\smile\!\!\!\!\!\!\mid}_{B,g} A$. But then $h \mathop{\smile\!\!\!\!\!\!\mid}_B A, g$, whence $h \mathop{\smile\!\!\!\!\!\!\mid}_A g$ and $hg \mathop{\smile\!\!\!\!\!\!\mid} A, h$. So $hg \mathop{\smile\!\!\!\!\!\!\mid} B, h$.

4. Suppose $h \models \mathrm{tp}(g/A)$ with $h \mathop{\smile\!\!\!\!\!\!\mid}_A g$. Then $gh \mathop{\smile\!\!\!\!\!\!\mid} A, g$, whence $gh \mathop{\smile\!\!\!\!\!\!\mid}_A g$. Hence g is left generic over $A \cup \{gh\}$, as is $(gh)^{-1}g = h^{-1}$. So $h^{-1} \mathop{\smile\!\!\!\!\!\!\mid} A \cup \{gh\}$ by genericity, whence $h^{-1} \mathop{\smile\!\!\!\!\!\!\mid}_A gh$ and $\mathrm{tp}(h^{-1}/A) = \mathrm{tp}(g^{-1}/A)$ is left generic.

5. If $\mathrm{tp}(g/A)$ is left generic and $h \mathop{\smile\!\!\!\!\!\!\mid}_A g$, then $\mathrm{tp}(g^{-1}/A)$ is left generic and $h^{-1} \mathop{\smile\!\!\!\!\!\!\mid}_A g^{-1}$. Hence $h^{-1}g^{-1} \mathop{\smile\!\!\!\!\!\!\mid} A, h^{-1}$, whence $gh \mathop{\smile\!\!\!\!\!\!\mid} A, h$, and $\mathrm{tp}(g/A)$ is right generic. The other direction is similar. ■

DEFINITION 4.1.3 Let π be a partial type over A extending "$x \in G$", and $g \in \mathrm{dcl}(A) \cap G$. We define

- the left translate $g\pi = \{\varphi(x) \in \mathfrak{L}(A) : \pi(x) \vdash \varphi(gx)\}$,
- the right translate $\pi g = \{\varphi(x) \in \mathfrak{L}(A) : \pi(x) \vdash \varphi(xg)\}$, and
- the inverse $\pi^{-1} = \{\varphi(x) \in \mathfrak{L}(A) : \pi(x) \vdash \varphi(x^{-1})\}$.

So if $h \models \pi$, then $gh \models g\pi$, $hg \models \pi g$, and $h^{-1} \models \pi^{-1}$. Furthermore, a complete type $p \in S_G(A)$ is generic if and only if gp, pg, and p^{-1} are generic. We now aim towards proving the existence of generic types.

DEFINITION 4.1.4 For any formula $\varphi(x, y)$ and any $k < \omega$, the *stratified* (φ, k)*-rank* $D^*(., \varphi, k)$ is defined on partial types π as follows

- $D^*(\pi, \varphi, k) \geq 0$ if π is consistent with G.

- $D^*(\pi, \varphi, k) \geq n+1$ if there are $g_i \in G$ and $a_i \in \mathfrak{C}$ for $i < \omega$, such that $\{\varphi(g_i x, a_i) : i < \omega\}$ is k-inconsistent and $D^*(\pi(x) \cup \{\varphi(g_i x, a_i)\}, \varphi, k) \geq n$ for all $i < \omega$.

We write $D^*(g/A, \varphi, k)$ for $D^*(\text{tp}(g/A), \varphi, k)$.

REMARK 4.1.5 It is clear that "$D^*(\pi(x, A), \varphi, k) \geq n$" is a type-definable condition on A. By compactness we may therefore require the sequence $(g_i, a_i : i < \omega)$ to be indiscernible over A. Furthermore, if $D^*(\pi, \varphi, k) \leq n$, then there is a finite part π_0 of π with $D^*(\pi_0, \varphi, k) \leq n$.

LEMMA 4.1.6 *1. If π is a partial type, then $D^*(\pi, \varphi, k) < \omega$ for all formulas φ and all $k < \omega$.*

2. *If π is a partial type over A, for any formula φ and any $k < \omega$ there is a completion p of π such that $D^*(\pi, \varphi, k) = D^*(p, \varphi, k)$.*

3. *For any $g \in G$ and $A \subseteq B$ we have $g \mathop{\smile\!\!\!\!\!\!\mid}_A B$ if and only if $D^*(g/A, \varphi, k) = D^*(g/B, \varphi, k)$ for all formulas φ and all $k < \omega$.*

4. *If $g \in G$ and $h \in \text{acl}(A)$, then $D^*(g/A, \varphi, k) = D^*(hg/A, \varphi, k)$ for all formulas φ and all $k < \omega$.*

PROOF:

1. $D^*(\pi(x), \varphi(x, y), k) \leq D(\pi(x), \varphi(y'x, y), k) < \omega$ by simplicity.

2. It is clear that if $D^*(\pi, \varphi, k) \geq n$, then for any formula $\psi(x)$ either $D^*(\pi \cup \{\psi\}, \varphi, k) \geq n$ or $D^*(\pi \cup \{\neg\psi\}, \varphi, k) \geq n$. As $D^*(., \varphi, k) \leq n$ is a local property, the assertion follows from Zorn's Lemma.

3. Suppose $g \not\mathop{\smile\!\!\!\!\!\!\mid}_A B$. Then there is a formula $\varphi(x, b) \in \text{tp}(g/B)$ which divides over A, so there is an A-indiscernible sequence $(b_i : i < \omega)$ of type $\text{tp}(b/A)$ and $k < \omega$ such that $\{\varphi(x, b_i) : i < \omega\}$ is k-inconsistent. Put $g_i = 1$ for all $i < \omega$. By definition of D^*
$$\begin{aligned} D^*(g/B, \varphi, k) &\leq D^*(\text{tp}(g/A) \cup \{\varphi(1x, b), \varphi, k) \\ &\leq D^*(g/A, \varphi, k) + 1 < \omega. \end{aligned}$$
Conversely, we show inductively that if $g \mathop{\smile\!\!\!\!\!\!\mid}_A B$ and $D^*(g/A, \varphi, k) \geq n$, then $D^*(g/B, \varphi, k) \geq n$. This is clear for $n = 0$. So suppose it holds for n and $D^*(g/A, \varphi, k) \geq n+1$. Then there is an A-indiscernible sequence $(g_i, a_i : i < \omega)$ and some $k < \omega$ such that $\{\varphi(g_i x, a_i) : i < \omega\}$ is k-inconsistent, and $D^*(\text{tp}(g/A) \cup \{\varphi(g_i x, a_i)\}, \varphi, k) \geq n$ for all $i < \omega$. Let q be a completion of $\text{tp}(g/A) \cup \{\varphi(g_0 x, a_0)\}$ to a complete type over $A \cup \{g_0, a_0\}$ with $D^*(q, \varphi, k) \geq n$. We may assume that $g \models q$, and

$B \mathop{\smile\!\!\!\!\!\!\;|}_{A,g}(g_i, a_i : i < \omega)$; as then $B \mathop{\smile\!\!\!\!\!\!\;|}_A (g_i, a_i : i < \omega)$, we may even move $(g_i, a_i : i > 0)$ over B, g_0, a_0 and assume that $(g_i, a_i : i < \omega)$ remains indiscernible over B by Proposition 2.2.7. Then $B \mathop{\smile\!\!\!\!\!\!\;|}_A g, g_0, a_0$, whence $g \mathop{\smile\!\!\!\!\!\!\;|}_{A,g_0,a_0} B$. By inductive hypothesis $D^*(\mathrm{tp}(g/B, g_0, a_0), \varphi, k) \geq n$, whence $D^*(\mathrm{tp}(g/B) \cup \{\varphi(g_i x, a_i)\}, \varphi, k) \geq n$ for all $i < \omega$ by indiscernibility. By definition

$$D^*(g/B, \varphi, k) \geq D^*(\mathrm{tp}(g/B) \cup \{\varphi(g_0 x, a_0)\}, \varphi, k) + 1 \geq n + 1.$$

4. For all formulas φ and all $k < \omega$ we show inductively on n that $D^*(g/A, \varphi, k) \geq n$ implies $D^*(hg/A, \varphi, k) \geq n$. The reverse implication will follow by translating by h^{-1}. Again the assertion is clear for $n = 0$ and we may assume it to hold for n. Suppose $D^*(g/A, \varphi, k) \geq n + 1$, as witnessed by a sequence $(g_i, a_i : i < \omega)$. Then $(g_i h^{-1}, a_i : i < \omega)$ witnesses $D^*(hg/A, \varphi, k) \geq n + 1$. ■

PROPOSITION 4.1.7 *There is a generic type for G over A.*

PROOF: Let $(\varphi_i, k_i : i < \alpha)$ be an enumeration of all pairs (φ, k), where φ is a formula and $k < \omega$. We define inductively partial types π_i and numbers $n_i < \omega$ for $i < \alpha$ such that $\pi_0(x)$ is "$x \in G$" and:

1. $n_i = D^*(\pi_i, \varphi_i, k_i)$,

2. $\pi_{i+1}(x) = \pi_i(x) \cup \{\neg\varphi(x) \in \mathfrak{L}(A) : D^*(\pi_i(x) \cup \{\varphi(x)\}, \varphi_i, k_i) < n_i$, and

3. $\pi_\lambda = \bigcup_{i<\lambda} \pi_i$ for limit ordinals $\lambda < \alpha$.

Since $D^*(., \varphi, k) \leq n$ is a local property, every π_i is consistent and we can extend $\bigcup_{i<\alpha} \pi_i$ to a complete type p over A, with $D^*(p, \varphi_i, k_i) = n_i$ for all $i < \alpha$.

CLAIM. For any $g \models p$ and $h \in G$ with $h \mathop{\smile\!\!\!\!\!\!\;|}_A g$ we have $hg \mathop{\smile\!\!\!\!\!\!\;|} A, h$.

PROOF OF CLAIM: For all $i < \omega$ we have

$$\begin{aligned} n_i = D^*(g/A, \varphi_i, k_i) &= D^*(g/A \cup \{h\}, \varphi_i, k_i) \\ &= D^*(hg/A \cup \{h\}, \varphi_i, k_i) \leq D^*(hg/\emptyset, \varphi_i, k_i). \end{aligned}$$

Inductively we see from the definition that $hg \models \pi_i$ for all $i < \alpha$. Hence

$$D^*(hg/\emptyset, \varphi_i, k_i) \leq D^*(\pi_i, \varphi_i, k_i) = n_i$$

for all $i < \alpha$, and $hg \mathop{\smile\!\!\!\!\!\!\;|} A, h$. ■

Thus p is a generic type. ■

DEFINITION 4.1.8 Let $\pi(x)$ be a partial type over A extending "$x \in G$". We call π *generic* for G if it is contained in some generic type for G.

LEMMA 4.1.9 *Let $\pi(x)$ be a partial type over A extending "$x \in G$". Then the following are equivalent:*

1. *π is generic for G.*
2. *$D^*(\pi, \varphi, k)$ is maximal possible among types in $S_G(A)$, for all formulas φ and all $k < \omega$.*
3. *For any $g \in G$ the partial type $g\pi$ does not fork over $\emptyset$.*
4. *For any $g \in G$ the partial type $g\pi$ does not fork over A.*

PROOF: 1. $\Rightarrow$ 2. Complete π to a generic type p with a realization $g \models p$. Fix a formula φ and some $k < \omega$. Choose $h \perp_A g$ in G such that $D^*(h^{-1}/A, \varphi, k)$ is maximal possible among types in $S_G(A)$. As $gh \perp_A h$ we get

$$\begin{aligned} D^*(h^{-1}/A, \varphi, k) &= D^*(h^{-1}/A \cup \{gh\}, \varphi, k) \\ &= D^*((gh)h^{-1}/A \cup \{gh\}, \varphi, k) \\ &= D^*(g/A \cup \{gh\}, \varphi, k) \\ &\leq D^*(g/A, \varphi, k) \leq D^*(\pi, \varphi, k). \end{aligned}$$

2. $\Rightarrow$ 3. Suppose $g\pi$ forks over $\emptyset$ for some $g \in G$. Then there is an indiscernible sequence of $\emptyset$-conjugates $(g_i\pi_i : i < \omega)$ of $g\pi$ such that $\bigwedge_{i<\omega} g_i\pi_i(x)$ is inconsistent. Hence there is a formula $\varphi(g^{-1}x, a) \in g\pi$ and some $k < \omega$ such that $\{\varphi(g_i^{-1}x, a_i) : i < \omega\}$ is k-inconsistent, where $\varphi(g_i^{-1}x, a_i)$ in $g_i\pi_i$ corresponds to $\varphi(g^{-1}x, a)$ in $g\pi$. But this shows

$$D^*(\pi, \varphi, k) = D^*(g\pi, \varphi, k) \leq D^*(x \in G, \varphi, k) + 1,$$

contradicting maximality.

3. $\Rightarrow$ 4. is trivial. 4. $\Rightarrow$ 1. Let g realize a generic type over A. By assumption there is $h \models \pi$ such that $gh \perp_A g$. Hence $\mathrm{tp}(g/A, gh)$ is generic, as is $\mathrm{tp}((gh)^{-1}g/A, gh) = \mathrm{tp}(h^{-1}/A, gh)$. So $\mathrm{tp}(h^{-1}/A)$ is generic, as is $\mathrm{tp}(h/A)$ and thus π. ■

DEFINITION 4.1.10 A type-definable subgroup H of G has *bounded index* in G if $|G : H| < \kappa$. For any set A the *A-connected component* of G, denoted G^0_A, is the intersection of all type-definable subgroups of G over A of bounded index. G is *connected over* A if it has no proper type-definable subgroup over A of bounded index.

LEMMA 4.1.11 *G^0_A is a normal subgroup of G of index at most $2^{|T(A)|}$.*

PROOF: Consider the map $\varphi : G \to \mathrm{Sym}(G/G^0_A)$ given by $g \mapsto \lambda_g$, where λ_g is the permutation of G/G^0_A mapping a coset hG^0_A to ghG^0_A. As $|G : G^0_A|$ is bounded, so is the image of φ, and hence the index of $\ker(\varphi)$ in G. But $\ker(\varphi)$ is a normal type-definable subgroup and A-invariant. Hence $\ker(\varphi) \geq G^0_A$. But for $g \in \ker(\varphi)$ we have $gG^0_A = G^0_A$, whence $\ker(\varphi) \leq G^0_A$, and equality holds. Thus G^0_A is normal in G.

For every definable superset $X \supseteq G^0_A$ there is some definable superset $Y \supseteq G$ and some $n < \omega$ such that for any n elements $g_1, \ldots, g_n \in Y$ we have $g_i g_j^{-1} \in X$ for at least one pair $i \neq j$, as otherwise we could find models in which G^0_A has arbitrarily large index in G by compactness. Now suppose $\{g_i : i < (2^{|T(A)|})^+\}$ were a set of representatives for different cosets of G^0_A in G. Then to any pair $\{i, j\}$ we could associate a definable superset $X \supseteq G^0_A$ such that $g_i g_j^{-1} \notin X$. By the Erdös-Rado Theorem 1.2.13 there is a subset $J \subset (2^{|T(A)|})^+$ of cardinality $|T(A)|^+$ such that all pairs from J are associated to the same formula. This contradicts the fact that we even have a finite bound on such sets. ∎

REMARK 4.1.12 A group G is *radicable* if for all $g \in G$ and all $n < \omega$ there is $h \in G$ with $h^n = g$. We shall see later that a *stable* or *supersimple* radicable group is connected, over every parameter set A. This holds in particular for the additive group of a stable or supersimple division ring in characteristic 0. It is unknown whether a radicable group in a simple theory must be connected.

EXAMPLE 4.1.13 Consider the circle $\mathbb{R}/\mathbb{Z}$ with addition and the induced order. It is divisible (whence radicable), and the elements infinitely close to 0 form the connected component.

EXAMPLE 4.1.14 Let $\mathbb{F}$ be the finite field with two elements, and $G = \mathbb{F}^{(\omega)}$. Let $\mathfrak{M}$ be the structure $(G, \mathbb{F}, 0, +, \cdot)$, where 0 is the neutral element, $+$ is addition on G, and $\cdot$ is the bilinear form $(a_i)_i \cdot (b_i)_i = \sum_i a_i b_i$ from G to $\mathbb{F}$. Then $\mathfrak{M}$ is simple; for any set $A \subset G^{\mathfrak{M}}$ we have $G^0_A = \{g \in G : \bigwedge_{a \in A} g \cdot a = 0\}$.

PROOF: We claim that $\mathfrak{M}$ eliminates quantifiers. Using bilinearity, a typical quantifier-free formula $\varphi(x, \bar{y})$ is a finite disjunction of formulas of the form

$$\psi(\bar{y}) \wedge x \cdot x = \epsilon \wedge \bigwedge_{i \in I_0} x = t^0_i(\bar{y}) \wedge \bigwedge_{i \in I_1} x \neq t^1_i(\bar{y}) \wedge \bigwedge_{i \in I_2} x \cdot t^2_i(\bar{y}) = \epsilon_i,$$

where the t^j_i are group terms and $\epsilon, \epsilon_i \in \mathbb{F}$ (where we have replaced φ by $(\varphi \wedge x \cdot x = 0) \vee (\varphi \wedge x \cdot x = 1)$ if necessary). If $I_0 \neq \emptyset$, it is obvious how to eliminate the quantifier $\exists x$. Suppose $I_0 = \emptyset$. We can always satisfy

$x \cdot x = \epsilon$ and $x \neq t$ by modifying x at some large co-ordinate; this will not affect the truth value of $x \cdot t_i^2(\bar{y}) = \epsilon_i$. Let $J_0 = \{i \in I_2 : \epsilon_i = 0\}$ and $J_1 = \{i \in I_2 : \epsilon_i = 1\}$. If $J_1 = \emptyset$, then $\bigwedge_{i \in I_2} x \cdot t_i^2(\bar{y}) = \epsilon_i$ is satisfied by $x = 0$. Otherwise, pick $j \in J_1$. Then $\exists x \bigwedge_{i \in I_2} x \cdot t_i^2(\bar{y}) = \epsilon_i$ if and only if $t_j^2(\bar{y}) \notin \langle t_{i_0}^2(\bar{y}) : i_0 \in J_0\,,\, t_{i_1}^2(\bar{y}) - t_j^2(\bar{y}) : i_1 \in J_1 \rangle$, which is a quantifier-free condition on $\bar{y}$.

Put $G_0 = \{g \in G : g \cdot g = 0\}$ and $G_1 = \{g \in G : g \cdot g = 1\}$, two $\emptyset$-definable subsets of G. We see from the last paragraph that every infinite definable subset of G is a finite union of cosets of subgroups of G of the form $\{g \in G : \bigwedge_i g \cdot g_i = 0\}$ (of finite index in G), intersected with G_0 or G_1; infinite indiscernible sequence of such sets alway have non-empty intersection. Simplicity of G follows; in fact $g \not\perp_A B$ if and only if $g \in \mathrm{acl}(B) - \mathrm{acl}(A)$. Furthermore, if C is such a coset, then

$$(C \cap G_0) + (C \cap G_0) = (C \cap G_1) + (C \cap G_1) = \langle C \rangle,$$

so every definable subgroup of G is a finite extension of some $\{g \in G : \bigwedge_i g \cdot g_i = 0\}$. ■

LEMMA 4.1.15 *Let H be a type-definable subgroup of G over A. Then the following are equivalent:*

1. *H is generic in G (as a partial type).*
2. *$H \geq G_A^0$.*
3. *H has bounded index in G.*
4. *$D^*(H, \varphi, k) = D^*(G, \varphi, k)$ for all formulas φ and all $k < \omega$.*

PROOF: The equivalence of 2. and 3. is obvious, and 1. ⇔ 4. follows from Lemma 4.1.9. Now consider an A-indiscernible sequence of cosets $(g_i H : i < \kappa)$. If H has bounded index in G, then all cosets must be the same by indiscernibility, so $g_0 H$ does not fork over A and H is generic by Lemma 4.1.9. Conversely, if H has unbounded index in G, the we can find such a sequence of distinct cosets, and $g_0 H$ forks over A. So H is not generic. ■

REMARK 4.1.16 If H is a type-definable subgroup of G, say over $\emptyset$, we can also define the stratified ranks inside H, say $D_H^*(.,\varphi,k)$. It is clear that generally $D^*(\pi,\varphi,k) \geq D_H^*(\pi,\varphi,k)$, and equality need not necessarily hold. Nevertheless, for type-definable subsets of H, the stratified D^*-ranks as well as the stratified D_H^*-ranks both witness forking, and a type-definable subset of H over A is generic for H if and only if it has maximal D^*-ranks, or D_H^*-ranks, among types in $S_H(A)$.

DEFINITION 4.1.17 Let $p(x) \in S_G(A)$. We put

$$S(p) = \{g \in G : gp \cup p \text{ does not fork over } \emptyset\}.$$

The *stabilizer* of p, denoted $\mathrm{stab}(p)$, is the set $S(p) \cdot S(p)$.

Similar notions exist for translations on the right rather than on the left.

REMARK 4.1.18 If $g \in S(p)$, then there is a realization $h \models p$ with $gh \mathop{\smile\!\!\!\!\!\!|}_A g$ and $gh \models p$. So for all formulas φ and all $k < \omega$

$$\begin{aligned} D^*(p, \varphi, k) &= D^*(gh/A, \varphi, k) = D^*(gh/A \cup \{g\}, \varphi, k) \\ &= D^*(h/A \cup \{g\}, \varphi, k) \leq D^*(h/A, \varphi, k) = D^*(p, \varphi, k). \end{aligned}$$

Hence we have equality, and $h \mathop{\smile\!\!\!\!\!\!|}_A g$. It follows that $g^{-1} \in S(p)$.

Note that by Lemma 2.3.15 both $S(p)$ and $\mathrm{stab}(p)$ are type-definable over A.

LEMMA 4.1.19 *Let X be a non-empty type-definable subset of G such that for independent $g, g' \in X$ we have $g^{-1}g' \in X$, and put $Y = X \cdot X$. Then Y is a type-definable subgroup of G, and X is generic in Y. In fact, X contains all generic types for Y.*

PROOF: Put $X' = \{g \in X : g^{-1} \in X\}$. If g, g', g'' are three independent elements of X, then $g^{-1}g' \in X'$ and $(g'^{-1}g)g'' \in X'$, whence $g'' \in X' \cdot X'$, so X' also satisfies the assumptions of the lemma and generates the same group. We may therefore assume that X is closed under inversion.

Enumerate all pairs of formulas and natural numbers as (φ_i, k_i), for $i < \alpha$. As in the proof of Lemma 4.1.7 there is a type p extending the partial type "$x \in X$", such that $(D^*(p, \varphi_i, k_i) : i < \alpha)$ is maximal possible in the lexicographic ordering on ω^α; put $n_i = D^*(p, \varphi_i, k_i)$. For any three elements g, g', g'' of X, choose $h \models p$ with $h \mathop{\smile\!\!\!\!\!\!|} g, g', g''$. Then $g'h \in X$; furthermore $g'h \mathop{\smile\!\!\!\!\!\!|}_{g'} g, g''$. Since for any $i < \alpha$

$$n_i = D^*(h/g', \varphi_i, k_i) = D^*(g'h/g', \varphi_i, k_i) \leq D^*(g'h, \varphi_i, k_i) \leq n_i$$

by maximality of n_i, equality holds all the way and thus $g'h \mathop{\smile\!\!\!\!\!\!|} g'$ by Lemma 4.1.6. Hence $g'h \mathop{\smile\!\!\!\!\!\!|} g, g'$ and $gg'h \in X$. Finally $h \mathop{\smile\!\!\!\!\!\!|} g''$, so $h^{-1}g'' \in X$, and $gg'g'' \in Y$. This shows that Y is a subgroup; it is clearly type-definable.

Now let g be a generic element of Y, say $g = g'g''$ for some $g', g'' \in X$. Let $h \models p$ independently of g', g''. Then $g''h \in X$, and by maximality of n_i again

$$D^*(g''h/g'', \varphi_i, k_i) = D^*(g''h, \varphi_i, k_i) = n_i$$

for all $i < \alpha$, whence $g''h \mathop{\smile\!\!\!\!\!\!\bigcup} g''$. As $g''h \mathop{\smile\!\!\!\!\!\!\bigcup}_{g''} g'$, we have $g''h \mathop{\smile\!\!\!\!\!\!\bigcup} g'$, and $gh = g'g''h$ is in X. But since g is generic independent of h, we have $gh \mathop{\smile\!\!\!\!\!\!\bigcup} h$, so $g = ghh^{-1}$ is in X as well. ■

REMARK 4.1.20 Note that the proof does not actually require G to be a group; it is enough if $g \cdot g' \cdot g''$ is defined and associative for all $g, g', g'' \in X$, every element in X has a two-sided inverse in X, and there is a unit $1 \in X$.

PROPOSITION 4.1.21 *Let $\mathfrak{M}$ be a model, and $p \in S_G(\mathfrak{M})$. If $g, g' \in S(p)$ with $g \mathop{\smile\!\!\!\!\!\!\bigcup}_{\mathfrak{M}} g'$, then $gg' \in S(p)$. Furthermore,* $\mathrm{stab}(p)$ *is a subgroup of G and $S(p)$ contains all generic types for* $\mathrm{stab}(p)$.

REMARK 4.1.22 The Proposition holds more generally if p is any Lascar strong type extending "$x \in G$".

PROOF: Suppose $g, g' \in S(p)$ with $g \mathop{\smile\!\!\!\!\!\!\bigcup}_{\mathfrak{M}} g'$. Then there are $h, h' \models p$ with $g \mathop{\smile\!\!\!\!\!\!\bigcup}_{\mathfrak{M}} h$, $g \mathop{\smile\!\!\!\!\!\!\bigcup}_{\mathfrak{M}} gh$, $g' \mathop{\smile\!\!\!\!\!\!\bigcup}_{\mathfrak{M}} h'$ and $g' \mathop{\smile\!\!\!\!\!\!\bigcup}_{\mathfrak{M}} g'h'$, such that $gh, g'h' \models p$. By the Independence Theorem, we may suppose $h = g'h'$ and $h \mathop{\smile\!\!\!\!\!\!\bigcup}_{\mathfrak{M}} g, g'$. So $gg'h' = gh \models p$; as $h \mathop{\smile\!\!\!\!\!\!\bigcup}_{\mathfrak{M},g} g'$ implies $gg'h' \mathop{\smile\!\!\!\!\!\!\bigcup}_{\mathfrak{M},g} g'$, transitivity yields $gg'h' \mathop{\smile\!\!\!\!\!\!\bigcup}_{\mathfrak{M}} g, g'$ and $gg' \in S(p)$. As $S(p)$ is closed under inversion, the rest follows from Lemma 4.1.19. ■

LEMMA 4.1.23 *Let $p \in S_G(\mathfrak{M})$ for some model $\mathfrak{M}$. Then* $\mathrm{stab}(p) \leq G^0_{\mathfrak{M}}$, *and equality holds if and only if p is generic for G.*

PROOF: The equivalence relation given by "$xy^{-1} \in G^0_{\mathfrak{M}}$" is type-definable over $\mathfrak{M}$ with boundedly many classes. Since $\mathfrak{M}$ is a model, every class is definable over $\mathfrak{M}$, and any type in $S_G(\mathfrak{M})$ must specify its equivalence class, i.e. its coset modulo $G^0_{\mathfrak{M}}$. Hence $\mathrm{stab}(p) \leq G^0_{\mathfrak{M}}$.

Now suppose p is generic for G, and let g, g' be independent realizations of p. Then $g'g^{-1}$ is generic and independent of g and of g' over $\mathfrak{M}$; moreover $(g'g^{-1})g = g'$. It follows that $g'g^{-1} \in S(p)$, so $S(p)$ is generic for G, as is $\mathrm{stab}(p)$. Hence $\mathrm{stab}(p) \geq G^0_{\mathfrak{M}}$, and we have equality.

Conversely, if $\mathrm{stab}(p) = G^0_{\mathfrak{M}}$, then there is $g \in S(p)$ which is generic for G. As $g \in S(p)$, there is $g' \models p$ with $g' \mathop{\smile\!\!\!\!\!\!\bigcup}_{\mathfrak{M}} g$, $gg' \mathop{\smile\!\!\!\!\!\!\bigcup}_{\mathfrak{M}} g$, and $gg' \models p$. As $\mathrm{tp}(g/\mathfrak{M})$ is generic, so is $\mathrm{tp}(g/\mathfrak{M}, g')$, whence also $\mathrm{tp}(gg'/\mathfrak{M}, g')$. It follows that $p = \mathrm{tp}(gg'/\mathfrak{M})$ is generic for G. ■

By symmetry $G^0_{\mathfrak{M}}$ is also the right stabilizer of the generic types for G over $\mathfrak{M}$.

REMARK 4.1.24 If $p \in S_G(\mathfrak{M})$ in a *stable* theory and q is the non-forking extension of p over a model $\mathfrak{N} \succ \mathfrak{M}$, then $\mathrm{stab}(p) = \mathrm{stab}(q) = S(p) = S(q) = G^0_{\mathfrak{M}} = G^0_{\mathfrak{N}}$. In particular the connected component $G^0_{\mathfrak{M}}$ does not depend on the model $\mathfrak{M}$ and is type-definable over the same parameters as G;

it is denoted by G^0. Moreover, if G is connected, then it has a unique generic type. More generally every coset of G^0 contains a unique generic type, there is a bijection between G/G^0 and the generic types of G, and G acts transitively by translation on the set of its generic types.

PROOF: Let $g, g' \in S(p)$. By definition there are $h, h' \models p$, such that $gh, g'h' \models p$ and $g \mathop{\smile\!\!\!\!\!\!\;|}_{\mathfrak{M}} h$, $g \mathop{\smile\!\!\!\!\!\!\;|}_{\mathfrak{M}} gh$, $g' \mathop{\smile\!\!\!\!\!\!\;|}_{\mathfrak{M}} h'$ and $g' \mathop{\smile\!\!\!\!\!\!\;|}_{\mathfrak{M}} g'h'$. As p has a unique non-forking extension to $\mathfrak{N}, g, g'$ (which extends q), we may assume that h, h' realize this extension, and in fact $h = h'$. An easy rank calculation shows that then $gh \mathop{\smile\!\!\!\!\!\!\;|}_{\mathfrak{M}} \mathfrak{N}, g, g'$ and $g'h \mathop{\smile\!\!\!\!\!\!\;|}_{\mathfrak{M}} \mathfrak{N}, g, g'$, so $g, g' \in S(q)$. Furthermore, $g'h \mathop{\smile\!\!\!\!\!\!\;|}_{\mathfrak{M}} g'g^{-1}$ and $g'h = (g'g^{-1})(gh) \models g'g^{-1}p$, whence $g'g^{-1} \in S(p)$ and $S(p) = \text{stab}(p)$. Hence $\text{stab}(p) = \text{stab}(q) = S(p) = S(q) = G^0_{\mathfrak{M}} = G^0_{\mathfrak{N}}$.

Thus $G^0_{\mathfrak{M}}$ does not depend on the choice of $\mathfrak{M}$ and is invariant under all automorphisms. So it is type-definable over $\emptyset$; as it has bounded index in G, it must equal $G^0_{\emptyset}$. It follows that $G^0_{\emptyset}$ is the minimal type-definable subgroup of bounded index in G.

Finally, if G is connected and g, h are generic independent over $\mathfrak{M}$, then g left-stabilizes $\text{tp}(h/\mathfrak{M})$ and h right-stabilizes $\text{tp}(g/\mathfrak{M})$. Hence $\text{tp}(g/\mathfrak{M}) = \text{tp}(gh/\mathfrak{M}) = \text{tp}(h/\mathfrak{M})$, and the generic type of G is unique.

The last assertion follows by translating G^0 to its cosets. ■

In contrast, if the theory is merely simple, then $G^0_{\mathfrak{M}}$ can contain more than one generic type. We should also note that we get a related result locally for a stable formula in a simple theory:

PROPOSITION 4.1.25 *Let G be a type-definable group in a simple theory, and suppose that $\varphi(x, \bar{y})$ is a stable formula (or merely that it is p-stable for all generic types p). Then there are only finitely many φ-definitions for generic types of G.*

PROOF: Suppose G is type-defined over a model $\mathfrak{M}$ which also contains all parameters of $\varphi(x, \bar{y})$. Since all extensions of a generic type $p \in S_G(\mathfrak{M})$ agree on instances of φ by p-stability, there are only boundedly many φ-types among the generic types of G over any parameter set containing $\mathfrak{M}$.

Suppose there are infinitely many generic φ-types $(p_i : i < \omega)$. Then by compactness for any I the partial type $\Gamma(a_i, \bar{b}_{ij} : i, j \in I, i \neq j)$ given by

$$\bigwedge_{i\in I}\{a_i \in G \wedge \bigwedge_{j\neq i} \neg[\varphi(a_i, \bar{b}_{ij}) \leftrightarrow \varphi(a_j, \bar{b}_{ij})] \wedge$$

$$\bigwedge_{\psi\in\mathfrak{L}, k<\omega} D(a_i/(a_j, \bar{b}_{jj'} : j' \neq j \neq i), \psi, k) \geq D(G, \psi, k)\}$$

is consistent, as any countable subset can be witnessed by a set of independent realizations of $(p_i : i < \omega)$ and corresponding parameters for the $\bar{b}_{ij}$. But any sequence $(a_i : i \in I)$ of realizations of Γ is a sequence of independent generic elements for G over $\mathfrak{M} \cup \{b_{ij} : i \neq j\}$ with different φ-types, contradicting boundedness. ■

Using the stratified ranks, we get a boundedness condition on descending chains of type-definable subgroups.

PROPOSITION 4.1.26 *Let $(H_i : i \leq \alpha)$ be a descending chain of type-definable subgroups of G with $H_\lambda = \bigcap_{i<\lambda} H_i$ for all limit ordinals $\lambda \leq \alpha$. Suppose $|G_i : G_{i+1}|$ is unbounded for all $i < \alpha$. Then $\alpha < |T|^+$.*

PROOF: For any $i < \alpha$ there is a formula φ_i and some $k_i < \omega$ such that $D^*(H_i, \varphi_i, k_i) > D^*(H_{i+1}, \varphi_i, k_i)$. If $\alpha \geq |T|^+$, then we could find a subset $J \subseteq \alpha$ with $|J| = |T|^+$, such that $(\varphi_i, k_i) = (\varphi, k)$ is constant for all $i \in J$. As $D^*(H_i, \varphi, k) > D^*(H_j, \varphi, k)$ for all $i < j$ in J, this contradicts finiteness of $D^*(G, \varphi, k)$. ■

4.2 RELATIVELY DEFINABLE GROUPS

DEFINITION 4.2.1 Let G be an arbitrary subgroup of a type-definable group.

- A subgroup H of G is *definable relative to G*, or just *relatively definable*, if there is some formula φ such that $H = \{g \in G :\models \varphi(g)\}$.
- A subgroup (or a subset) of G is *(relatively) φ-definable* if it is definable (relative to G) by a formula $\varphi(x, \bar{a})$ for some $\bar{a} \in \mathfrak{C}$.
- A family of *uniformly (relatively) definable* subgroups of G is a family of (relatively) φ-definable subgroups of G, for some fixed formula φ.
- A family of (relatively) definable subgroups of G is *type-definable* if there are a formula $\varphi(x, \bar{y})$ and a partial type $\pi(\bar{y})$ such that the family is given by $\{\{g \in G :\models \varphi(g, \bar{a})\} :\models \pi(\bar{a})\}$.

REMARK 4.2.2 We usually fix a model with an ambient group G, and view a family $\mathfrak{H}$ of subgroups as just a collection of subgroups of G. However, when G and $\mathfrak{H}$ are both type-definable, for any model $\mathfrak{M}$ we obtain a family $\mathfrak{H}^{\mathfrak{M}}$ of subgroups of $G^{\mathfrak{M}}$; we can then ask about properties which are independent of the particular choice of model.

DEFINITION 4.2.3 Let G be an arbitrary group. Two subgroups H_1 and H_2 of G are *commensurable* if their intersection has finite index in both H_1 and H_2. A subgroup K of G is *uniformly commensurable* to a family $\mathfrak{H}$ of subgroups of G if the index of $K \cap H$ in K and in H is finite and bounded independently of $H \in \mathfrak{H}$. If G is type-definable, we call two type-definable subgroups H_1 and H_2 of G *commensurate* if the index of their intersection in both H_1 and H_2 is bounded independently of the ambient model.

By compactness, if a relatively definable group has bounded index in a type-definable group, then the index must be finite. In particular, two commensurate relatively definable subgroups of a type-definable group are commensurable.

The following theorem does not need simplicity:

THEOREM 4.2.4 *Let G be an arbitrary group and $\mathfrak{H}$ a family of subgroups of G such that there is some $n < \omega$ bounding the index $|H : H \cap H'|$ for any $H, H' \in \mathfrak{H}$. Then there is a subgroup N which is uniformly commensurable to $\mathfrak{H}$ and invariant under all automorphisms of G which stabilize $\mathfrak{H}$ setwise. We have $\bigcap \mathfrak{H} \leq N \leq \langle \mathfrak{H} \rangle$; in fact N is a finite extension of a finite intersection of groups in $\mathfrak{H}$. In particular, if $\mathfrak{H}$ consists of relatively definable subgroups, then N is relatively definable.*

PROOF: Let I be a finite intersection of elements of $\mathfrak{H}$. Then $I \leq H'$ for some $H' \in \mathfrak{H}$, so for any $H \in \mathfrak{H}$ we get $|HI : H| \leq |HH' : H| = |H' : H \cap H'| \leq n$. Therefore there is a maximal value $m(I) = \max\{|HI : H| : H \in \mathfrak{H}\}$. We call I *strong* if $m(I)$ takes the minimal possible value m (which is at most n); for strong I we define $\mathfrak{H}(I) = \{H \in \mathfrak{H} : |HI : H| = m\}$. Clearly, if J is another intersection of groups in $\mathfrak{H}$ with $J \leq I$, then J is also strong, $\mathfrak{H}(J) \subseteq \mathfrak{H}(I)$, and for $H \in \mathfrak{H}(J)$ we have $|HI : H| = |HJ : H| = m$, whence $HI = HJ$.

LEMMA 4.2.5 *If H and I are subgroups of G, then $\bigcap_{i \in I}(HI)^i$ is a subgroup of G containing I.*

PROOF: Put $I_H = \bigcap_{i \in I}(HI)^i$. Then $I \subseteq I_H$ is obvious. Now suppose $x, y \in I_H$. Then for any $k \in I$ we can write $x^k = hi$ with $h \in H$ and $i \in I$, and $y^{ki^{-1}} = h'i'$ with $h' \in H$ and $i' \in I$. Then $(xy)^k = hi(h'i')^i = (hh')(i'i) \in HI$, so I_H is closed under multiplication. And $(x^{-1})^{ki^{-1}} = (i^{-1}h^{-1})^{i^{-1}} = h^{-1}i^{-1} \in HI$; since ki^{-1} runs through I when k runs through I, we get $x^{-1} \in I_H$. Thus I_H is a subgroup of G. ∎

Put $N(I) = \bigcap_{H \in \mathfrak{H}(I)} I_H$. This is a subgroup of G containing I; if J is another intersection of groups in $\mathfrak{H}$ with $J \leq I$, then $I_H \leq J_H$ for any $H \in \mathfrak{H}(J) \subseteq \mathfrak{H}(I)$, so $N(I) \leq N(J)$. Furthermore, for $H \in \mathfrak{H}(J)$ we have $|N(J) : I| \leq |HJ : I| = |HI : I| = |H : H \cap I| \leq kn$, where k is the number of groups in $\mathfrak{H}$ whose intersection is I.

As $I \cap J$ is strong for any strong intersections I and J of groups in $\mathfrak{H}$, there is a unique maximal group N of the form $N(I_0)$, with strong I_0. Clearly, N is invariant under all automorphisms of G which stabilize $\mathfrak{H}$ setwise. Furthermore, $|H : N(I_0) \cap H| \leq |H : I_0 \cap H| \leq k_0 n$ for any $H \in \mathfrak{H}$, where k_0 is the number of groups in $\mathfrak{H}$ whose intersection is I_0. If $H' \in \mathfrak{H}(I_0 \cap H)$, then $N(I_0) = N(I_0 \cap H) \leq H'(I_0 \cap H)$, so

$$|N(I_0) : N(I_0) \cap H| = |N(I_0)H : H| \leq |H'H : H| = |H' : H' \cap H| \leq n.$$

So N is uniformly commensurable to $\mathfrak{H}$. As $I_0 \leq N \leq \langle\mathfrak{H}\rangle$, we are done. ■

From now on, the ambient theory will be simple again.

LEMMA 4.2.6 *Let G_0 be a type-definable group with a type-definable subgroup G. Suppose π is a partial type such that for any $\bar{a} \models \pi$ the formula $\varphi(x, \bar{a})$ defines a subgroup $H(\bar{a})$ relative to G_0. Then there is some integer $n < \omega$ and a definable superset X of G such that whenever $H(\bar{a}')$ intersects G in a subgroup of finite index for some $\bar{a}' \models \pi$, the index is bounded by n and X is covered by n cosets of $H(a')$.*

PROOF: By compactness, there is a definable superset Y of G_0 such that on Y multiplication is defined (but may go outside Y) and associative (whenever all partial products are also in Y), every $y \in Y$ has an inverse $y^{-1} \in Y$, and for any $x, y \in Y$ and $\bar{a} \models \pi$ with $\models \varphi(x, \bar{a}) \wedge \varphi(y, \bar{a})$ also $\models \varphi(x^{-1}y, \bar{a})$. Suppose $x \in Y$ with $xG \subseteq Y$, and consider $\{g \in xG : \models \varphi(g, \bar{a})\}$. If g and h are in this set, then $g^{-1}h \in G$ and $\models \varphi(g^{-1}h, \bar{a})$, so this set must be a left coset of $G \cap H(\bar{a})$. Therefore, if there is a definable superset X of G and $1 = x_0, x_1, \ldots, x_n$ in Y with $\bigcup_{i \leq n} x_i X \subseteq Y$ and $\models \forall x \in X \bigvee_{i<n} \varphi(x_i x, \bar{a})$, then $G \cap H(\bar{a})$ has index at most $n+1$ in G. Conversely, if $G \cap H(\bar{a})$ has index at most $n+1$ in G, by compactness there must be such elements $x_0, \ldots, x_n$ (which we may even choose to come from G) and such a set X (which we may take definable over the parameters used to type-define G). It follows that the condition on $\bar{a}'$ that $H(\bar{a}')$ intersects G in a subgroup of finite index can be expressed as an infinite disjunction. However, it is also type-definable, since it is equivalent to $D^*(G \cap H(\bar{a}'), \varphi, k) = D^*(G, \varphi, k)$ for all $\mathfrak{L}$-formulas φ and all $k < \omega$. By compactness, it is definable, and there is a bound n and a definable $X \supseteq G$ as required. ■

PROPOSITION 4.2.7 *Let G be type-definable over $\emptyset$, and $\mathfrak{H}$ a type-definable family of relatively definable subgroups of G. Then there is a finite intersection $N_0 = \bigcap_{i<n} H_i$ of elements in $\mathfrak{H}$, and a finite extension N of N_0 (which is thus also relatively definable) such that N is uniformly commensurable with $H \cap N_0$ for all $H \in \mathfrak{H}$, and N is invariant under all automorphisms of $\mathfrak{C}$ stabilizing $\mathfrak{H}$ setwise.*

In particular, any intersection of a family of uniformly relatively definable subgroups is commensurate with a finite subintersection.

PROOF: By Proposition 4.1.26 there is an intersection K of groups in $\mathfrak{H}$ of size at most $|T|$, such that any bigger intersection of this form yields a subgroup of bounded index. By relative definability the index $|K : K \cap H|$ is finite for all $H \in \mathfrak{H}$, and by Lemma 4.2.6 there is $k < \omega$ and a finite subintersection $N_1 = \bigcap_{i<n} H_i$ of groups in $\mathfrak{H}$ containing K, such that $|N_1 : N_1 \cap H| \leq k$ for all $H \in \mathfrak{H}$.

Now consider the family of conjugates of N_1 under automorphisms of $\mathfrak{C}$ stabilizing $\mathfrak{H}$ setwise. For any such conjugate N_1', the index $|N_1 : N_1 \cap N_1'|$ is bounded by k^n. By Theorem 4.2.4, there is a finite intersection N_0 of these conjugates (which is thus a finite intersection of groups in $\mathfrak{H}$), and a finite extension N of N_0, such that N is invariant under all these automorphisms. It is clear that N is uniformly commensurable with $N_0 \cap H$ for all $H \in \mathfrak{H}$. ■

REMARK 4.2.8 If $\mathfrak{H}$ is invariant under all automorphisms fixing some parameter set A in a $|T(A)|^+$-saturated and -homogeneous model $\mathfrak{M}$, then N is A-definable. This happens in particular if the family $\mathfrak{H}$ is type-definable over A, say of the form $\{\varphi(x, \bar{a}) :\models \pi(\bar{a})\}$ where π is a partial type over A. In this case, for every model $\mathfrak{M}$ containing A, we obtain a family $\mathfrak{H}^{\mathfrak{M}}$ of uniformly $\mathfrak{M}$-definable subgroups of $G^{\mathfrak{M}}$. Choosing $\mathfrak{M}$ sufficiently saturated, we obtain a relatively A-definable subgroup N such that $N^{\mathfrak{M}}$ is uniformly commensurable to $\mathfrak{H}^{\mathfrak{M}}$; as this is a first-order property, N is uniformly commensurable to $\mathfrak{H}$ in any model.

DEFINITION 4.2.9 Let G be type-definable over $\emptyset$. We call a type-definable subgroup H of G *locally connected* if every commensurate G-conjugate or automorphic image of H is equal to H.

This definition is similar to the stable case. However, while a connected group in a stable theory is locally connected, a connected (over some parameters) group in a simple theory need not be locally connected. For instance, in Example 4.1.14 only G itself is locally connected, while over any non-empty parameter set A the connected component G_A^0 is a proper subgroup of G. Note further that local connectivity of H depends on the ambient group G (and the parameters in the language); every type-definable group is trivially locally connected as a subgroup of itself!

COROLLARY 4.2.10 *Let G be type-definable over $\emptyset$ and H a relatively definable subgroup of G. If $\mathfrak{H}$ denotes the family of images of H under $\emptyset$-automorphisms, then the equivalence relation on $\mathfrak{H}$ given by commensurability is definable. Furthermore, there is a locally connected relatively definable subgroup H^c of G commensurable with H.*

Note that H^c need not be unique.

PROOF: Since H is relatively definable, commensurability and commensurativity are the same here. By Lemma 4.2.6 commensurability is a type-definable realtion on $\mathrm{tp}(\bar{a})$, where $\bar{a}$ is a parameter over which H is relatively defined. Let $\mathfrak{H}_0$ be the subfamily of those G-conjugates of automorphic images of H which are commensurable with H; this is a type-definable family of uniformly relatively definable subgroups of G. Theorem 4.2.4 yields a relatively definable subgroup H^c commensurable with H, such that H^c is stabilized by all automorphisms stabilizing $\mathfrak{H}_0$. In particular, any commensurable automorphic image of H^c is equal to H^c, as it must arise from the same family $\mathfrak{H}_0$. Similarly, any commensurable group-theoretic conjugate of H^c must be equal to H^c. Hence H^c is locally connected. ■

REMARK 4.2.11 Let G be a type-definable group over $\emptyset$, and H a relatively $\bar{a}$-definable locally connected subgroup of G. If E is the type-definable equivalence relation on $\mathrm{tp}(\bar{a})$ induced by commensurability, then any automorphism of $\mathfrak{C}$ fixes $\bar{a}_E$ if and only if it stabilizes H. Hence $\bar{a}_E$ is a canonical parameter for H.

We also have a chain condition on intersections of uniformly relatively definable subgroups, but only up to commensurability.

THEOREM 4.2.12 *Let G be type-definable over $\emptyset$ and $\mathfrak{H}$ a type-definable family of relatively definable subgroups of G. Then there are integers $k, k' < \omega$ such that there is no sequence $\{H_i : i \leq k\} \subset \mathfrak{H}$ with $|\bigcap_{i<j} H_i : \bigcap_{i\leq j} H_i| > k'$ for all $j \leq k$. In particular, any intersection N of elements in $\mathfrak{H}$ is commensurate with a subintersection N_0 of size at most k and $|N_0 : N_0 \cap H| \leq k'$ for any $H \in \mathfrak{H}$ with $H \geq N$. Furthermore, any chain of intersections of elements of $\mathfrak{H}$, each of unbounded index in its predecessor, has length at most k.*

PROOF: If the first assertion does not hold, then by compactness the condition

$$\{H_i \in \mathfrak{H} : i < \omega\} \cup \{|\bigcap_{i<k} H_i : \bigcap_{i\leq k} H_i| > k' : k, k' < \omega\}$$

is a consistent first-order condition on the parameters needed to define the groups $(H_i : i < \omega)$. But a realization of it yields a family $\{H_i : i < \omega\} \subset \mathfrak{H}$ such that for any finite subset $I \subset \omega$ the index

$$|\bigcap_{i\in I} H_i : H_{i_0} \cap \bigcap_{i\in I} H_i|$$

is infinite for some $i_0 \leq \max(I) + 1$. This contradicts Proposition 4.2.7.

If N is an intersection of elements of $\mathfrak{H}$, put $\mathfrak{H}_0 = \{H \in \mathfrak{H} : H \geq N\}$. Choose successively $H_0, H_1, \dots \in \mathfrak{H}_0$ such that $|\bigcap_{i<j} H_i : \bigcap_{i\leq j} H_i| > k'$ for all j, if possible. By the first assertion this sequence must stop after k such choices, and $N_0 = H_0 \cap \dots \cap H_{k-1}$ will do.

Finally, consider a chain $C_0 > C_1 > \dots > C_k$ of intersections of groups in $\mathfrak{H}$ such that every C_i has unbounded index in its predecessor. As the index of C_{i+1} in C_i is unbounded, there must be some element H_{i+1} in the intersection which forms C_{i+1}, such that $C_i \cap H_{i+1}$ has infinite index in C_i. Pick arbitrary $H_0 \in \mathfrak{H}$ with $H_0 \geq C_0$. Then the indices $|\bigcap_{i<j} H_i : \bigcap_{i\leq j} H_i|$ are infinite for all $j \leq k$, contradicting the first assertion. ■

4.3 HYPERDEFINABLE GROUPS

DEFINITION 4.3.1 A hyperdefinable group is given by a hyperdefinable set G and a hyperdefinable binary function $*$, such that $(G, *)$ forms a group.

In other words, G is given by a type-definable equivalence relation $E(x, y)$, a partial E-type $\Gamma(x)$, and a type-definable E-relation $\mu(x, y, z)$, such that $G = \Gamma/E$ and $* = \mu/E$. We interpret the group law as a ternary relation "$xy = z$"; if a_E, b_E and c_E are in G, we write $a_E b_E = c_E$ instead of $a_E * b_E = c_E$, or $\models \mu(a, b, c)$. From now on, we fix a hyperdefinable group $G = \langle \Gamma(x), E, \mu \rangle$ over $\emptyset$, with unit 1_E. Clearly, we may assume that E refines the type-definable equivalence relation $\bigwedge_{\gamma \in \Gamma}[\gamma(x) \leftrightarrow \gamma(y)]$.

DEFINITION 4.3.2 Let A be a set of parameters. $S_G(A)$ is the space of all E-types in G over A extending $\Gamma(x)$.
A type $p \in S_G(A)$ is *left generic* if for all b_E realizing a type in $S_G(A)$ and $a_E \models p$ with $a_E \mathop{\smile\!\!\!\!\!\!|}_A b_E$, we have $b_E a_E \mathop{\smile\!\!\!\!\!\!|} A, b_E$. A *right generic* type is defined analogously.

DEFINITION 4.3.3 If $\pi(x)$ is a partial type over A extending $\Gamma(x)$ and $a_E \in \mathrm{dcl}(A) \cap G$, put

- $a_E\pi(x) = \Gamma(x) \wedge \exists y\, [\pi(y) \wedge \mu(a, y, x)]$, the *left translate* of π by a_E,
- $\pi a_E(x) = \Gamma(x) \wedge \exists y\, [\pi(y) \wedge \mu(y, a, x)]$, the *right translate* of π by a_E,
- $\pi^{-1}(x) = \Gamma(x) \wedge \exists y\, [\pi(y) \wedge \mu(x, y, 1)]$, the *inverse* of π.

Hence if $b_E \models \pi$, then $a_E b_E \models a_E\pi$, $b_E a_E \models \pi a_E$, and $b_E^{-1} \models \pi^{-1}$. If π is a complete E-type over A, so are $a_E\pi$, πa_E, and π^{-1}. Note that since μ is E-invariant, $a_E\pi$ and πa_E do not depend on the particular choice of a in its E-class. If $a \notin \mathrm{dcl}(A)$, we can consider π as a partial type over Aa_E, in order to form $a_E\pi$ and πa_E, which will be partial types over Aa_E.

Note that $\pi(x)$ and $\exists x' [E(x, x') \wedge \pi(x')]$ have the same translates.

LEMMA 4.3.4 *Let $p \in S_G(A)$.*

1. *If p is left generic and $a_E \in G \cap \mathrm{dcl}(A)$, then $a_E p$ is left generic.*
2. *If q is a non-forking extension of p, then p is left generic if and only if q is left generic.*
3. *If p is left generic, so is p^{-1}.*
4. *A type in $S_G(A)$ is left generic if and only if it is right generic.*

PROOF: As the proof of Lemma 4.1.2. ■

So we can simply speak of a *generic* type of G. In order to prove the existence of generic types, we have to introduce a suitable stratified rank. Suppose $E(x, y) = \{\epsilon(x, y) : \epsilon \in E\}$. For any formula $\varphi(x, y)$ and $\epsilon \in E$ put

$$\varphi_\epsilon^k(y_1, \dots, y_k) = \neg \exists x \bigwedge_{s=1}^{k} \exists y \, [\epsilon(x, y) \wedge \varphi(y, y_s)].$$

DEFINITION 4.3.5 The *stratified* (φ, ϵ, k)*-rank* $D_G(., \varphi(x, z), \epsilon, k)$ is defined inductively on partial types $\pi(x)$ extending $\Gamma(x)$ as follows:

- $D_G(\pi(x), \varphi, \epsilon, k) \geq 0$ if $\pi(x)$ is consistent.
- $D_G(\pi(x), \varphi, \epsilon, k) \geq n + 1$ if there is $b \models \Gamma$ and a sequence $(b_i : i < \omega)$, such that
$$D_G(b_E \pi(x) \cup \{\varphi(x, b_i)\}, \varphi, \epsilon, k) \geq n$$
for all $i < \omega$, and $\varphi_\epsilon^k(b_i : i \in I)$ holds for all subsets $I \subset \omega$ of size k.

If $D_G(\pi, \varphi, \epsilon, k) \geq n$ for all $n < \omega$, we put $D_G(\pi, \varphi, \epsilon, k) = \infty$.

Note that $D_G(a/A, \varphi, \epsilon, k) = D_G(a_E/A, \varphi, \epsilon, k)$ for all formulas φ, all $\epsilon \in E$ and $k < \omega$, as $b_E \operatorname{tp}(a/A) = b_E \operatorname{tp}(a_E/A)$ (considered as partial types over Ab), and $D_G(a/A, \varphi, \epsilon, k) \geq 0$ if and only if $a \models \Gamma$. So D_G is also defined on E-types.

LEMMA 4.3.6 *1.* *D_G is left translation invariant on G: if $c_E \in G \cap \operatorname{dcl}(A)$, then $D_G(a_E/A, \varphi, \epsilon, k) = D_G(c_E a_E/A, \varphi, \epsilon, k)$ for all formulas φ, all $\epsilon \in E$ and all $k < \omega$.*

2. *For every formula φ, $\epsilon \in E$ and $k < \omega$, the rank $D_G(., \varphi, \epsilon, k)$ is closed and continuous: $D_G(\pi, \varphi, \epsilon, k) = \min\{D_G(\psi, \varphi, \epsilon, k) : \pi \vdash \psi\}$. Furthermore, if $\pi(x) = \pi(x, A)$ is a partial type with parameters A, then for every $n < \omega$ there is a partial type $\nu_n^\pi(X)$ such that $\models \nu_n^\pi(A')$ if and only if $D_G(\pi(x, A'), \varphi, \epsilon, k) \geq n$.*

3. *In the definition of D_G we may require the sequence $(b_i : i < \omega)$ to be indiscernible over b and the parameters used in π.*

4. *$D_G(\Gamma, \varphi, \epsilon, k) < \omega$.*

5. *D_G witnesses forking on G: if $a \in \Gamma$ and $A \subseteq B$, then $a_E \mathop{\smile\!\!\!\!\!\!\;|}_A B$ if and only if $D_G(a_E/A, \varphi, \epsilon, k) = D_G(a_E/B, \varphi, \epsilon, k)$ for all $\mathfrak{L}$-formulas φ, all $\epsilon \in E$ and all $k < \omega$.*

PROOF:

1. By induction on n: if b and $(b_i : i < \omega)$ witness $D_G(a_E/A, \varphi, \epsilon, k) \geq n+1$, then some pre-image of $b_E c_E^{-1}$ and $(b_i : i < \omega)$ witness $D_G(c_E a_E/A, \varphi, \epsilon, k) \geq n+1$.

2. We show inductively on n that $D_G(\psi, \varphi, \epsilon, k) \geq n$ for all ψ provable from π implies $D_G(\pi, \varphi, \epsilon, k) \geq n$, and there is a partial type ν_n^π such that $\models \nu_n^\pi(A')$ if and only if $D_G(\pi(x, A'), \varphi, \epsilon, k) \geq n$. Clearly, this holds for $n = 0$, where $\nu_0^\pi(X)$ is the partial type $\exists x\, \pi(x, X)$. So assume the assertion holds for n, and $D_G(\psi, \varphi, \epsilon, k) \geq n+1$ for all $\psi(x, A)$ implied by $\pi(x, A)$. Then for every such ψ there is $b^\psi \in \Gamma$ and a sequence $(b_i^\psi : i < \omega)$, such that
$$D_G(b_E^\psi \psi(x, A) \cup \{\varphi(x, b_i^\psi)\}, \varphi, \epsilon, k) \geq n$$
for all $i < \omega$, and $\models \varphi_\epsilon^k(b_i^\psi : i \in I)$ for all subsets $I \subset \omega$ of size k. By inductive hypothesis, there is a partial type $\nu_n^\psi(X, x, y)$ true of (A, b^ψ, b_i^ψ) for all $i < \omega$, such that $D_G(b_E' \psi(x, A') \cup \{\varphi(x, b'')\}, \varphi, \epsilon, k) \geq n$ if and only if $\models \nu_n^\psi(A', b', b'')$. Clearly, if $\psi \vdash \psi'$, then $\nu_n^\psi \vdash \nu_n^{\psi'}$. By compactness, there are $(b', b_i' : i < \omega)$ such that $\models \nu_n^\psi(A, b', b_i')$ for all ψ which are implied by π and all $i < \omega$, and $\models \varphi_\epsilon^k(b_i' : i \in I)$ for all subsets $I \subset \omega$ of size k. By inductive hypothesis, $D_G(b_E' \pi(x) \cup \{\varphi(x, b_i')\}, \varphi, \epsilon, k) \geq n$ for all $i < \omega$, whence $D_G(\pi, \varphi, \epsilon, k) \geq n+1$ by definition. Finally, we may define $\nu_{n+1}^\pi(X)$ as
$$\exists x\, \exists (y_i : i < \omega) [\bigwedge_{i<\omega, \pi \vdash \psi} \nu_n^\psi(X, x, y_i) \wedge \bigwedge_{I \subset \omega, |I|=k} \varphi_\epsilon^k(y_i : i \in I)].$$

3. Compactness, since $D_G(b_E \pi(x) \cup \{\varphi(x, b_i)\}, \varphi, \epsilon, k) \geq n$ is a closed condition on b_i over the domain of π and b_E by part 2.

4. Suppose $D_G(\Gamma, \varphi, \epsilon, k) = \infty$. For a large ordinal α consider the following type $\Phi_\alpha(x^\beta, y_i^\beta : \beta < \alpha, i < \omega)$:

   (a) $(y_i^\beta : i < \omega)$ is indiscernible over $\{x^\beta, x^\gamma, y_0^\gamma : \gamma < \beta\}$, for all $\beta < \alpha$.

   (b) $\varphi_\epsilon^k(y_i^\beta : i \in I)$ for all $I \subset \omega$ of size k and all $\beta < \alpha$.

   (c) $\Gamma(x^\beta)$ for all $\beta < \alpha$.

   (d) $\{x_E^\beta \varphi(x, y_0^\beta) : \beta < \alpha\}$ is consistent.

   If Φ_0 is a finite bit of Φ_α, it is contained in Φ_n for some $n < \omega$ (after renaming variables). But since $D_G(\Gamma, \varphi, \epsilon, k) \geq n+1$, we can inductively find $b_E^t \in G$ and $(b_i^t : i < \omega)$ for $t \leq n$, such that for all $t \leq n$:

- $(b_i^t : i < \omega)$ is indiscernible over $(b_E^t, b_E^j, b_0^j : j < t)$.
- $\models \varphi_\epsilon^k(b_i^t : i \in I)$ for all $I \subset \omega$ of size k.
- $D_G(\bigwedge_{j \leq t} b_E^t b_E^{t-1} \cdots b_E^{j+1} \varphi(x, b_0^j), \varphi, \epsilon, k) \geq n - t$.

Realizing y_i^t by b_i^t and x_E^t by $(b_E^t \cdots b_E^0)^{-1}$, we see that (a), (b) and (c) in Φ_n are satisfied. Furthermore, if $a \models \bigwedge_{j \leq n} b_E^n b_E^{n-1} \cdots b_E^{j+1} \varphi(x, b_0^j)$, then

$$(b_E^n b_E^{n-1} \cdots b_E^0)^{-1} a \models \bigwedge_{j \leq n} (b_E^j b_E^{j-1} \cdots b_E^0)^{-1} \varphi(x, b_0^j) = \bigwedge_{j \leq n} x_E^j \varphi(x, y_0^t),$$

so (d) is satisfied as well. Thus Φ_n and hence Φ_α are consistent.

If $(b_E^\beta, b_i^\beta : \beta < \alpha, i < \omega)$ realizes Φ_α and a realizes $\bigwedge_{\beta < \alpha} b_E^\beta \varphi(x, b_0^\beta)$, then

$$\exists y\, [E(x,y) \wedge \varphi(y, b_0^\beta)] \in \mathrm{tp}((b_E^\beta)^{-1} a_E / b_E^\gamma, b_0^\gamma : \gamma \leq \beta),$$

$(b_i^\beta : i < \omega)$ is indiscernible over $(b_E^\beta, b_E^\gamma, b_0^\gamma : \gamma < \beta)$, and

$$\bigwedge_{i < \omega} \exists y\, [E(x,y) \wedge \varphi(y, b_i^\beta)]$$

is k-inconsistent (as $\models \varphi_\epsilon^k(b_i^\beta : i \in I)$ for all $I \subset \omega$ of size k). Hence

$$(b_E^\beta)^{-1} a_E \mathop{\not\perp}_{(b_E^\beta, b_E^\gamma, b_0^\gamma : \gamma < \beta)} b_0^\beta, \quad \text{whence} \quad a_E \mathop{\not\perp}_{(b_E^\gamma, b_0^\gamma : \gamma < \beta)} b_E^\beta, b_0^\beta$$

for all $\beta < \alpha$, contradicting simplicity for big α.

5. Suppose $a_E \not\perp_A B$. Put $p(x, B) = \mathrm{tp}(a_E/B)$. If $(B_i : i < \omega)$ is a Morley sequence in $\mathrm{tp}(B/A)$ with $B_0 = B$, then $\bigcup_{i<\omega} p(x, B_i)$ is inconsistent. It follows that there is some formula $\varphi_0(x, B_0) \in p(x, B_0)$ and some $k < \omega$ such that $(\varphi_0(x, B_i) : i < \omega)$ is k-inconsistent. Since p is invariant under E, there is a formula $\varphi(x, b_0) \in p(x, B_0)$ and $\epsilon \in E$ such that $\exists y\, [\varphi(y, b_0) \wedge \epsilon(x, y)] \vdash \varphi_0(x, B_0)$. Clearly $\varphi_\epsilon^k(b_i : i \in I)$ holds for all $I \subseteq \omega$ of size k, where $b_i \in B_i$ corresponds to $b_0 \in B_0$. By definition of D_G we get $D_G(a_E/A, \varphi, \epsilon, k) \geq D_G(a_E/B, \varphi, \epsilon, k) + 1$ (where the additional parameter b satisfies $b_E = 1$).

   For the converse, we show by induction on n that if $a_E \perp_A B$ and $D_G(a_E/A, \varphi, \epsilon, k) \geq n$, then $D_G(a_E/B, \varphi, \epsilon, k) \geq n$, for all $\mathfrak{L}$-formulas φ, all $\epsilon \in E$ and $k < \omega$. This is clear for $n = 0$. So suppose $D_G(a_E/A, \varphi, \epsilon, k) \geq n + 1$. Then there are b and a sequence $(b_i : i < \omega)$ indiscernible over A, b with $D_G(\mathrm{tp}(b_E a_E/Ab) \cup \{\varphi(x, b_0)\}, \varphi, \epsilon, k) \geq n$

and $\models \varphi_\epsilon^k(b_i : i \in I)$ for all $I \subset \omega$ of size k. It is easy to see that we can extend $\text{tp}(b_E a_E/Ab) \cup \{\exists y\,[E(x,y) \wedge \varphi(y, b_0)]\}$ to a complete E-type $p \in S_G(A, b, b_0)$ with $D_G(p, \varphi, \epsilon, k) \geq n$; we may suppose that this is realized by $b_E a_E$. We can choose $b, b_0 \mathop{\smile\!\!\!\!\!\!|}_{A,a_E} B$, whence $B \mathop{\smile\!\!\!\!\!\!|}_A a_E, b, b_0$ and $b_E a_E \mathop{\smile\!\!\!\!\!\!|}_{A,b,b_0} B$. By inductive hypothesis $D_G(b_E a_E/A, b, b_0, \varphi, \epsilon, k) \geq n$ implies $D_G(b_E a_E/B, b, b_0, \varphi, \epsilon, k) \geq n$.

As $b_0 \mathop{\smile\!\!\!\!\!\!|}_{A,b} B$, we may assume that $(b_i : i < \omega)$ is indiscernible over B, b by Lemma 3.2.4. Then the sequence $(b_i : i < \omega)$ witnesses that $D_G(b_E a_E/B, b, \varphi, \epsilon, k) \geq n + 1$. Therefore

$$\begin{aligned} D_G(a_E/B, \varphi, \epsilon, k) &\geq D_G(a_E/B, b, \varphi, \epsilon, k) \\ &= D_G(b_E a_E/B, b, \varphi, \epsilon, k) \geq n + 1. \quad \blacksquare \end{aligned}$$

THEOREM 4.3.7 *There are generic types for G.*

PROOF: Having done all the work in the previous lemma, this is now easy. Enumerate all triples (φ, ϵ, k) of an $\mathfrak{L}$-formula φ, an $\epsilon \in E$ and a $k < \omega$ as $(\varphi_\alpha, \epsilon_\alpha, k_\alpha : \alpha < \beta)$ for some ordinal β. Put $\bigcup_{j<0} \Gamma_j = \Gamma$, and let $n_\alpha = D_G(\bigcup_{j<\alpha} \Gamma_j, \varphi_\alpha, \epsilon_\alpha, k_\alpha)$ and

$$\Gamma_\alpha = \bigcup_{j<\alpha} \Gamma_j \cup \{\neg\psi(x) : D_G(\psi, \varphi_\alpha, \epsilon_\alpha, k_\alpha) < n_\alpha\}$$

for all $\alpha < \beta$. Then by continuity there is some type $p \in S_G(A)$ extending $\bigcup_{\alpha<\beta} \Gamma_\alpha$ with $D_G(p, \varphi_\alpha, \epsilon_\alpha, k_\alpha) = n_\alpha$ for all $\alpha < \beta$. We claim that p is generic.

So consider any $a_E \models p$ and $b_E \in G$ with $a_E \mathop{\smile\!\!\!\!\!\!|}_A b_E$. Then for all $\alpha < \beta$

$$\begin{aligned} n_\alpha = D_G(a_E/A, \varphi_\alpha, \epsilon_\alpha, k_\alpha) &= D_G(a_E/A, b_E, \varphi_\alpha, \epsilon_\alpha, k_\alpha) \\ = D_G(b_E a_E/A, b_E, \varphi_\alpha, \epsilon_\alpha, k_\alpha) &\leq D_G(b_E a_E, \varphi_\alpha, \epsilon_\alpha, k_\alpha); \end{aligned}$$

since $D_G(b_E a_E, \varphi_\alpha, \epsilon_\alpha, k_\alpha) \leq n_\alpha$ inductively for all $\alpha < \beta$ by our choice of n_α, we get equality all the way through. Hence $b_E a_E \mathop{\smile\!\!\!\!\!\!|} A, b_E$, and p is generic. ■

PROPOSITION 4.3.8 *The following are equivalent for a type $p \in S_G(A)$:*

1. *p is generic for G.*
2. *$a_E b_E \mathop{\smile\!\!\!\!\!\!|}_A b_E$ for all $a_E \models p$ and $b_E \mathop{\smile\!\!\!\!\!\!|}_A a_E$.*
3. *$D_G(p, \varphi, \epsilon, k) = D_G(\Gamma, \varphi, \epsilon, k)$ for all $\mathfrak{L}$-formulas φ, all $\epsilon \in E$ and all $k < \omega$.*

4. *$p\lceil_\emptyset$ is generic over $\emptyset$ and p does not fork over $\emptyset$.*

PROOF: 1. ⇒ 2. Obvious, since left-generic is the same as right-generic.

2. ⇒ 3. Suppose p satisfies 2. and $a_E \models p$. Choose $b_E \downarrow_A a_E$ such that $D_G(b_E/A, \varphi, \epsilon, k) = D_G(\Gamma, \varphi, \epsilon, k)$ for some formula φ, some $\epsilon \in E$, and $k < \omega$. The assumption implies $a_E b_E^{-1} \downarrow_A b_E$, whence

$$\begin{aligned} D_G(\Gamma, \varphi, \epsilon, k) &= D_G(b_E/A, \varphi, \epsilon, k) = D_G(b_E/A \cup \{a_E b_E^{-1}\}, \varphi, \epsilon, k) \\ &= D_G(a_E/A \cup \{a_E b_E^{-1}\}, \varphi, \epsilon, k) \leq D_G(a_E/A, \varphi, \epsilon, k) \\ &= D_G(p, \varphi, \epsilon, k). \end{aligned}$$

3. ⇒ 4. If $D_G(p, \varphi, \epsilon, k)$ is maximal possible for all φ, ϵ, k, then p cannot fork over $\emptyset$ since D_G witnesses forking. Now let $a_E \models p\lceil_\emptyset$ and $b_E \in G$ with $a_E \downarrow b_E$. Then

$$\begin{aligned} D_G(\Gamma, \varphi, \epsilon, k) &= D_G(a_E, \varphi, \epsilon, k) = D_G(a_E/b_E, \varphi, \epsilon, k) \\ &= D_G(b_E a_E/b_E, \varphi, \epsilon, k) \leq D_G(b_E a_E, \varphi, \epsilon, k) \\ &\leq D_G(\Gamma, \varphi, \epsilon, k) \end{aligned}$$

for all $\mathfrak{L}$-formulas φ, all $\epsilon \in E$ and all $k < \omega$. Hence equality must hold all the way through, and $b_E a_E \downarrow b_E$. Therefore $p\lceil_\emptyset$ is generic.

4. ⇒ 1. Clear, since a non-forking extension of a generic type is again generic. ■

DEFINITION 4.3.9 A partial type $\pi(x)$ is *generic* for G if it is contained in some generic type for G.

PROPOSITION 4.3.10 *Let π be a partial type over A extending Γ. Then the following are equivalent:*

1. *$\pi(x)$ is generic over A.*

2. *For all $b_E \in G$ the partial type $b_E\pi$ does not fork over $\emptyset$.*

3. *For all $b_E \in G$ the partial type $b_E\pi$ does not fork over A.*

PROOF: 1. ⇒ 2. Since π contains a generic type over any $b_E \in G$, there is a generic $a_E \models \pi$ with $a_E \downarrow_A b_E$. By genericity, $b_E a_E \downarrow A, b_E$. Then $b_E a_E \models b_E\pi$, and $b_E\pi$ does not fork over $\emptyset$.

2. ⇒ 3. is trivial.

3. ⇒ 1. Let b_E be generic over A. Then $b_E\pi$ does not fork over A, and there is $b_E c_E \models b_E\pi$ with $b_E c_E \downarrow_A b_E$. Hence b_E and therefore b_E^{-1} is generic over $A, b_E c_E$, as is $b_E^{-1}(b_E c_E) = c_E$. Therefore $\mathrm{tp}(c_E/A)$ is generic and extends π. ■

REMARK 4.3.11 Suppose H is a subgroup of G such that a coset aH is hyperdefinable over A (this does not imply that a is definable over A). We shall say that $p \in S_G(A)$ is *generic* for aH if for some non-forking extension q of p over A, a the translate $a^{-1}q$ is generic for H. Then p is generic for aH if and only if $D_G(p, \varphi, \epsilon, k)$ takes the maximal possible value for all φ, all $\epsilon \in E$ and all $k < \omega$, namely $D_G(H, \varphi, \epsilon, k)$. The definition for right cosets is analogous; as the right coset Ha of H is equal to the left coset aH^a of H^a, one has to check that the two definitions agree (which follows from the fact that p is generic if and only if p^{-1} is generic).

A type is generic for the coset space G/H if it is the image of a generic type of G; this happens if and only if it has maximal D_G-ranks.

LEMMA 4.3.12 *Let H a hyperdefinable subgroup of G over A, and g generic for G over A. Let $\bar{g}$ be the class of g under the hyperdefinable equivalence relation $x^{-1}y \in H$. Then the coset gH is hyperdefinable over $A, \bar{g}$ and g is generic for gH over $A, \bar{g}$.*

PROOF: Hyperdefinability of gH over $A, \bar{g}$ is obvious. Let g' realize a generic type for gH over A, g, so $g' = gh$ for some generic $h \in H$ over A, g. Hence $h \perp_A g$, and $gh \perp A, h$ by genericity of g. As $ghH = gH$, it follows that $\bar{g} \in \mathrm{dcl}(gh, A)$ and $gh \perp_{A,\bar{g}} h$. Hence $\mathrm{tp}(gh/A, \bar{g}, h)$ is generic for gH, as is $\mathrm{tp}(g/A, \bar{g})$. ■

Note that $\bar{g}$ is a canonical parameter for gH.

4.4 CHAIN CONDITIONS AND COMMENSURATIVITY

As in the last section, the G will be a hyperdefinable group over $\emptyset$ of type E (i.e. the domain consists of classes modulo E) in a simple theory. However, from now on we shall usually denote elements of G by $g, h, \ldots$ rather than $a_E, b_E, \ldots$.

As in the case of type-definable groups, we say that a hyperdefinable subgroup H of G has *bounded index* in G if $|G : H| < \kappa$; if H is hyperdefined over A, an immediate generalization of Lemma 4.1.11 yields $|G : H| \leq 2^{|T(A)|}$. Similarly, if X is a hyperdefinable subset of G which is a union of less than κ cosets of H, we say that H has *bounded index* in X.

Two hyperdefinable subgroups H_1 and H_2 of G are again called *commensurate* if their intersection has bounded index both in H_1 and in H_2.

PROPOSITION 4.4.1 *Suppose H is a hyperdefinable subgroup of G, and X is a hyperdefinable subset. Then the index of H in XH is bounded if and only if $D_G(H, \varphi, \epsilon, k) = D_G(XH, \varphi, \epsilon, k)$ for all formulas φ, all $\epsilon \in E$, and all $k < \omega$.*

PROOF: Let A be a set of parameters such that H and X are hyperdefinable over A. The index of H in XH is unbounded if and only if there is an infinite indiscernible sequence of distinct cosets $(g_iH : i < \omega)$ of H in XH over A. This happens if and only if gH forks over A for some $g \in XH - H$.

Suppose first that gH forks over A for some $g \in X$. Then $gh \not\perp_A g$ for all $h \in H$; choosing $h \perp_A g$ generic for H, there are a formula φ, some $\epsilon \in E$ and $k < \omega$ with

$$\begin{aligned} D_G(XH, \varphi, \epsilon, k) &\geq D_G(gh/A, \varphi, \epsilon, k) > D_G(gh/A, g, \varphi, \epsilon, k) \\ &= D_G(h/A, g, \varphi, \epsilon, k) = D_G(H, \varphi, \epsilon, k). \end{aligned}$$

Conversely, if gH does not fork over A for any $g \in XH$, consider $gh \in gH$ with $gh \perp_A g$. Then

$$\begin{aligned} D_G(H, \varphi, \epsilon, k) &\geq D_G(h^{-1}/A, gh, \varphi, \epsilon, k) \\ &= D_G(g/A, gh, \varphi, \epsilon, k) = D_G(g/A, \varphi, \epsilon, k); \end{aligned}$$

since $g \in XH$ was arbitrary, $D_G(H, \varphi, \epsilon, k) \geq D_G(XH, \varphi, \epsilon, k)$. ■

COROLLARY 4.4.2 *If H is a hyperdefinable subgroup of G, then H has bounded index in G if and only if it has the same local D_G-ranks, or equivalently if and only if it is generic in G (as a partial E-type). If $H = H(a)$ is hyperdefinable over a, then the equivalence relation F on $\mathrm{tp}(a)$ given by $a'Fa''$ if $H(a')$ and $H(a'')$ are commensurate is type-definable.*

PROOF: The first assertion is immediate from Lemma 4.4.1 and Proposition 4.3.8, putting $X = G$. Secondly, F is type-defined by the conditions "$D_G(H(a') \cap H(a''), \varphi, \epsilon, k) \geq D_G(H, \varphi, \epsilon, k)$" for all formulas φ, all $\epsilon \in E$ and all $k < \omega$. ■

PROPOSITION 4.4.3 *Let $(H_i : i \leq \alpha)$ be a descending chain of hyperdefinable subgroups of G, continuous at limits, such that each successor group has unbounded index its predecessor. Then $\alpha < |T|^+$.*

PROOF: For any $j < \alpha$ let $(\varphi_j, \epsilon_j, k_j)$ be a triple such that $D(H_j, \varphi_j, \epsilon_j, k_j) > D(H_{j+1}, \varphi_j, \epsilon_j, k_j)$. If the chain had length $|T|^+$, then we could find a subchain J of length $|T|^+$, an $\mathfrak{L}$-formula φ, some $\epsilon \in E$ and $k < \omega$, such that $(\varphi_j, \epsilon_j, k_j) = (\varphi, \epsilon, k)$ is constant for all $j \in J$. As $D(H_j, \varphi, \epsilon, k)$ is strictly descending for $j \in J$, this contradicts finiteness of $D(G, \varphi, \epsilon, k)$. ■

DEFINITION 4.4.4 For any set A, the *A-connected component* G_A^0 of G is the smallest hyperdefinable subgroup of bounded index.

PROPOSITION 4.4.5 *1. G_A^0 is normal in G, of index at most $2^{|T(A)|}$.*

2. *A subgroup H which is hyperdefinable over A is generic (as a partial type) if and only if $H \geq G_A^0$.*

PROOF: Normality follows from the fact that the kernel of the action of G on G/G_A^0 by left multiplication is a hyperdefinable normal subgroup of bounded index in G. It therefore contains in G_A^0, and must be equal to it. The bound on the index follows as in Lemma 4.1.11 from the Erdös-Rado Theorem 1.2.13.

The second assertion is obvious. ■

DEFINITION 4.4.6 A generic type of G_A^0 over A is called a *principal* generic type of G over A.

REMARK 4.4.7 Note that a principal generic type may have a non-principal generic non-forking extension (namely, whenever $G_B^0 < G_A^0$ for some $A \subset B$). However, we shall see later that a principal generic type always has a principal generic extension to any superset. The restriction of a principal generic type is obviously principal generic.

The fact that the chain condition holds only up to commensurativity has some consequences for the hyperdefinability of certain subgroups. In particular, for a hyperdefinable subgroup H of G, neither the centre $Z(H)$ nor the normalizer $N_G(H)$ are necessarily hyperdefinable, and the centralizer $C_G(A)$ of a subset A need only be hyperdefinable if A is finite. However, we may define approximations to those subgroups. Let us first state the analogue of Lemma 4.1.19 for hyperdefinable groups.

LEMMA 4.4.8 *Let X be a hyperdefinable subset of G such that for independent $x, x' \in X$ we have $x^{-1}x' \in X$. Put $H = X \cdot X$. Then H is a hyperdefinable subgroup of G, and X is generic in H. In fact, X contains all generic types for H.*

PROOF: As the proof of Lemma 4.1.19, but this time we enumerate all triples $(\varphi_\alpha, \epsilon_\alpha, k_\alpha : \alpha < \beta)$ of an $\mathfrak{L}$-formula φ, an $\epsilon \in E$ and $k < \omega$. ■

DEFINITION 4.4.9 Let H and K be subgroups of G, both hyperdefinable over a model $\mathfrak{M}$, and $p \in S_G(\mathfrak{M})$.

1. The *approximate centralizer* of p in H is

$$\tilde{C}_H(p) = \{h \in H : \exists x \models p[x \underset{\mathfrak{M}}{\perp} h \wedge hx = xh]\}^2.$$

2. The *approximate centralizer* of K in H is

$$\tilde{C}_H(K) = \{h \in H : |K/C_K(h)| \text{ is bounded}\}.$$

3. The *approximate centre* of G is defined as $\tilde{Z}(G) = \tilde{C}_G(G)$.

4. The *approximate iterated centres of* G, denoted $\tilde{Z}_i(G)$ for $i < \omega$, are defined inductively as $\tilde{Z}_0(G) = 1$, and $\tilde{Z}_{i+1}(G)$ is the pre-image in G of $\tilde{Z}(G/\tilde{Z}_i(G))$.

5. The *approximate normalizer* of K in H is

$$\tilde{N}_H(K) = \{h \in H : K \text{ and } K^h \text{ are commensurate}\}.$$

Clearly, $\tilde{Z}(G) = \tilde{Z}_1(G)$.

For the next Proposition, recall that for two subgroups H, K of a group G the iterated commutator group $[H,_i K]$ is defined inductively as follows: $[H,_0 K] = H$, and $[H,_{i+1} K] = [[H,_i K], K]$.

PROPOSITION 4.4.10 *Let H, K, and p be as in Definition 4.4.9. Then*

1. $\tilde{C}_H(p)$ is a hyperdefinable subgroup of H. For any $x \models p$, the centralizer of x intersects $\tilde{C}_H(p)$ in a subgroup of bounded index. If q is a non-forking extension of p (over some model $\mathfrak{N} \succeq \mathfrak{M}$), then $\tilde{C}_H(p) = \tilde{C}_H(q)$.

2. $\tilde{C}_H(K)$ is a hyperdefinable subgroup of H. If L is another hyperdefinable subgroup of G commensurate with K, then $\tilde{C}_H(K) = \tilde{C}_H(L)$. Furthermore, $\tilde{C}_K(\tilde{C}_H(K))$ has bounded index in K, and $\tilde{C}_H(\tilde{C}_K(\tilde{C}_H(K))) = \tilde{C}_H(K)$.

3. $\tilde{Z}_i(G)$ is a characteristic hyperdefinable subgroup of G for all $i < \omega$, and $[\tilde{Z}_{i+1}(G), G^0_\emptyset]/\tilde{Z}_i(G)$ is bounded and centralized by $G^0_\emptyset$. In particular, $[\tilde{Z}_i(G),_{2i} G^0_\emptyset] = 1$, and $\tilde{Z}_i(G)^0_\emptyset$ is nilpotent of class $2i$.

4. $\tilde{N}_H(K)$ is hyperdefinable.

PROOF:

1. $\tilde{C}_H(p)$ is clearly hyperdefinable by Remark 3.2.9. Suppose $h, h' \in H$ are such that there are $x, x' \models p$ with $h \mathop{\smile\!\!\!\!\!\!\!\;\mid}_{\mathfrak{M}} x$, $h' \mathop{\smile\!\!\!\!\!\!\!\;\mid}_{\mathfrak{M}} x'$, $xh = hx$ and $x'h' = h'x'$. If $h \mathop{\smile\!\!\!\!\!\!\!\;\mid}_{\mathfrak{M}} h'$, we may assume that $x = x'$ and $x \mathop{\smile\!\!\!\!\!\!\!\;\mid}_{\mathfrak{M}} h, h'$ by the Independence Theorem. Then $x \mathop{\smile\!\!\!\!\!\!\!\;\mid}_{\mathfrak{M}} h^{-1}h'$, and $x(h^{-1}h') = (h^{-1}h')x$. It follows from Lemma 4.4.8 that $\tilde{C}_H(p)$ is a subgroup of H, and for generic $h \in \tilde{C}_H(p)$ there is $x \models p$ with $x \mathop{\smile\!\!\!\!\!\!\!\;\mid}_{\mathfrak{M}} h$ and $xh = hx$. So the centralizer of x is generic in $\tilde{C}_H(p)$ and has bounded index. As p is a complete type, the centraliser of any realization of p has bounded index in $\tilde{C}_H(p)$.

Lastly, if $h \in \tilde{C}_H(p)$, then there are generic $g, g' \in \tilde{C}_H(p)$ over $\mathfrak{N}$ with $h = gg'$. Since g is generic, there is $x \models p$, with $x \mathop{\smile\!\!\!\!\!\!\!\;\mid}_{\mathfrak{M}} g$ and $xg = gx$. As

$g \mathop{\smile\!\!\!\!\!\!\,|}_{\mathfrak{M}} \mathfrak{N}$, by the Independence Theorem we may assume that $x \models q$ and $x \mathop{\smile\!\!\!\!\!\!\,|}_{\mathfrak{M}} \mathfrak{N}, g$, whence $x \mathop{\smile\!\!\!\!\!\!\,|}_{\mathfrak{N}} g$. Thus $g \in \tilde{C}_H(q)$. Similarly, $g' \in \tilde{C}_H(q)$, so $h \in \tilde{C}_H(q)$ and $\tilde{C}_H(p) \le \tilde{C}_H(q)$. Conversely, suppose h is such that there is $x \models q$ with $x \mathop{\smile\!\!\!\!\!\!\,|}_{\mathfrak{N}} h$ and $hx = xh$. Then $x \mathop{\smile\!\!\!\!\!\!\,|}_{\mathfrak{M}} \mathfrak{N}$, whence $x \mathop{\smile\!\!\!\!\!\!\,|}_{\mathfrak{M}} h$ and $x \models p$. It follows that $\tilde{C}_H(q) \le \tilde{C}_H(p)$.

2. We have $\tilde{C}_H(K) = \{h \in H : D_G(C_K(h), \varphi, \epsilon, k) \ge D_G(K, \varphi, \epsilon, k)$ for all formulas φ, all $\epsilon \in E$ and all $k < \omega\}$, so $\tilde{C}_H(K)$ is hyperdefinable; it is obvious from the definition that this is a subgroup of H. Clearly $\tilde{C}_H(K) = \tilde{C}_H(L)$ for commensurate K and L.

   For the next assertion, consider a generic $h \in \tilde{C}_H(K)$. Then $C_K(h)$ has bounded index in K, and there is a generic $k \in K$ with $k \mathop{\smile\!\!\!\!\!\!\,|}_{\mathfrak{M}} h$ and $kh = hk$. Hence $C_H(k)$ intersects $\tilde{C}_H(K)$ in a subgroup of bounded index, and $k \in \tilde{C}_K(\tilde{C}_H(K))$. So $\tilde{C}_K(\tilde{C}_H(K))$ has bounded index in K. The last assertion follows from commensurativity of K and $\tilde{C}_K(\tilde{C}_H(K))$.

3. If $\tilde{Z}_i(G)$ is a hyperdefinable characteristic subgroup of G, then $G/\tilde{Z}_i(G)$ is a hyperdefinable group. Now if $C_G(h)$ has bounded index in G and σ is an automorphism of G, then $C_G(h^\sigma) = C_G(h)^\sigma$ has bounded index in G. Hence $\tilde{Z}(G)$ is characteristic in G, and hyperdefinable by part 2. Therefore $\tilde{Z}(G/\tilde{Z}_i(G))$ is a characteristic hyperdefinable subgroup of $G/\tilde{Z}_i(G)$, and $\tilde{Z}_{i+1}(G)$ is a characteristic hyperdefinable subgroup of G. The first assertion now follows by induction, the case $i = 0$ being trivial.

   In order to prove boundedness and centrality of $[\tilde{Z}_{i+1}(G), G^0_\emptyset]/\tilde{Z}_i(G)$, it is clearly enough to consider the case $i = 0$. Now $C_G(h)$ has bounded index in G for any $h \in \tilde{Z}(G)$; since $[h, G]$ is in bijection with $G/C_G(h)$, it is bounded as well. As $G^0_\emptyset \le \tilde{C}_G(\tilde{C}_G(G)) = \tilde{C}_G(\tilde{Z}(G))$, for any $g \in G^0_\emptyset$ the index of $C_{\tilde{Z}(G)}(g)$ in $\tilde{Z}(G)$ is also bounded, as is $[\tilde{Z}(G), g]$. In particular, for independent $h \in \tilde{Z}(G)$ and $g \in G^0_\emptyset$,

   $$[h, g] \in \mathrm{bdd}(h) \cap \mathrm{bdd}(g) = \mathrm{bdd}(\emptyset).$$

   As any element h of $\tilde{Z}(G)$ can be written as the product of two generic elements h' and h'', both independent of any given $g \in G^0_\emptyset$, we obtain

   $$[h, g] = [h'h'', g] = [h', g]^{h''}[h'', g] = [h', g][[h', g], h''][h'', g];$$

   so $[h', g], [h'', g] \in \mathrm{bdd}(\emptyset)$, whence $[h', g] \mathop{\smile\!\!\!\!\!\!\,|} h''$ and $[[h', g], h''] \in \mathrm{bdd}(\emptyset)$ as well. It follows that $[\tilde{Z}(G), G^0_\emptyset]$ is bounded; as it is normal in G, its centralizer has bounded index in G and contains $G^0_\emptyset$.

   By induction on i, we see that $[\tilde{Z}_i(G),_{2i} G^0_\emptyset] = 1$, and $\tilde{Z}_i(G)^0_\emptyset$ is nilpotent of class $2i$ for all $i < \omega$.

4. $\tilde{N}_H(K) = \{h \in H : D_G(K^h \cap K, \varphi, \epsilon, k) \geq D_G(K, \varphi, \epsilon, k)$ for all formulas φ, all $\epsilon \in E$ and all $k < \omega\}$, clearly a hyperdefinable set. ∎

REMARK 4.4.11 A similar proof shows that $\tilde{Z}(G)$ is bounded-by-abelian.

PROBLEM 4.4.12 Can we improve Proposition 4.4.10.3. above? Is it true that $[\tilde{Z}_i(G),_i G^0_\emptyset]$ is bounded for all $i < \omega$? If $G = \tilde{Z}_i(G)$, is $G^0_\emptyset$ (bounded central)-by-(nilpotent of class i)?

4.5 STABILIZERS

Again, G is a hyperdefinable group of type E over $\emptyset$ in a simple theory.

DEFINITION 4.5.1 For $p \in S_G(A)$ put

$$S(p) = \{g \in G : p \cup gp \text{ does not fork over } A\}.$$

The *stabilizer* $\mathrm{stab}(p)$ of p is defined as $S(p) \cdot S(p)$.

We can similarly define stabilizers on the right: $S(p)^* = \{g \in G : p \cup pg$ does not fork over $A\}$, and $\mathrm{stab}(p)^* = S(p)^* \cdot S(p)^*$. It is clear from the definition that $S(p)^* = S(p^{-1})^{-1}$ and $\mathrm{stab}(p)^* = \mathrm{stab}(p^{-1})^{-1}$.

LEMMA 4.5.2 *1. $S(p)$ and $\mathrm{stab}(p)$ are hyperdefinable over A.*

2. *If $g \in S(p)$, there are $h, h' \models p$, each independent of g over A, with $gh = h'$. In particular, $g^{-1} \in S(p)$.*

3. *If g and g' are independent elements of $S(p)$ and p is over a model $\mathfrak{M}$ (or if p is a Lascar strong type), then $gg' \in S(p)$.*

PROOF:

1. Immediate from Remark 3.2.9.

2. By definition there is $h' \models p \cup gp$ with $h' \perp_A g$. Put $h = g^{-1}h'$. Then $h \models p$ and

$$\begin{aligned} D_G(p, \varphi, \epsilon, k) &= D_G(h/A, \varphi, \epsilon, k) \geq D_G(h/A, g, \varphi, \epsilon, k) \\ &= D_G(gh/A, g, \varphi, \epsilon, k) = D_G(h'/A, g, \varphi, \epsilon, k) \\ &= D_G(h'/A, \varphi, \epsilon, k) = D_G(p, \varphi, \epsilon, k) \end{aligned}$$

for all $\mathfrak{L}$-formulas φ, all $\epsilon \in E$ and all $k < \omega$. Hence we have equality all the way through, and $h \perp_A g$.

3. There are h and h' realizing p with $g \perp_{\mathfrak{M}} h$ and $g' \perp_{\mathfrak{M}} h'$, such that gh and $g'h'$ realize p, and $gh \perp_{\mathfrak{M}} g$ and $g'h' \perp_{\mathfrak{M}} g'$. By the Independence

Theorem we may assume that $h = g'h'$, and $h \mathop{\smile\!\!\!\!\!\!\,|}_{\mathfrak{M}} g, g'$. Then $(gg')h'$ realizes $p \cup (gg')p$. Furthermore, $h \mathop{\smile\!\!\!\!\!\!\,|}_{\mathfrak{M}} g, g'$ implies $gh \mathop{\smile\!\!\!\!\!\!\,|}_{\mathfrak{M},g} g'$, whence $gh \mathop{\smile\!\!\!\!\!\!\,|}_{\mathfrak{M}} g, g'$. Thus $gg' \in S(p)$. ■

Corollary 4.5.3 *If $\mathfrak{M}$ is a model and $p \in S_G(\mathfrak{M})$, or p is a Lascar strong type, then* $\mathrm{stab}(p)$ *is a hyperdefinable subgroup of G, and $S(p)$ is a hyperdefinable subset of* $\mathrm{stab}(p)$ *containing all generic types.*

Proof: Immediate, using Lemmas 4.5.2.3 and 4.4.8. ■

Proposition 4.5.4 *Let $p \in S_G(\mathfrak{M})$. Then the following are equivalent:*

1. *p is generic for G.*
2. $\mathrm{stab}(p)$ *is generic in G.*
3. $\mathrm{stab}(p)$ *has bounded index in G.*
4. $\mathrm{stab}(p) = G^0_{\mathfrak{M}}$.

Proof: Suppose p is generic for G, and g and h are two independent realizations of p. Then by genericity $g \mathop{\smile\!\!\!\!\!\!\,|} gh^{-1}$ and gh^{-1} is also generic for G. But $g \models p \cup gh^{-1}p$ and $gh^{-1} \in S(p)$, so $S(p)$ and hence $\mathrm{stab}(p)$ are generic in G.

Conversely, if $\mathrm{stab}(p)$ and therefore $S(p)$ are generic in G, we may choose generic $g \in S(p)$ and $h \models p \cup gp$ with $g \mathop{\smile\!\!\!\!\!\!\,|}_{\mathfrak{M}} h$. As g is generic, $g^{-1}h$ is generic, as is p.

The equivalence of 2. and 3. follows from Proposition 4.4.5.

Finally, since p must specify its coset modulo $G^0_{\mathfrak{M}}$, we get $\mathrm{stab}(p) \leq G^0_{\mathfrak{M}}$ for any type $p \in S_G(\mathfrak{M})$. Hence 3. and 4. are equivalent. ■

Since non-forking extensions need not be unique in a simple theory, we have to link the stabilizer of a type with the stabilizer of its non-forking extensions.

Lemma 4.5.5 *Let $p \in S_G(\mathfrak{M})$ and q be a non-forking extension of p over some $\mathfrak{N} \succ \mathfrak{M}$. Then* $\mathrm{stab}(q)$ *is a subgroup of bounded index in* $\mathrm{stab}(p)$.

Proof: By definition, $S(p) \supseteq S(q)$, so $\mathrm{stab}(p) \geq \mathrm{stab}(q)$. Conversely, suppose g is generic for $\mathrm{stab}(p)$ over $\mathfrak{N}$. Then $g \mathop{\smile\!\!\!\!\!\!\,|}_{\mathfrak{M}} \mathfrak{N}$ and $g \in S(p)$; by the Independence Theorem 3.2.15 and Lemma 4.5.2 there is $x \mathop{\smile\!\!\!\!\!\!\,|}_{\mathfrak{M}} \mathfrak{N}, g$ with $x \models q$ and $gx \models p$; since $gx \mathop{\smile\!\!\!\!\!\!\,|}_{\mathfrak{M}} g$ we get $gx \mathop{\smile\!\!\!\!\!\!\,|}_{\mathfrak{M}} \mathfrak{N}, g$. Now suppose $g' \models \mathrm{tp}(g/\mathfrak{N})$ with $g' \mathop{\smile\!\!\!\!\!\!\,|}_{\mathfrak{N}} g$. By the Independence Theorem we may assume $gx = g'x'$ for some $x' \models q$, and $g'x' \mathop{\smile\!\!\!\!\!\!\,|}_{\mathfrak{N}} g, g'$. Then $g^{-1}g'x' \models q$; moreover $g^{-1}g'x' \mathop{\smile\!\!\!\!\!\!\,|}_{\mathfrak{N},g} g'$ and $g^{-1}g'x' \mathop{\smile\!\!\!\!\!\!\,|}_{\mathfrak{N}} g$ yield $g^{-1}g'x' \mathop{\smile\!\!\!\!\!\!\,|}_{\mathfrak{N}} g^{-1}g'$. Therefore $g^{-1}g' \in S(q)$. But g and g' are two independent generic elements for $\mathrm{stab}(p)$

over $\mathfrak{N}$, so $g^{-1}g'$ is generic for $\mathrm{stab}(p)$ over $\mathfrak{N}$, and $\mathrm{stab}(q)$ has bounded index in $\mathrm{stab}(p)$ by Corollary 4.4.2. ■

The following Proposition can be seen as an analogue to the uniqueness of the principal generic type in a stable theory (Remark 4.1.24).

PROPOSITION 4.5.6 *Let p, p' and p'' be three principal generic types of G over a model $\mathfrak{M}$. Then there are independent realizations $g \models p$ and $g' \models p'$ such that $gg' \models p''$.*

PROOF: Let $h' \models p'$ and $h'' \models p''$ with $h' \perp_{\mathfrak{M}} h''$, and put $a = h''h'^{-1}$. Then $ah' = h''$, a is principal generic over $\mathfrak{M}$, and $\{a, h', h''\}$ is pairwise independent over $\mathfrak{M}$. We can similarly find a principal generic b over $\mathfrak{M}$, such that $ab \models p$ and $\{a, b, ab\}$ is pairwise independent over $\mathfrak{M}$. Now $S(p')$ contains all principal generics and in particular b^{-1}; as $a \perp_{\mathfrak{M}} h'$ we may assume by the Independence Theorem that $b \perp_{\mathfrak{M}} a, h'$ and $b^{-1}h'$ realises p'. Put $g = ab$ and $g' = b^{-1}h'$; we only have to check that $g \perp_{\mathfrak{M}} g'$. But $\{a, b, h'\}$ are independent over $\mathfrak{M}$, whence $ab \perp_{\mathfrak{M},b} b^{-1}h'$. Since $ab \perp_{\mathfrak{M}} b$ (as a and b are independent generic elements), we get $ab \perp_{\mathfrak{M}} b^{-1}h'$, as required. ■

COROLLARY 4.5.7 *Let K be a hyperdefinable field in a simple theory, and T a hyperdefinable multiplicative subgroup of bounded index. Then $K^\times = aT+bT$ for any two cosets aT and bT of T.*

PROOF: For $c \in K^\times$ choose a model $\mathfrak{M}$ containing a, b, c, together with a system of representatives for all the cosets of T in K. Let p be a principal additive generic type over $\mathfrak{M}$. As T has bounded index, there is some $d \in K^\times$ with $p \vdash x \in dT$; since $d(K^+)^0 = (K^+)^0$, the translate $d^{-1}p$ is also a principal additive generic, and we may assume $p \vdash x \in T$. Similarly, ap, bp and cp are principal additive generic types over $\mathfrak{M}$. By Proposition 4.5.6 there are three pairwise independent realizations x, y and z of p such that $ax + by = cz$. Then $a(x/z) + b(y/z) = c$, and $x/z, y/z \in T$. ■

A simple field, even a supersimple field, need not be connected additively or multiplicatively; in fact it may be connected one way but not the other. For instance, a pseudofinite field of characteristic 0 is supersimple, divisible, and additively connected, but $(K^\times)^n$ has index n in $K^\times$ for all $n < \omega$. In the stable case, things are different:

REMARK 4.5.8 An infinite stable division ring D is additively and multiplicatively connected.

PROOF: Consider the additive connected component $(D^+)^0$; it is independent of the parameters by Remark 4.1.24, and invariant under left and right translation by elements in $D^\times$. Hence it is an ideal; since it has bounded index, it is non-trivial, and $(D^+)^0 = D$. So D has a unique additive generic type p by Remark 4.1.24; since dp is also additive generic for any $d \in D^\times$, it follows that p is the unique multiplicative generic type, and D is multiplicatively connected. ■

We can now prove the existence of a principal generic extension for a principal generic type over a model.

COROLLARY 4.5.9 *Let $p \in S_G(\mathfrak{M})$ be principal generic, and $A \supseteq \mathfrak{M}$. Then p has a principal generic extension to A.*

PROOF: Extending A, we may assume that it is a model. By Proposition 4.5.6 we find two independent realizations $x, y \models p$ such that $xy \models p$. We choose $y \perp_{\mathfrak{M}} A$; as $xy \perp_{\mathfrak{M}} y$ by genericity, we may assume by the Independence Theorem that $\mathrm{tp}(xy/A) = \mathrm{tp}(y/A)$ and $xy \perp_{\mathfrak{M}} A, y$. Then xy and y are independent generic over A, as are $x = (xy)y^{-1}$ and xy. So $x \in S(y/A) \subseteq G_A^0$. Thus $\mathrm{tp}(x/A)$ extends p and is principal generic for G over A. ■

The following lemma establishes the connection between the approximate centralizers of a group and its principal generic types.

LEMMA 4.5.10 *Let H be a hyperdefinable subgroup of G, and p a principal generic type of H. Then $\tilde{C}_G(p) = \tilde{C}_G(H)$.*

PROOF: If $g \in \tilde{C}_G(H)$, let h realize a principal generic extension of p to $\mathfrak{M}, g$. As $C_H(g)$ has bounded index in H, we get $h \in H^0_{\mathfrak{M},g} \leq C_H(g)$, and $gh = hg$. Since $h \models p$ and $h \perp_{\mathfrak{M}} g$, we have $g \in \tilde{C}_G(p)$.

Conversely, if $g \in G$ and $x \models p$ with $x \perp_{\mathfrak{M}} g$ and $xg = gx$, then $C_H(g)$ has bounded index in H by genericity of p, whence $g \in \tilde{C}_G(H)$. ■

In fact, this also shows that for a principal generic type p over $\mathfrak{M}$ we have $\tilde{C}_G(p) = \{g \in G : \exists x \models p\,[x \perp_{\mathfrak{M}} g \wedge xg = gx]\}$.

Stabilizers can be helpful in showing the existence of hyperdefinable subgroups.

LEMMA 4.5.11 *Let $p \in S_G(\mathfrak{M})$ for some model $\mathfrak{M}$, such that for independent realizations a and b of p the product $a^{-1}b$ is independent of a. Then a left translate of (some non-forking extension of) p is a generic type of* $\mathrm{stab}(p)^*$.

PROOF: If a and b are independent realizations of p, then $b^{-1}a$ is independent of b; as $b^{-1}a = (a^{-1}b)^{-1}$, we have that $a^{-1}b$ is independent both of a and of b. By definition, $a^{-1}b \in S(p)^*$, so

$$\begin{aligned} D_G(S(p)^*, \varphi, \epsilon, k) &\geq D_G(a^{-1}b/\mathfrak{M}, \varphi, \epsilon, k) = D_G(a^{-1}b/\mathfrak{M}, a, \varphi, \epsilon, k) \\ &= D_G(b/\mathfrak{M}, a, \varphi, \epsilon, k) = D_G(b/\mathfrak{M}, \varphi, \epsilon, k) \\ &= D_G(p, \varphi, \epsilon, k) \end{aligned}$$

for all $\mathfrak{L}$-formulas φ, all $\epsilon \in E$ and all $k < \omega$. On the other hand, for any $g \in S(p)^*$ there is by definition some $x \models p$ with $x \perp_{\mathfrak{M}} g$, $xg \perp_{\mathfrak{M}} g$ and $xg \models p$. Hence

$$\begin{aligned} D_G(g/\mathfrak{M}, \varphi, \epsilon, k) &= D_G(g/\mathfrak{M}, x, \varphi, \epsilon, k) = D_G(xg/\mathfrak{M}, x, \varphi, \epsilon, k) \\ &\leq D_G(xg/\mathfrak{M}, \varphi, \epsilon, k) = D_G(p, \varphi, \epsilon, k) \end{aligned}$$

for all $\mathfrak{L}$-formulas φ, all $\epsilon \in E$ and all $k < \omega$; whence $D_G(S(p)^*, \varphi, \epsilon, k) \leq D_G(p, \varphi, \epsilon, k)$. Thus

$$\begin{aligned} D_G(\mathrm{stab}(p)^*, \varphi, \epsilon, k) &= D_G(S(p)^*, \varphi, \epsilon, k) = D(a^{-1}b/\mathfrak{M}, \varphi, \epsilon, k) \\ &= D_G(p, \varphi, \epsilon, k). \end{aligned}$$

Proposition 4.3.8.3 implies that $\mathrm{tp}(a^{-1}b/\mathfrak{M})$ is generic for $\mathrm{stab}(p)^*$; it has a non-forking extension $\mathrm{tp}(a^{-1}b/\mathfrak{M}, a)$, whose left translate $\mathrm{tp}(b/\mathfrak{M}, a)$ is a non-forking extension of p. ∎

REMARK 4.5.12 Of course, since $\mathrm{stab}(p)^* = \mathrm{stab}(p^{-1})$, the lemma implies that if ab^{-1} is independent of a for any two independent realizations of a Lascar strong type p, then p is a right translate of a generic type for $\mathrm{stab}(p)$.

We shall now use stabilizers to prove the existence of certain components of a hyperdefinable group G. In analogy to Definition 4.2.1 we call a family $\mathfrak{H}$ of subgroups of G *hyperdefinable over* A if it consists of subgroups of the form $\Theta(x, B)/E$, where $\Theta(x, B)$ is a partial E-type over B and B runs through all realizations of some partial type over A.

THEOREM 4.5.13 *Let $\mathfrak{H}$ be a family of subgroups of G, and suppose $\mathfrak{H}$ is hyperdefinable over A. Then there are a hyperdefinable subgroup N of G over A contained in $\langle \mathfrak{H} \rangle$ and containing some bounded intersection of G-conjugates of groups in $\mathfrak{H}$, and a bounded intersection L of groups in $\mathfrak{H}$, such that N is commensurate with $L \cap H$ for all $H \in \mathfrak{H}$. Furthermore, for any hyperdefinable group Γ over A acting on G and stabilizing $\mathfrak{H}$ setwise (under its induced action on the collection of subgroups of G), there is some hyperdefinable N_Γ with the same properties as N, which in addition is Γ-invariant.*

Note that the subgroup of all $g \in G$ which stabilize $\mathfrak{H}$ setwise (under their action by conjugation) may not be hyperdefinable.

PROOF: Suppose $\mathfrak{H}$ is hyperdefined over A as $\{H(a) : \models \pi(a)\}$. We may assume that $A = \mathrm{bdd}(A)$ and π is a complete type p over A, simply by obtaining groups N_p for the families $\{H(a) : a \models p\}$ for every completion p of π over $\mathrm{bdd}(A)$, and then taking N to be the intersection of all A-conjugates of $\bigcap_{p \in [\pi]} N_p$ (which will be hyperdefined over A). Put

$$X = \{x \in G : \exists y \models p \, [y \underset{A}{\perp} x \wedge x \in H(y)]\}.$$

Suppose $x, x' \in X$ with $x \perp_A x'$, as witnessed by some $y, y' \models p$ with $x \in H(y), x' \in H(y'), x \perp_A y$ and $x' \perp_A y'$. By the Independence Theorem, we may assume that $y = y'$ and $y \perp_A x, x'$, so $x^{-1}x' \in H(y)$ and $x^{-1}x' \in X$. Put $K = X \cdot X$; by Remark 3.2.9 both X and K are hyperdefinable over A, and K is a subgroup of G by Lemma 4.4.8. Clearly $K \leq \langle \mathfrak{H} \rangle$.

By Proposition 4.4.3 there are $\alpha < |T|^+$ and a sequence $(a_i : i < \alpha)$ of independent realizations of p, such that for any $a \models p$ independent of $(a_i : i < \alpha)$, the group $H(a)$ intersects $\bigcap_{i<\alpha} H(a_i)$ in a subgroup of bounded index. Put $L = \bigcap_{i<\alpha} H(a_i)$.

CLAIM. L is commensurate with K.

PROOF OF CLAIM: First, let $a \models p$ be independent of $(a_i : i < \alpha)$, and x generic in $L \cap H(a)$. Since $L \cap H(a)$ has bounded index in L, the element x is generic in L. Therefore $x \perp_{A \cup (a_i : i<\alpha)} a$; as $a \perp_A (a_i : i < \alpha)$, we get $a \perp_A x$ and $x \in X \subseteq K$. Thus $L \cap K$ has bounded index in L.

Next consider a generic type q of K. As $q(x) \vdash x \in X$, by the Independence Theorem there is a realization x of q such that $x \in H(a_i)$ for all $i < \alpha$, with $x \perp_A (a_i : i < \alpha)$. Hence $K \cap L$ has bounded index in K. ■

Now, put

$$N = \bigcap_{k \in K} (\bigcap_{a \models p} H(a) \cdot K)^k.$$

This is invariant under A-automorphisms; by Lemma 4.2.5 it is a subgroup of G containing K. Since N lies between K and LK, the intersection in its definition is in fact a bounded one, so N is hyperdefinable over A and contains a bounded intersection of G-conjugates of groups in $\mathfrak{H}$. Because K is commensurate with L, it follows that N is commensurate with L. Clearly $N \leq \langle \mathfrak{H} \rangle$. If $a \models p$, let L_0 be an A-automorphic conjugate of L whose parameters are independent of a. Then $H(a) \cap L_0$ is commensurate with L_0, and L_0 is commensurate with N and therefore L, so $H(a) \cap L$ is commensurate with N.

For the last assertion, suppose Γ is a hyperdefinable group over A acting on G and stabilizing $\mathfrak{H}$. Again we may assume that $A = \mathrm{bdd}(A)$. Let p be generic for Γ over A, and put

$$Y = \{x \in G : \exists \gamma \models p \, [\gamma \perp_A x \wedge x \in N^\gamma\},$$

and $Z = Y \cdot Y$. Then both Y and Z are hyperdefinable over A, and Z is a subgroup of G commensurate with some intersection of Γ-conjugates of N (apply the first half of the proof to the family of Γ-conjugates of N). But a Γ-conjugate of N is commensurate with a Γ-conjugate of L; since Γ stabilizes $\mathfrak{H}$, it is commensurate with N. Therefore Z is commensurate with N.

CLAIM. If $x \in Y$ and $\gamma_0 \in \Gamma_A^0$ is generic with $x \perp_A \gamma_0$, then $x^{\gamma_0} \in Y$.

PROOF OF CLAIM: Since γ_0 stabilizes p on the right, by the Independence Theorem we can find some $\gamma \models p$ such that $\gamma\gamma_0 \models p$, $x \in N^\gamma$, and $\gamma \perp_A x, \gamma_0$. Then $x^{\gamma_0} \in N^{\gamma\gamma_0}$; since $\gamma\gamma_0 \perp_{A,\gamma_0} x^{\gamma_0}$ and $\gamma\gamma_0 \perp_A \gamma_0$ by genericity, we get $x^{\gamma_0} \perp_A \gamma\gamma_0$ and $x^{\gamma_0} \in Y$. ■

CLAIM. Z is normalized by Γ_A^0.

PROOF OF CLAIM: Since any element in Γ_A^0 is the product of two generic elements, it is sufficient to show $Z^\gamma = Z$ for generic $\gamma \in \Gamma_A^0$. But for any $x \in Z$ there are $a, b \in Y$, both independent of γ over A, with $x = ab$. As $a^\gamma \in Y$ and $b^\gamma \in Y$, we get $x^\gamma \in Z$. ■

Put $N_0 = \bigcap_{\gamma\in\Gamma} Z^\gamma$. If Γ_0 is a set of representatives for Γ/Γ_A^0, then $N_0 = \bigcap_{\gamma\in\Gamma_0} Z^\gamma$, so N_0 is a hyperdefinable group over A stabilized by Γ. As any Γ-conjugate of Z is commensurate with N, so is N_0.

Finally, we put $N_\Gamma = \bigcap_{n\in N_0} (\bigcap_{\gamma\in\Gamma} N^\gamma \cdot N_0)^n$. This is a subgroup of G by Lemma 4.2.5; as it lies between N_0 and N_0N, it is commensurate with N and therefore hyperdefinable (over A) as a bounded intersection of hyperdefinable groups. It is clearly Γ-invariant and contains some bounded intersection of conjugates of N, whence of G-conjugates of groups in $\mathfrak{H}$. ■

COROLLARY 4.5.14 *Let H be a hyperdefinable subgroup of G. Then there is a hyperdefinable normal subgroup K of G, which contains and is commensurate with sufficiently big intersections of G-conjugates of H.*

PROOF: Obvious from Theorem 4.5.13, taking $\mathfrak{H} = \{H^g : g \in G\}$ and $\Gamma = G$. ■

DEFINITION 4.5.15 A hyperdefinable subgroup H of G is *locally connected* if for any G-conjugate or automorphic image H^* of H, either $H = H^*$ or $H \cap H^*$ has unbounded index in H.

Note that if H^* is a G-conjugate of an automorphic image of H, then it has the same local D_G-ranks. Hence if $H \cap H^*$ has unbounded index in H, it has smaller $D_G(., \varphi, \epsilon, k)$-rank for some φ, some $\epsilon \in E$ and $k < \omega$. So $H \cap H^*$ must have infinite index in H^* as well.

COROLLARY 4.5.16 *Let H be a hyperdefinable subgroup of G. Then there is a unique minimal hyperdefinable locally connected subgroup H^c commensurate with H.*

We call H^c the *locally connected component* of H. However, H^c is not necessarily a bounded intersection of commensurate conjugates of H. Clearly, if H is commensurate with all its G-conjugates, then H^c is normal in G.

PROOF: Suppose H is hyperdefined over a and indicate this by writing $H = H(a)$. Let $\mathfrak{H}$ be the family of G-conjugates of automorphic images of H

commensurate with H. Since $H(a')^g$ is commensurate with H for $a' \models \mathrm{tp}(a)$ and $g \in G$ if and only if $D_G(H \cap H(a')^g, \varphi, \epsilon, k) \geq D_G(H, \varphi, \epsilon, k)$ for all $\varphi, i \in I$ and $k < \omega$, the family $\mathfrak{H}$ is hyperdefinable. Furthermore, the group $\Gamma = \{g \in G : H \text{ and } H^g \text{ are commensurate}\}$ is hyperdefinable; if $g \in \Gamma$ and $H(a')^h \in \mathfrak{H}$, then $H(a')^h$ and $H(a')^{hg}$ are also commensurate. Therefore Γ stabilizes $\mathfrak{H}$.

By Corollary 4.4.2 the equivalence relation F on $\mathrm{tp}(a)$ induced by commensurativity of $H(a)$ and $H(a')$ is type-definable. It is obvious that an automorphism stabilizes $\mathfrak{H}$ setwise if and only if it fixes a_F, so a_F is a canonical parameter for $\mathfrak{H}$, and $\mathfrak{H}$ is hyperdefinable over a_F. It follows that Γ is hyperdefinable over a_F as $\exists\, a' \models \mathrm{tp}(a)\, [a'Fa \wedge H(a')^g \in \mathfrak{H}]$. By Theorem 4.5.13 there is a subgroup H_0 of G which is hyperdefinable over a_F, commensurate with a sufficiently big intersection of groups in $\mathfrak{H}$, and normalized by Γ. But any (bounded) intersection of groups in $\mathfrak{H}$ is commensurate with H, so H_0 is commensurate with H. If σ is an automorphism of the monster model, then either H and $\sigma(H)$ are commensurate, or not. In the first case σ fixes a_F and stabilizes H_0; in the second case H_0 and $\sigma(H_0)$ cannot be commensurate. Similarly, if $g \in G$, then either H and H^g are commensurate, $g \in \Gamma$ and $H_0 = H_0^g$, or neither H and H^g nor H_0 and H_0^g are commensurate. So H_0 is locally connected.

Finally, let H^c be the intersection of all hyperdefinable locally connected groups commensurate with H. If H_1 is hyperdefinable, locally connected and commensurate with H, then H_1 is stabilized under all automorphisms stabilizing H, so H_1 is hyperdefinable over the parameters needed for H. In particular, H^c is a bounded intersection and thus a hyperdefinable group commensurate with H; it is clearly locally connected and minimal with these properties. ∎

REMARK 4.5.17 If H_0 and H_1 are commensurate, then $H_0^c = H_1^c$. In particular, if q is a non-forking extension of p, then $\mathrm{stab}(p)^c = \mathrm{stab}(q)^c$.

LEMMA 4.5.18 *For a locally connected hyperdefinable subgroup H of G, the normalizer $N_G(H)$ is hyperdefinable. If H and H^* are two commensurate locally connected groups, then $N_G(H) = N_G(H^*)$.*

PROOF: By local connectivity, any $g \in G$ normalizes H if and only if H and H^g are commensurate, so $N_G(H) = \tilde{N}_G(H)$, which is hyperdefinable by Lemma 4.4.10.4.

Now if H and H^* are two commensurate groups, then $\tilde{N}_G(H) = \tilde{N}_G(H^*)$; the second assertion follows. ∎

LEMMA 4.5.19 *A locally connected hyperdefinable subgroup H of G has a canonical parameter u. (That is, any automorphism stabilizes H setwise if* *and only if it fixes u.) Any coset gH of H also has a canonical parameter v_g. Moreover, if p is generic for H and q is generic for gH, then* $u \in \mathrm{dcl}(\mathrm{Cb}(p)) \subseteq \mathrm{bdd}(u)$ *and* $v_g \in \mathrm{dcl}(\mathrm{Cb}(q)) \subseteq \mathrm{bdd}(v_g)$.

PROOF: If H is hyperdefinable over a, say $H = H(a)$, then the equivalence relation F on $\mathrm{tp}(a)$ induced by commensurativity of $H(a')$ and $H(a'')$ is type-definable by Corollary 4.4.2. By local connectivity, a_F is a canonical parameter for H. Similarly, the equivalence relation F' on $G \times \mathrm{tp}(a)$ given by $(g', a')F'(g'', a'')$ if and only if $g'H(a') = g''H(a'')$ is type-definable as "$D_G(g'H(a') \cap g''H(a''), \varphi, \epsilon, k) \geq D_G(H, \varphi, \epsilon, k)$ for all formulas φ, all $\epsilon \in E$ and all $k < \omega$", since $g'H(a') \cap g''H(a'')$ is either empty or a coset of $H(a') \cap H(a'')$. Hence the canonical parameter of gH is $(g, a)_{F'}$.

Now let p be a generic type for H. As H is hyperdefinable over u, by genericity p does not fork over u, whence $\mathrm{Cb}(p) \subseteq \mathrm{bdd}(u)$. On the other hand, if p is over u and p^* is an automorphic image of p over some u^* with $p \sim_0 p^*$, we may choose a common non-forking realization g of p and p^*. Then

$$D_G(g/u, u^*, \varphi, \epsilon, k) = D_G(p, \varphi, \epsilon, k) = D_G(H, \varphi, \epsilon, k)$$

for all formulas φ, all $\epsilon \in E$ and all $k < \omega$. If H^* is the automorphic image of H corresponding to u^*, then $g \in H \cap H^*$. Therefore H and H^* must be commensurate, whence equal by local connectivity. So u is invariant under all automorphisms stabilizing the $\sim$-class of p, whence $u \in \mathrm{dcl}(\mathrm{Cb}(p))$.

Similarly, let q be generic for gH over u, g. Then $D_G(q, \varphi, \epsilon, k) = D_G(H, \varphi, \epsilon, k)$ for all formulas φ, all $\epsilon \in E$ and all $k < \omega$. But these ranks are maximal possible in gH, so q cannot fork over v_g and $\mathrm{Cb}(q) \subseteq \mathrm{bdd}(v_g)$. Finally, $v_g \in \mathrm{dcl}(\mathrm{Cb}(q))$ is proved in the same way as $u \in \mathrm{dcl}(\mathrm{Cb}(p))$ in the preceding paragraph. ∎

4.6 QUOTIENT GROUPS AND ANALYSABILITY

Again, G will denote a hyperdefinable group of type E over $\emptyset$ in a simple theory. We shall begin with an improvement of Proposition 3.4.9 linking internality and finite generation in the case of groups.

PROPOSITION 4.6.1 *If $p \in S_G(A)$ is generic for G and internal in some A-invariant family Σ of types, then G is finitely generated over Σ.*

PROOF: As a non-forking extension of a generic type is again generic, we may assume that A is a model, and there are types $\bar\sigma$ in Σ over A such that for every realization x of p there are $\bar y \models \bar\sigma$, with $x \in \mathrm{dcl}(A, \bar y)$.

Suppose $a \in S(p)$. Then there are $b, b' \models p$ with $ab = b'$. By assumption there are realizations $\bar c$ and $\bar c'$ of $\bar\sigma$ with $b \in \mathrm{dcl}(A, \bar c)$ and $b' \in \mathrm{dcl}(A, \bar c')$. It follows that $a \in \mathrm{dcl}(A, \bar c, \bar c')$. But every element in $\mathrm{stab}(p) = G^0_A$ is the product of two elements in $S(p)$, so every element in G_A^0 is in the definable closure of A and some realizations of $\bar{\sigma}$. Finally, if $B \supseteq A$ contains a set of representatives for G/G_A^0, then every element of G is in the definable closure of B and some realizations of $\bar{\sigma}$. Hence G is finitely generated over Σ. ■

We are now aiming for the existence of normal subgroups with particular properties.

LEMMA 4.6.2 *Suppose $\mathfrak{M}$ is a model, $p \in S_G(\mathfrak{M})$ is generic for G, $a \models p$, and $a_0 \in \mathrm{dcl}(\mathfrak{M}, a) - \mathrm{dcl}(\mathfrak{M})$ is q-internal for some $q \in S(\mathfrak{M})$. Put*

$$X = \{h \in G : \exists x \models p\,[x \mathop{\smile\!\!\!\!\!\!|}_{\mathfrak{M}} h \wedge xh \models p \wedge \exists x_0\, \mathrm{tp}(x, x_0/\mathfrak{M}) = \mathrm{tp}(xh, x_0/\mathfrak{M}) = \mathrm{tp}(a, a_0/\mathfrak{M})]\},$$

and $H = X \cdot X$. Then H is a hyperdefinable subgroup of unbounded index in G. There is a hyperdefinable normal subgroup N of G containing and commensurate with some intersection of G-conjugates of H, such that G/N is q-internal. Furthermore, if Γ is a hyperdefinable group acting on G, then we may choose N to be Γ-invariant.

PROOF: Clearly X is hyperdefinable.

CLAIM. If $h, h' \in X$ with $h \mathop{\smile\!\!\!\!\!\!|}_{\mathfrak{M}} h'$, then $h^{-1}h' \in X$.

PROOF OF CLAIM: The Independence Theorem yields the existence of $x \models p$ with $x \mathop{\smile\!\!\!\!\!\!|}_{\mathfrak{M}} h, h'$ and $xh, xh' \models p$, and of some x_0, such that

$$\mathrm{tp}(xh, x_0, \mathfrak{M}) = \mathrm{tp}(x, x_0/\mathfrak{M}) = \mathrm{tp}(xh', x_0/\mathfrak{M}) = \mathrm{tp}(a, a_0/\mathfrak{M}).$$

Put $y = xh$. Then $y \mathop{\smile\!\!\!\!\!\!|}_{\mathfrak{M}} h$ by genericity of p, and $x \mathop{\smile\!\!\!\!\!\!|}_{\mathfrak{M},h} h'$ implies $y \mathop{\smile\!\!\!\!\!\!|}_{\mathfrak{M}} h^{-1}h'$. Furthermore both $y = xh$ and $y(h^{-1}h') = xh'$ realize p, and

$$\mathrm{tp}(y, x_0, \mathfrak{M}) = \mathrm{tp}(y(h^{-1}h'), x_0, \mathfrak{M}) = \mathrm{tp}(a, a_0/\mathfrak{M}).$$

Therefore $h^{-1}h' \in X$. ■

So H is a hyperdefinable subgroup of G over $\mathfrak{M}$ by Lemma 4.4.8, and X contains all generic types of H. Consider the coset aH, a hyperimaginary element with respect to the type-definable equivalence relation $x^{-1}y \in H$.

CLAIM. The type $\mathrm{tp}(aH/\mathfrak{M})$ is q-internal.

PROOF OF CLAIM: As $S(p)$ is generic in G, there is an infinite generic Morley sequence $I = (x_i : i < \alpha)$ independent of a, such that $x_i a \models p$ for all $i < \alpha$; furthermore for every $i < \alpha$ there is some $(x_i a)_0$ with $\mathrm{tp}(x_i a, (x_i a)_0/\mathfrak{M}) = \mathrm{tp}(a, a_0/\mathfrak{M})$. Put $Y = \{x_i, (x_i a)_0 : i < \alpha\}$.

We claim that $aH \in \mathrm{dcl}(Y \cup \mathfrak{M})$. So suppose $\mathrm{tp}(a'/Y, \mathfrak{M}) = \mathrm{tp}(a/Y, \mathfrak{M})$. As I is independent and independent of a over $\mathfrak{M}$, it is independent over $\mathfrak{M}, a$.

So $(x_i a : i < \alpha)$ is independent over $\mathfrak{M}$, a, and $x_i a \mathop{\smile\!\!\!\!\!\mid}_{\mathfrak{M},a} (x_j a : j \neq i)$ for all $i < \alpha$. Since $x_i a \mathop{\smile\!\!\!\!\!\mid}_{\mathfrak{M}} a$ by genericity of x_i, we see that $(x_i a : i < \alpha)$ is independent over $\mathfrak{M}$. By simplicity, as α is big, there must be some $i < \alpha$ with $a^{-1}a' \mathop{\smile\!\!\!\!\!\mid}_{\mathfrak{M}} x_i a$. But then $x_i a$, $x_i a'$ and $(x_i a)_0$ witness $a^{-1}a' \in X$, whence $aH = a'H$.

As $a \mathop{\smile\!\!\!\!\!\mid}_{\mathfrak{M}} I$ and $\mathrm{tp}((x_i a)_0/\mathfrak{M})$ is q-internal for all $i < \alpha$, the claim follows. ■

CLAIM. G/H is q-internal.

PROOF OF CLAIM: For any $g \in G$ there is $a \models p$ with $a \mathop{\smile\!\!\!\!\!\mid}_{\mathfrak{M}} g$; as $\mathrm{tp}(a^{-1}/\mathfrak{M})$ is generic, $ga^{-1} \mathop{\smile\!\!\!\!\!\mid}_{\mathfrak{M}} g$, whence $ga^{-1} \mathop{\smile\!\!\!\!\!\mid}_{\mathfrak{M}} gH$. Now $gH \in \mathrm{dcl}(ga^{-1}, aH)$, so $\mathrm{tp}(gH/\mathfrak{M})$ is $\mathrm{tp}(aH/\mathfrak{M})$-internal. Since $\mathrm{tp}(aH/\mathfrak{M})$ is q-internal, the claim follows from Lemma 3.4.8. ■

By Corollary 4.5.14 there is a hyperdefinable normal subgroup N of G, which contains and is commensurate with some intersection $\bigcap_{i<\alpha} H^{g_i}$. As $G/H^{g_i} = (G/H)^{g_i}$ is q-internal for all $i < \alpha$, so is $G/\bigcap_{i<\alpha} H^{g_i}$, and hence G/N.

If a hyperdefinable group Γ acts on G, we apply Corollary 4.5.14 to H as a subgroup of $G \rtimes \Gamma$. As $G/H^\gamma = (G/H)^\gamma$ for any $\gamma \in \Gamma$, we thus obtain a Γ-invariant subgroup N_Γ such that G/N_Γ is q-internal. ■

THEOREM 4.6.3 *If $p \in S_G(A)$ is generic for G and almost internal in some A-invariant family Σ of types, then there is a hyperdefinable bounded normal subgroup N of G such that G/N is finitely generated over Σ. Furthermore, if G is normal in a hyperdefinable group Γ, we may choose N to be Γ-invariant.*

PROOF: Since p is almost internal in Σ, there is $a \models p$, some $B \mathop{\smile\!\!\!\!\!\mid}_A a$ and realizations $\bar{c}$ of types $\bar{\sigma}$ from Σ over B, such that $a \in \mathrm{bdd}(B, \bar{c})$. We add B to the language, and assume $A = B$ is a model. Let a_1 be the canonical base of $\mathrm{lstp}(\bar{c}/A, a)$, so $\bar{c} \mathop{\smile\!\!\!\!\!\mid}_{a_1} A, a$, whence $\bar{c} \mathop{\smile\!\!\!\!\!\mid}_{A,a_1} a$ and $a \in \mathrm{bdd}(A, a_1)$. Clearly $a_1 \in \mathrm{bdd}(A, a)$; since a_1 is in the definable closure of any Morley sequence in $\mathrm{tp}(\bar{c}/A, a)$ by Corollary 3.3.13, we see that $\mathrm{tp}(a_1/A)$ is $\bar{\sigma}$-internal. If a_0 denotes the hyperimaginary set of (A, a)-conjugates of a_1, then $a_0 \in \mathrm{dcl}(A, a)$ and $\mathrm{tp}(a_0/A)$ is $\bar{\sigma}$-internal. Furthermore, $a_1 \in \mathrm{bdd}(a_0)$ and $a \in \mathrm{bdd}(A, a_0)$.

Let X and H be as in Lemma 4.6.2. So there is a normal hyperdefinable subgroup N of G, containing and commensurate with an intersection of G-conjugates of H, such that G/N is $\bar{\sigma}$-internal. Hence G/N is finitely generated over $\bar{\sigma}$, whence over Σ, by Lemma 4.6.1.

CLAIM. X is bounded.

PROOF OF CLAIM: Consider any $h \in X$. There are $x \models p$ and a $\bar{\sigma}$-internal $x_0 \in \mathrm{dcl}(A, x)$ with $x \in \mathrm{bdd}(A, x_0)$, such that $h \mathop{\smile\!\!\!\!\!\mid}_A x$ with $\mathrm{tp}(xh, x_0/A) = \mathrm{tp}(a, a_0/A)$. But $x, xh \in \mathrm{bdd}(A, x_0)$, whence $h \in \mathrm{bdd}(A, x_0) \subseteq \mathrm{bdd}(A, x)$; as $h \mathop{\smile\!\!\!\!\!\mid}_A x$, we get $h \in \mathrm{bdd}(A)$. ■

It follows that H and hence N are bounded. The last assertion follows from the "furthermore" in Lemma 4.6.2. ■

THEOREM 4.6.4 *Suppose a generic type p of G is not foreign to some partial type π. Then there is a normal hyperdefinable subgroup N of unbounded index in G such that the quotient G/N is π-internal. Furthermore, if G is normal in a hyperdefinable group Γ, then we may choose N to be Γ-invariant.*

PROOF: There is a set A on which π is based, a realization a of a non-forking extension of p to A, and a realization c of π with $a \not\perp_A c$. By Remark 3.4.12 there is some π-internal $a_0 \in \mathrm{dcl}(A,a) - \mathrm{bdd}(A)$. Let again X and H be as in Lemma 4.6.2, and N a normal hyperdefinable subgroup of G containing and commensurate with an intersection of G-conjugates of H, such that G/N is π-internal.

CLAIM. X is not generic for G.

PROOF OF CLAIM: Suppose X is generic for G. Then there are a generic $h \in X$, some $x \models p$ with $x \perp_A h$, and a π-internal $x_0 \in \mathrm{dcl}(A,x) - \mathrm{bdd}(A)$, such that $\mathrm{tp}(x, x_0/A) = \mathrm{tp}(xh, x_0/A)$. But $xh \perp_A x$ by genericity of h, and $x_0 \in \mathrm{dcl}(A,x) \cap \mathrm{dcl}(A,xh) \subseteq \mathrm{bdd}(A)$, a contradiction. ■

Hence neither X, nor H or N, are generic in G, so N has unbounded index. The last assertion follows again from the "furthermore" in Lemma 4.6.2. ■

COROLLARY 4.6.5 *G is analysable in some $\emptyset$-invariant family Σ if and only if there is a sequence $G = G_0 > G_1 > \cdots > G_\alpha = \{1\}$ of hyperdefinable normal subgroups of G, continuous at limits, such that for all $i < \alpha$ the section G_i/G_{i+1} is Σ-internal.*

PROOF: Suppose G is Σ-analysable, and let N be a hyperdefinable normal unbounded subgroup of G, over some parameters A, say. Let g be generic for G over A. If $\bar{g}$ is a canonical parameter for the coset gN over A, then g is generic for gN over $A, \bar{g}$ by Lemma 4.3.12; as $\mathrm{tp}(g)$ is Σ-analysable, $\mathrm{tp}(g/A, \bar{g})$ is not foreign to some type $\sigma \in \Sigma$ by Lemma 3.4.14. But the generic type of N is a translate of the generic type of gN, and hence not foreign to Σ. By Theorem 4.6.4 there is a normal subgroup N_1 of G such that N/N_1 is unbounded and σ-internal.

Iterating this process, we obtain a chain of non-commensurate normal subgroups $G = G_0 > G_1 > \cdots > G_\alpha$ such that G_i/G_{i+1} is Σ-internal for all $i < \alpha$, and $G_\lambda = \bigcap_{i<\lambda} G_i$ for limit ordinals λ. By Proposition 4.1.26 this sequence must stop for some α. But then G_α must be bounded; putting $G_{\alpha+1} = \{1\}$, we obtain the desired chain.

For the converse, suppose every group in the chain is defined over A, and let g be generic for G over A. If $\bar{g}_i$ denotes the canonical parameter of the coset gG_i over A, then $\bar{g}_{i+1}$ is generic for G/G_{i+1}, and hence also generic for the coset $\bar{g}_{i+1}(G_i/G_{i+1})$ over $A, \bar{g}_i$. Furthermore, since $\mathrm{tp}(\bar{g}_{i+1}/A, \bar{g}_i)$ is a translate of the generic type of G_i/G_{i+1}, it is Σ-internal. It follows that $(\bar{g}_i : i < \alpha)$ is a Σ-analysis of g over A. ■

DEFINITION 4.6.6 Let Σ be a family of partial types. A hyperdefinable group G is *Σ-connected* if one of its generic types is foreign to Σ (and then they all are).

If N is a hyperdefinable normal subgroup of G and G is Σ-connected, then G/N is Σ-connected, as a generic element gN of G/N is definable in a generic element g of G.

COROLLARY 4.6.7 *Let Σ be an $\emptyset$-invariant family of types. Then there is a unique minimal locally connected Σ-connected hyperdefinable normal subgroup G^Σ of G with Σ-analysable quotient G/G^Σ.* ■

We call G^Σ the *Σ-connected component* of G.

PROOF: By Proposition 4.1.26 there is a maximal α such that there is a chain of non-commensurate normal hyperdefinable subgroups $G = G_0 > G_1 > \cdots > G_\alpha$ of G with G_i/G_{i+1} internal in Σ for all $i < \alpha$. Then the generic types of G_α must be foreign to Σ by Theorem 4.6.4. If G^* is another hyperdefinable normal Σ-connected subgroup of G such that G/G^* is Σ-analysable, then $G_\alpha/(G_\alpha \cap G^*)$ and $G^*/(G_\alpha \cap G^*)$ are both Σ-analysable and Σ-connected, whence bounded. So G and G^* are commensurate. In particular, G_α^c is uniquely determined, and we may take $G^\Sigma = G_\alpha^c$. ■

4.7 GENERICALLY GIVEN GROUPS

In this section, we shall reconstruct a hyperdefinable group from generically given data. Again, we shall work in a simple theory.

THEOREM 4.7.1 *Let π be a partial (hyperimaginary) type and $*$ a partial hyperdefinable function defined on pairs of independent realizations of π, both over $\emptyset$, such that*

1. GENERIC INDEPENDENCE *for independent realizations a and b of π the product $a * b$ realizes π and is independent of a and of b,*
2. GENERIC ASSOCIATIVITY *for three independent realizations a, b and c of π we have $(a * b) * c = a * (b * c)$, and*
3. GENERIC SURJECTIVITY *for any independent a and b realizing π there are c and c' independent of a and of b, with $a * c = b$ and $c' * a = b$.*

Then there are a hyperdefinable group G and a definable bijection from π to the generic types of G, such that generically $*$ is mapped to the group multiplication. G is unique up to definable isomorphism.

Hrushovski calls the first condition *generic cancellation.*

REMARK 4.7.2 If π consists of a single Lascar strong type, generic surjectivity follows from 1. and 2.

PROOF: Given any two independent realizations a and b of π, if $x \models \pi$ independently of a, then $a * x \models \pi$ independently of a and of x by generic independence. Furthermore there are independent realizations x' and x'' of π independent of b with $x' * x'' = b$. By the Independence Theorem we may assume $a * x = x'$ and $x' \perp a, b$. We may choose $x \perp_{a,x'} b$ and $x'' \perp_{b,x'} a, x$. Then $b \perp a, x'$, whence $b \perp_{x'} a, x$ and $a, x \perp_{x'} b, x''$, yielding $x'' \perp a, x$. Thus $\{a, x, x''\}$ is a triple of independent realizations of π, and $a * (x * x'') = (a * x) * x'' = x' * x'' = b$. So we may take $c = x * x''$; the existence of c' is proven similarly. ■

PROOF OF THEOREM 4.7.1: On pairs of realizations of π consider the reflexive and symmetric relation $(a, b)R(a', b')$ given by

$$\exists\, x, y \models \pi\,[x \perp a, b, a', b' \wedge y \perp a, b, a', b' \wedge a * x = a' * y \wedge b * x = b' * y].$$

CLAIM. R is hyperdefinable.

PROOF OF CLAIM: The problem lies in type-defining the condition "$\exists\, x \models \pi\; x \perp a$"; if π is a complete type, this is just Remark 3.2.9. For the general case, fix a complete (hyperimaginary) type $p \vdash \pi$. It is enough to show that for any partial type $\Phi(\bar{a}, x)$ and $\bar{a}_0 \subseteq \bar{a}$ the condition "$\exists x \models \pi\,[x \perp \bar{a}_0 \wedge \Phi(\bar{a}, x)]$" is type-definable as

$$\Psi(\bar{a}) := \exists\, x \models \pi\; \exists\, x' \models p\,[x' \perp \bar{a}_0, x \wedge \Phi(\bar{a}, x * x')].$$

So assume there is $x \models \pi$ with $x \perp \bar{a}_0$ and $\models \Phi(\bar{a}, x)$. If $x' \models p$ independently of $\bar{a}_0, x$, there is $x'' \models \pi$ independent of x and of x' with $x'' * x' = x$ by generic surjectivity; we may assume $x'' \perp_{x,x'} \bar{a}_0$. Then $\bar{a}_0 \perp x, x', x''$, whence $x' \perp_{x''} \bar{a}_0$ and $x' \perp \bar{a}_0, x''$. Therefore $\Psi(\bar{a})$ holds.

Conversely, suppose x and x' witness $\models \Psi(\bar{a})$. So $x' \perp \bar{a}_0, x$, whence $x * x' \perp_x \bar{a}_0$; as $x * x' \perp x$ by generic independence, we get $x * x' \perp \bar{a}_0$. ■

CLAIM. If $(a, b)R(a', b')$ and $(a, b)R(a'', b'')$, then $(a', b')R(a'', b'')$.

PROOF OF CLAIM: Choose elements x, y witnessing $(a, b)R(a', b')$ and elements x', y' witnessing $(a, b)R(a'', b'')$, such that $x, y \perp_{a,b,a',b'} a'', b''$ and $x', y' \perp_{a,b,a'',b''} a', b', x, y$. Then any one of x, y, x' or y' is independent of a, b, a', b', a'', b''. Take $u \models \pi$ independent of everything. Then $x * u \perp a, b, a', b', x, x', y, y'$ by generic independence; by generic surjectivity there is $u' \models \pi$ independent of x' and of $x * u$ with $x * u = x' * u'$; we may choose

$u' \mathop{\smile\!\!\!\!\!\mid}_{x*u,x'} a, b, a', b', a'', b'', y'$. Therefore $u', x * u \mathop{\smile\!\!\!\!\!\mid}_{x'} a, b, a', b', a'', b'', y'$, whence $u' \mathop{\smile\!\!\!\!\!\mid} a, b, a', b', a'', b'', x', y'$. Then

$$\begin{aligned} a' * (y * u) &= (a' * y) * u = (a * x) * u = a * (x * u) \\ &= a * (x' * u') = (a * x') * u' = (a'' * y') * u' = a'' * (y' * u') \end{aligned}$$

by generic associativity (all triples in question are independent); similarly $b' * (y * u) = b'' * (y' * u')$. Since $y * u \mathop{\smile\!\!\!\!\!\mid} y$ by generic independence and $u \mathop{\smile\!\!\!\!\!\mid}_y a', b', a'', b''$ implies $y * u \mathop{\smile\!\!\!\!\!\mid}_y a', b', a'', b''$, we have $y * u \mathop{\smile\!\!\!\!\!\mid} a', b', a'', b''$; similarly $y' * u' \mathop{\smile\!\!\!\!\!\mid} a', b', a'', b''$. Hence $(a', b')R(a'', b'')$. ■

Thus R is an equivalence relation. Let $[a, b]$ denote the class of (a, b) modulo R.

CLAIM. If $a * x = b * x$ for some $x \models \pi$ independent of a and of b, then $a * y = b * y$ for all $y \models \pi$ independent of a and of b.

PROOF OF CLAIM: Suppose $z \models \pi$ independently of a, x and of b, x. Then there is $c \models \pi$ independent of x and of z, with $x * c = z$; we may assume that $c \mathop{\smile\!\!\!\!\!\mid}_{x,z} a, b$. Then both a and b are independent of c, x, z; since $c \mathop{\smile\!\!\!\!\!\mid} x$, both $\{a, x, c\}$ and $\{b, x, c\}$ are independent triples. Thus

$$a * z = a * (x * c) = (a * x) * c = (b * x) * c = b * (x * c) = b * z.$$

Now choose $z \mathop{\smile\!\!\!\!\!\mid} a, b, x, y$. Then $a * z = b * z$, and hence $a * y = b * y$. ■

By symmetry, the same holds for multiplication on the left.

CLAIM. If $x \models \pi$ independently of a, b, then $[a, b] = [a * x, b * x]$. If $c \models \pi$ is independent of a, b, then there are $d, d' \models \pi$ independent of a, b with $[a, b] = [c, d] = [d', c]$.

PROOF OF CLAIM: Let $y \mathop{\smile\!\!\!\!\!\mid} a, b, x$ realize π. So $y \mathop{\smile\!\!\!\!\!\mid} a, b, a * x, b * x$; moreover $x * y$ realizes π independently of $a, b, a * x, b * x$ by generic independence. Finally, $a * (x * y) = (a * x) * y$ and $b * (x * y) = (b * x) * y$ by generic associativity. Thus $(a, b)R(a * x, b * x)$ holds.

Now consider $c \models \pi$ independent of a, b. By generic surjectivity there is $x \models \pi$ independent of a and of c with $a * x = c$; if we choose $x \mathop{\smile\!\!\!\!\!\mid}_{a,c} b$, then $b \mathop{\smile\!\!\!\!\!\mid}_a c, x$, whence $a, b \mathop{\smile\!\!\!\!\!\mid} x$ and $[a, b] = [c, b * x]$. The existence of d' follows by symmetry. ■

Define a multiplication $\circ$ on the classes modulo R by $[a, b] \circ [b, c] = [a, c]$.

CLAIM. The multiplication $\circ$ is well-defined.

PROOF OF CLAIM: Suppose $[a', b'] = [a, b]$ and $[b', c'] = [b, c]$. So there are realizations x and y of π independent of a, a', b, b', and realizations x' and y' of π independent of b, b', c, c', with $a * x = a' * y$, $b * x = b' * y$, $b * x' = b' * y'$ and $c * x' = c' * y'$. We may assume that $x', y' \mathop{\smile\!\!\!\!\!\mid}_{b,b',c,c'} a, a', x, y$, so in particular $x \mathop{\smile\!\!\!\!\!\!\perp} x'$. By generic surjectivity there is $z \models \pi$ independent of x and of x' with $x * z = x'$; we may assume that z is independent over x, x' of everything else. Then $x' \mathop{\smile\!\!\!\!\!\!\perp} a, a', b, b', x, y$, whence $x' \mathop{\smile\!\!\!\!\!\!\perp}_x a, a', b, b', y$. As $a, a', b, b', y \mathop{\smile\!\!\!\!\!\!\perp}_{x,x'} z$, we get $a, a', b, b', y \mathop{\smile\!\!\!\!\!\!\perp}_x x', z$; now $z \mathop{\smile\!\!\!\!\!\!\perp} x$ implies $z \mathop{\smile\!\!\!\!\!\!\perp} a, a', b, b', x, y$. It follows that the triples $\{b, x, z\}$, $\{b', y, z\}$, $\{a, x, z\}$ and $\{a', y, z\}$ are all independent. Therefore

$$b' * y' = b * x' = b * (x * z) = (b * x) * z = (b' * y) * z = b' * (y * z).$$

As b' is independent of y' and of $y * z$, we get $u * y' = u * (y * z)$ for all $u \models \pi$ independent of y' and of $y * z$. In particular

$$a' * y' = a' * (y * z) = (a' * y) * z = (a * x) * z = a * (x * z) = a * x',$$

whence $[a', c'] = [a' * y', c' * y'] = [a * x', c * x'] = [a, c]$. ■

CLAIM. $[a, a] = [b, b]$ for any $a, b \models \pi$.

PROOF OF CLAIM: Choose $c \models \pi$ independent of a, b. Then there are realizations $x \mathop{\smile\!\!\!\!\!\!\perp} a$ and $y \mathop{\smile\!\!\!\!\!\!\perp} b$ of π with $a * x = c$ and $b * y = c$, whence $[a, a] = [c, c] = [b, b]$. ■

Clearly, $[a, a]$ is a (left and right) unit for multiplication. Any class $[a, b]$ has a (left and right) inverse $[b, a]$.

CLAIM. Multiplication is associative.

PROOF OF CLAIM: Consider three classes $[a, b]$, $[a', b']$ and $[a'', b'']$. Replacing $[a', b']$ by $[a' * x, b' * x]$ for some $x \models \pi$ independently of a, b, a', b', a'', b'', we may assume $a', b' \mathop{\smile\!\!\!\!\!\!\perp} a, b, a'', b''$. Then there are realizations c and c' of π with $[a, b] = [c, a']$ and $[a'', b''] = [b', c']$. We get

$$\begin{aligned}([a, b] \circ [a', b']) \circ [a'', b''] &= [c, b'] \circ [b', c'] = [c, c'] \\ &= [c, a'] \circ [a', c'] = [a, b] \circ ([a', b'] \circ [a'', b'']). \quad ■\end{aligned}$$

Thus the set of R-classes of π^2 under $\circ$ forms a hyperdefinable group G, with unit $1_G = [a, a]$ (for any $a \models \pi$) and inverse $[a, b]^{-1} = [b, a]$.

CLAIM. $[a * x, x] = [a * y, y]$ for any x and y independent of a.

PROOF OF CLAIM: Choose $u \models \pi$ independently of a, x, y. As before there are realizations z and z' of π independent of a, x, y with $x * z = u = y * z'$. Then $[a * x, x] = [a * x * z, x * z] = [a * y * z', y * z'] = [a * y, y]$. ■

Consider the map $\sigma : a \mapsto [a * x, x]$ (for any $x \models \pi$ independent of a). This is clearly a hyperdefinable map from π to G.

CLAIM. $G = \text{im}(\sigma)^2$.

PROOF OF CLAIM: If $a, b \models \pi$, pick $x \models \pi$ independently of a, b. Then there are $y \models \pi$ independently of a and of x with $y * x = a$, and $y' \models \pi$

independently of x and of b with $y' * b = x$. Therefore

$$[a, b] = [a, x][x, b] = [y * x, x][y' * b, b] \in \mathrm{im}(\sigma)^2. \quad \blacksquare$$

CLAIM. σ is injective.

PROOF OF CLAIM: Suppose $\sigma(a) = \sigma(b)$ for two realizations a and b of π. If x realizes π independently of a, b, then $[a * x, x] = [b * x, x]$, whence

$$[a, b] = [a * x, b * x] = [a * x, x][b * x, x]^{-1} = 1_G = [a, a].$$

So there are $x \mathop{\smile\!\!\!\!\!\mid}\, a, b$ and $y \mathop{\smile\!\!\!\!\!\mid}\, a, b$ with $a * x = a * y = b * x$; therefore $a * z = b * z$ for all $z \models \pi$ independently of a and of b.

Take $c \models \pi$ independently of a, b. Choose $a_1 \models \pi$ independently of a and of c with $a = a_1 * c$ and $b_1 \models \pi$ independently of b and of c with $b = b_1 * c$. For any $y \models \pi$ independently of a, a_1, b, b_1, c the triples $\{a_1, c, y\}$ and $\{b_1, c, y\}$ are independent; furthermore

$$a_1 * (c * y) = (a_1 * c) * y = a * y = b * y = (b_1 * c) * y = b_1 * (c * y).$$

Therefore $a_1 * x = b_1 * x$ for all $x \models \pi$ independently of a_1 and of b_1. Thus

$$a = a_1 * c = b_1 * c = b. \quad \blacksquare$$

CLAIM. σ generically preserves multiplication, and maps π to the generic types of G.

PROOF OF CLAIM: Let a and b be two independent realizations of π, and consider $x \models \pi$ independently of a, b. Then

$$\sigma(a * b) = [a * b * x, x] = [a * b * x, b * x] \circ [b * x, x] = \sigma(a) \circ \sigma(b),$$

so σ preserves multiplication.

Take $c \models \pi$ and consider any $a, b \models \pi$ independently of c. Then $\sigma(c) = [c * a, a]$ and $\sigma(c) \circ [a, b] = [c * a, b]$. As $c * a \mathop{\smile\!\!\!\!\!\mid}_a b$ and $c * a \mathop{\smile\!\!\!\!\!\mid}\, a$, we have $c * a \mathop{\smile\!\!\!\!\!\mid}\, a, b$ and there is $d \models \pi$ independently of $c * a$ and of b with $c * a = d * b$ and $d \mathop{\smile\!\!\!\!\!\mid}_{c*a,b} a$. So $\sigma(c) \circ [a, b] = [d * b, b] = \sigma(d)$; since $a \mathop{\smile\!\!\!\!\!\mid}_b c * a, d$ and $d \mathop{\smile\!\!\!\!\!\mid}\, b$, we get $d \mathop{\smile\!\!\!\!\!\mid}\, a, b$. Hence $\sigma(c) \circ [a, b] \mathop{\smile\!\!\!\!\!\mid}\, [a, b]$ and $\sigma(c)$ is generic.

Since $\sigma(c) \circ [a, b] = \sigma(d) \in \mathrm{im}(\sigma)$, we see that $\mathrm{im}(\sigma)$ is invariant under multiplication by (independent) elements of G. But these act transitively on the generic types, so $\mathrm{im}(\sigma)$ contains all generic types of G. $\blacksquare$

Finally, suppose G_1 is another group such that there is an injective map σ_1 from π to the generic types of G_1 generically preserving multiplication. Then there is a bijection τ from the generic types of G to the generic types of G_1 generically preserving multiplication. So if x, y, x', y' are generic in G with $xy = x'y'$ and $x, y \mathop{\smile\!\!\!\!\!\mid}\, x'$ (but otherwise not necessarily independent), then there is generic $z \in G$ independent of x and of x' with $xz = x'$. Hence $y = zy'$; as $y \mathop{\smile\!\!\!\!\!\!\,|}_x x^{-1}x'$ implies $y, x \mathop{\smile\!\!\!\!\!\!\,|} z$, genericity of $\mathrm{tp}(y)$ yields $z^{-1}y \mathop{\smile\!\!\!\!\!\!\,|} z$, whence $y' \mathop{\smile\!\!\!\!\!\!\,|} z$. So

$$\tau(x)\tau(y) = \tau(x)[\tau(z)\tau(y')] = [\tau(x)\tau(z)]\tau(y') = \tau(x')\tau(y');$$

it follows that τ induces an isomorphism of G and G_1. ■

4.8 LOCALLY MODULAR GROUPS

In this section, G will be a hyperdefinable group over $\emptyset$ in a simple theory, and P an $\emptyset$-invariant locally modular family of types.

DEFINITION 4.8.1 Let Σ be the family of all partial types which are co-foreign to P. We shall call a hyperdefinable group *P-connected* if it is Σ-connected; similarly the Σ-connected component G^Σ of G will also be called the *P-connected component*, and be denoted by G^P.

So P is foreign to G/G^P, and G^P is P-connected. Moreover, a hyperdefinable group is P-connected if and only if one of its generic types is P-minimal (and then all of them are).

REMARK 4.8.2 Since we want the Σ- and P-connected component of G to be a subgroup of G, we have only defined them for the ambient group. So in order to obtain H^Σ or H^P for a hyperdefinable subgroup $H \leq G$, we have to add the parameters used for the definition, and then consider H as the ambient group. However, this will affect the notion of local connectivity: a locally connected subgroup of H need not be locally connected in G.

PROPOSITION 4.8.3 *Let G be P-internal. If H is a P-connected hyperdefinable subgroup of G, then its locally connected component H^c is hyperdefinable over* $\mathrm{cl}_P(\emptyset)$.

PROOF: Let u be the canonical parameter for H^c, and consider a generic element h for H^c over u and a generic element g for G over u, h. So hg is generic for H^cg over g, u, and hg is generic for G over u, h. If v is the canonical parameter for the coset H^cg, then $v \in \mathrm{Cb}(hg/g, u)$ by Lemma 4.5.19. On the other hand, since $\mathrm{tp}(hg/g, u)$ is P-minimal as a translate of a generic type of H, we get $\mathrm{Cb}(hg/g, u) \subset \mathrm{cl}_P(hg)$ by Lemma 3.5.13.

As $H^c = (H^cg)(H^cg)^{-1}$, the canonical parameter u is definable over v, whence $u \in \mathrm{cl}_P(hg)$. But $hg \mathop{\smile\!\!\!\!\!\!\,|} u$, so $u \mathop{\smile\!\!\!\!\!\!\,|}_{\mathrm{cl}_P(\emptyset)} \mathrm{cl}_P(hg)$ by Lemma 3.5.5, and $u \in \mathrm{cl}_P(\emptyset)$. ■

THEOREM 4.8.4 *Let G be P-internal P-connected. Then $\tilde{Z}(G) \geq G^0_\emptyset$, and $G^0_\emptyset$ is (bounded central)-by-Abelian.*

PROOF: Given $g \in G$, consider the subgroup $H_g = \{(h, h^g) : h \in G\}$ of G^2. Then H_g is P-connected, so H_g^c is definable over $\mathrm{cl}_P(\emptyset)$ by Proposition 4.8.3. For any $g, g' \in G$, the groups H_g and $H_{g'}$ are commensurate if and only if $C_G(g'g^{-1})$ has bounded index in G. Thus the cosets of $\tilde{Z}(G)$ in G are in bijection with the commensurativity classes among $\{H_g : g \in G\}$, whence $G/\tilde{Z}(G) \subset \mathrm{cl}_P(\emptyset)$ by Lemma 4.8.3. Therefore $G/\tilde{Z}(G)$ is co-foreign to P and $\tilde{Z}(G) \geq G^0_\emptyset$ by P-connectivity. The last assertion follows from Proposition 4.4.10.3. ■

REMARK 4.8.5 1. We do not claim that the bounded central subgroup $\tilde{Z}(G)'$ is hyperdefinable.

2. In a one-based theory (where P is the family of all types), every hyperdefinable group is P-internal P-connected.

3. If G is type-definable and finitary, then by compactness the set of commutators of G is bounded if and only if it is finite. But in that case G' is finite and definable.

4. If G is type-definable and finitary and if $G^0_\emptyset$ is Abelian, then by compactness there is some definable superset X of $G^0_\emptyset$ such that $xy = yx$ for all $x, y \in X$. We may choose X and a definable superset Y of G small enough such that $C_G(C_Y(X)) \cap C_G(X)$ is a relatively definable Abelian subgroup of G containing $G^0_\emptyset$; by compactness G is Abelian-by-finite.

Hence a P-internal P-connected type-definable finitary group is finite-by-Abelian-by-finite.

DEFINITION 4.8.6 The *descending central series* of G, denoted $(\gamma_i(G) : i < \omega)$, is defined inductively by $\gamma_1(G) = G$, $\gamma_{\alpha+1}(G) = [\gamma_\alpha(G), G]$, and $\gamma_\lambda(G) = \bigcap_{\alpha<\lambda} \gamma_\alpha(G)$ for a limit ordinal λ. A group G is *hypernilpotent* if $\gamma_\alpha(G) = \{1\}$ for some ordinal α.

REMARK 4.8.7 G is hypernilpotent if and only if there is a continuous series $(G_i : i \leq \alpha)$ of normal subgroups of G with $G_0 = G$, $[G_i : G] \leq G_{i+1}$ for all $i < \alpha$, and $G_\alpha = \{1\}$.

THEOREM 4.8.8 *Let G be P-analysable P-connected. Then G is hypernilpotent-by-bounded.*

PROOF: By Corollary 4.6.5 there is a descending sequence $(G_i : i \leq \alpha)$ of normal subgroups of G, hyperdefinable over $\emptyset$ and continuous at limits, with $G = G_0$ and $G_\alpha = \{1\}$, such that every quotient G_i/G_{i+1} is P-internal. Replacing G by the quotients G/G_i, it is enough to show that if H is a P-internal normal subgroup of $G^0_\emptyset$ which is hyperdefinable over $\emptyset$, then $H \leq Z_4(G^0_\emptyset)$.

If K is the P-connected component of H (inside H), then K is $\emptyset$-definable and normal in G. For $g \in G$, put $K_g = \{(h, h^g) : h \in K\} < K^2$. As K_g is P-internal P-connected, K_g^c is hyperdefinable over $\mathrm{cl}_P(\emptyset)$ by Proposition 4.8.3; since again K_g and $K_{g'}$ are commensurate if and only if $C_K(g'g^{-1})$ has bounded index in K, the commensurativity classes among $\{K_g : g \in G\}$ are in hyperdefinable bijection with the cosets of $\tilde{C}_G(K)$ in G. Therefore $G/\tilde{C}_G(K) \subset \mathrm{cl}_P(\emptyset)$ by Proposition 4.8.3, so $\tilde{C}_G(K) \geq G^P$ and $\tilde{C}_G(K)$ has bounded index in G by P-connectivity. Hence there are a generic element of G and an independent generic element of K which commute; it follows that $\tilde{Z}(G)$ intersects K in a subgroup of bounded index.

Put $\bar{G} = G/\tilde{Z}(G)$. Then $\bar{G}$ is again P-connected, and

$$\bar{G}/C_{\bar{G}}(h\tilde{Z}(G)) \in \mathrm{dcl}(h^G/\tilde{Z}(G)) \subseteq \mathrm{dcl}(H\,\tilde{Z}(G)/\tilde{Z}(G)) \subseteq \mathrm{bdd}(H/K) \subset \mathrm{cl}_P(\emptyset)$$

for all $h \in H$, so $C_{\bar{G}}(h\tilde{Z}(G)) \geq \bar{G}^0_{\{h\}}$ by P-connectivity. Since principal generic types have principal generic extensions by Lemma 4.5.9, for any generic $g \in G^0_\emptyset$ there is generic $h \in H$ with $[g, h] \in \tilde{Z}(G)$. It follows that for generic $g \in G$ and independent $h \in H$ the coset $[g, h]\tilde{Z}(G)$ is in $\mathrm{bdd}(g) \cap \mathrm{bdd}(h) = \mathrm{bdd}(\emptyset)$. Hence $[H, G^0_\emptyset]\tilde{Z}(G)/\tilde{Z}(G)$ is bounded, and $H \leq Z_4(G^0_\emptyset)$. ■

REMARK 4.8.9 If in addition G is type-definable and finitary, then G is nilpotent-by-finite.

PROOF: Let $(G_i : i \leq \alpha)$ be a descending continuous sequence of type-definable (over $\emptyset$) normal subgroups of $G^0_\emptyset$ with $G^0_\emptyset = G_0$ and $G_\alpha = \{1\}$, such that G_i/G_{i+1} is P-internal for all $i < \alpha$. Fix $i < \omega$, and suppose $\beta > 0$ is minimal such that $G_\beta \leq Z_i(G^0_\emptyset)$. By compactness there are definable supersets X of $G^0_\emptyset$ and Y of G_β such that $[Y,_i X] = 1$ (which just means that $[y, x_1, \ldots, x_n] = 1$ for all $y \in Y$ and $x_1, \ldots, x_n \in X$). Then any subgroup of $G^0_\emptyset$ contained in Y is in $Z_i(G^0_\emptyset)$, so β cannot be a limit ordinal by compactness.

Thus by the proof of Theorem 4.8.8 we get $G_{\beta-1} \leq Z_{i+4}(G^0_\emptyset)$. Starting with $G_\alpha = Z_0(G^0_\emptyset)$, we obtain a sequence of descending ordinals $(\alpha_i : i < \omega)$ such that $G_{\alpha_i} \leq Z_k(G^0_\emptyset)$ for some $k < \omega$. It must break off at some point; this happens only if $G_0 \leq Z_k(G^0_\emptyset)$ for some $k < \omega$, and $G^0_\emptyset$ is nilpotent of class k. By compactness, there is a definable superset X of $G^0_\emptyset$ such that $[X,_k X] = 1$. Since X is definable, there are only finitely many $g_i \in G$ such that $g_i g_j^{-1} \notin X$ for all $i \neq j$. But then $\langle X \rangle \cap G$ is a subgroup of finite index in G, and is nilpotent of class k. ■

PROPOSITION 4.8.10 *Suppose G is P-internal, and $p \in S_G(\mathfrak{M})$ is P-minimal. Then there is a locally connected P-connected subgroup H of G*

which is hyperdefinable over $\mathrm{cl}_P(\emptyset)$, *such that* p *is a generic type of some* $\mathfrak{M}$*-definable coset of* H.

PROOF: Let $x \models p$ and h be generic for G over $\mathfrak{M}$ with $h \mathop{\smile\!\!\!\!\!\!|}_{\mathfrak{M}} x$. Put $C = \mathrm{cl}_P(\mathrm{Cb}(hx/\mathrm{cl}_P(\mathfrak{M},h))\cup\mathfrak{M})$ and choose $h' \mathop{\smile\!\!\!\!\!\!|}_C h$ with $\mathrm{tp}(h'/C) = \mathrm{tp}(h/C)$; by the Independence Theorem we may re-choose x and assume $hx \mathop{\smile\!\!\!\!\!\!|}_C h, h'$ and $hx = h'x'$ for some $x' \models p$. Then $x \mathop{\smile\!\!\!\!\!\!|}_{C,h} h'^{-1}h$ and $x \mathop{\smile\!\!\!\!\!\!|}_{\mathfrak{M}} h$; since $C \subseteq \mathrm{cl}_P(\mathfrak{M},h)$ we get $x \mathop{\smile\!\!\!\!\!\!|}_{\mathrm{cl}_P(\mathfrak{M})} C, h$ by Lemma 3.5.5. As $x \mathop{\smile\!\!\!\!\!\!|}_{\mathfrak{M}} \mathrm{cl}_P(\mathfrak{M})$ by P-minimality, $x \mathop{\smile\!\!\!\!\!\!|}_{\mathfrak{M}} C, h$, whence $x \mathop{\smile\!\!\!\!\!\!|}_{\mathfrak{M}} h'^{-1}h$. As $h'^{-1}hx = x' \models p$, we obtain $h'^{-1}h \in \mathrm{stab}(p)$, and $h\,\mathrm{stab}(p) = h'\,\mathrm{stab}(p)$.

Let H be the P-connected component of $\mathrm{stab}(p)$ (inside $\mathrm{stab}(p)$ as ambient group). Since $\mathrm{stab}(p)/H \subset \mathrm{cl}_P(\mathfrak{M})$, the coset $u = h\,H^c$ lies in $\mathrm{cl}_P(\mathfrak{M},h) \cap \mathrm{cl}_P(\mathfrak{M},h') \subseteq C$. As $\mathrm{tp}(hx/C)$ is P-minimal, $\mathrm{Cb}(hx/C) \subseteq \mathrm{cl}_P(hx)$ by Lemma 3.5.13, and $u \in \mathrm{cl}_P(\mathfrak{M},hx)$.

Let v be the canonical parameter of hH^cx. As $hx \mathop{\smile\!\!\!\!\!\!|}_{\mathfrak{M}} x$ by genericity of h, Lemma 4.3.12 implies that hx is generic for $hx(x^{-1}H^cx) = hH^cx$ over $\mathfrak{M}, v$, so $v \in \mathrm{Cb}(hx/\mathfrak{M},v)$ by Lemma 4.5.19. As H^c is P-connected, $\mathrm{tp}(hx/\mathfrak{M},v)$ is P-minimal, so Lemma 3.5.13 implies $\mathrm{Cb}(hx/\mathfrak{M},v) \subseteq \mathrm{cl}_P(hx)$, whence $v \in \mathrm{cl}_P(\mathfrak{M},hx)$.

If w is the canonical parameter w for H^cx, then $H^cx = (hH^c)^{-1}(hH^cx)$ yields $w \in \mathrm{dcl}(u,v) \subseteq \mathrm{cl}_P(\mathfrak{M},hx)$. Now $hx \mathop{\smile\!\!\!\!\!\!|}_{\mathfrak{M}} x$; as $\mathrm{tp}(x/\mathfrak{M})$ is P-minimal, so is $\mathrm{tp}(x/\mathfrak{M},hx)$, whence $x \mathop{\smile\!\!\!\!\!\!|}_{\mathfrak{M},hx} \mathrm{cl}_P(\mathfrak{M},hx)$ and $x \mathop{\smile\!\!\!\!\!\!|}_{\mathfrak{M}} w$. Since $w \in \mathrm{dcl}(\mathfrak{M},x)$, we have $w \in \mathrm{dcl}(\mathfrak{M})$, and H^cx is hyperdefinable over $\mathfrak{M}$. Moreover, H^c is hyperdefinable over $\mathrm{cl}_P(\emptyset)$ by Proposition 4.8.3.

Finally, if $g \in \mathrm{stab}(p)$ is generic and $x \models p$ with $x \mathop{\smile\!\!\!\!\!\!|}_{\mathfrak{M}} g$ and $gx \models p$, then gx is generic for $\mathrm{stab}(p)\,x$ over $\mathfrak{M}, x$, whence

$$\begin{aligned} D_G(H^cx,\varphi,\epsilon,k) &\le D_G(\mathrm{stab}(p)\,x,\varphi,\epsilon,k) = D_G(gx/\mathfrak{M}\cup\{x\},\varphi,\epsilon,k) \\ &\le D_G(gx/\mathfrak{M},\varphi,\epsilon,k) = D_G(p,\varphi,\epsilon,k) \end{aligned}$$

for all formulas φ, all $\epsilon \in E$ and all $k < \omega$; since $x \in H^cx$ for $x \models p$, we have equality, and p is generic for H^cx by Lemma 4.4.2. ∎

In particular, H^c and $\mathrm{stab}(p)$ are commensurable, and $\mathrm{stab}(p)$ is P-connected itself.

Proposition 4.8.10 yields a characterization of one-basedness for groups.

PROPOSITION 4.8.11 *A hyperdefinable group* G *in a simple theory is one-based if and only if every type is a generic type for some coset of a hyperdefinable subgroup of* G^n *(for all* $n < \omega$*).*

REMARK 4.8.12 We may require these subgroups to be locally connected, and hence hyperdefinable over $\mathrm{bdd}(\emptyset)$ by Proposition 4.8.3.

PROOF: Left to right follows from Proposition 4.8.10. For the converse, suppose every type is a generic type for some coset of a hyperdefinable subgroup of G^n (for the appropriate n).

CLAIM. A locally connected hyperdefinable subgroup H of G^n is hyperdefinable over $\text{bdd}(\emptyset)$.

PROOF OF CLAIM: Let v be the canonical parameter of H. By Proposition 4.5.6 applied to H^2 there are principal generic elements x_0, y_0, x_1, y_1 of H independent over v, with

$$\text{lstp}(x_0, y_0/v) = \text{lstp}(x_1, y_1/v) = \text{lstp}(x_0x_1, y_0y_1/v).$$

By assumption $\text{tp}(x_0, y_0/\emptyset)$ is a generic type of some coset kK of a hyperdefinable subgroup K of G^{2n} (whose parameters A are independent of x_0, y_0, v); similarly there is A' with $\text{lstp}(A', x_0x_1, y_0y_1/v) = \text{lstp}(A, x_0, y_0/v)$. As $x_0, y_0 \perp_v x_0x_1, y_0y_1$, we may assume $A = A'$ and $A \perp_v x_0, y_0, x_1, y_1$ by the Independence Theorem. Then $A \perp x_0, y_0, x_1, y_1, v$, and $(x_1, y_1) \in K$. Since

$$\begin{aligned} D_G(x_1, y_1/A, \varphi, \epsilon, k) &= D_G(x_1, y_1, \varphi, \epsilon, k) \\ &= D_G(x_0, y_0, \varphi, \epsilon, k) = D_G(x_0, y_0/A, \varphi, \epsilon, k) \end{aligned}$$

for all formulas φ, all $\epsilon \in E$ and $k < \omega$, the type $\text{tp}(x_1, y_1/A)$ must be generic for K.

Since we might just as well have started with H^n instead of H for any $n < \omega$, this yields independent principal generic elements $\bar{x}$ and $\bar{y}$ of H^ω over v, a set B of parameters with $B \perp \bar{x}, \bar{y}, v$ and a hyperdefinable group N over B, such that $(\bar{x}, \bar{y})$ is generic for N. Note that by local connectivity, $v \in \text{dcl}(\bar{x}) \cap \text{dcl}(\bar{y})$.

By Proposition 4.5.6 there are principal generic $(\bar{x}^*, \bar{y}^*)$ in N such that $(\bar{x}, \bar{y})$, $(\bar{x}^*, \bar{y}^*)$ and $(\bar{x}^* \cdot \bar{x}, \bar{y}^* \cdot \bar{y})$ are pairwise independent over B, with

$$\text{tp}(\bar{x}, \bar{y}/B) = \text{tp}(\bar{x}^* \cdot \bar{x}, \bar{y}^* \cdot \bar{y}/B).$$

Then there is $v^* \in \text{dcl}(\bar{x}^* \cdot \bar{x}) \cap \text{dcl}(\bar{y}^* \cdot \bar{y})$, the canonical parameter of a group H^*, which corresponds to $\bar{x}^* \cdot \bar{x}$ (or $\bar{y}^* \cdot \bar{y}$) in the same way H corresponds to $\bar{x}$ (or $\bar{y}$). As $v, v^* \in \text{dcl}(\bar{y}, \bar{y}^*)$, over $B, \bar{y}, \bar{y}^*$ we have:

- $\bar{x}$, $\bar{x}^*$ and $\bar{x}^* \cdot \bar{x}$ are pairwise independent,
- $\bar{x}$ is generic for H^ω, and
- $\bar{x}^* \cdot \bar{x}$ is generic for $(H^*)^\omega$.

Furthermore $v \perp_B v^*$; since $v \perp B$ we get $v \perp v^*$. But $\bar{x}^* \cdot \bar{x}$ is also generic for $\bar{x}^* H^\omega$ over $B, \bar{y}, \bar{y}^*, \bar{x}^*$. It follows that $H^\omega \cap (H^*)^\omega$ has bounded index in H^ω, so $H \cap H^*$ has bounded index in H, and $H = H^*$ by local connectivity. Since $v \perp v^*$, we have $v \in \text{bdd}(\emptyset)$. ■

Now consider a model $\mathfrak{M}$ and a type $p \in S_{G^n}(\mathfrak{M})$ realized by a. Let H be a hyperdefinable subgroup of G^n and $g \in G^n$ be such that p is generic for gH; clearly we may assume that H is hyperdefinable over $\mathfrak{M}$ and $g \in \mathfrak{M}$. Furthermore, as gHH^c contains only boundedly many cosets of H^c, we may replace H by its locally connected component, possibly modifying g inside $\mathrm{bdd}(\mathfrak{M}) = \mathrm{dcl}(\mathfrak{M})$. So we can assume that H is locally connected, and hence hyperdefinable over $\mathrm{bdd}(\emptyset)$. Since $gH = aH$, it follows that the canonical parameter v of gH is in $\mathrm{bdd}(a)$. As $\mathrm{tp}(a/\mathfrak{M})$ is generic for gH, we have $a \perp_v \mathfrak{M}$; since $v \in \mathrm{bdd}(a) \cap \mathrm{bdd}(\mathfrak{M})$, we are done. ■

Under suitable circumstances, we may interpret these subgroups as endomorphisms and use them to describe the structure of forking on generic elements. We shall first generalize the notion of an endomorphism.

DEFINITION 4.8.13 Let A be a bounded-by-abelian group hyperdefinable over $\emptyset$, and P an invariant family of types. A *P-endomorphism* of A, or *P-endogeny*, is a partial endomorphism of $A/\mathrm{cl}_P(\emptyset)$, whose domain is a subgroup of bounded index in $A/\mathrm{cl}_P(\emptyset)$ and which is induced by a hyperdefinable subgroup of $A \times A$ over $\mathrm{cl}_P(\emptyset)$.

Next, a notion which will be of particular interest in connection with supersimple theories in the next chapter.

DEFINITION 4.8.14 A type is *regular* if it is unbounded and foreign to all its forking extensions.

REMARK 4.8.15 Regularity is clearly preserved under non-forking extensions and definable bijections. We shall see in Lemma 5.2.13 that if a Lascar srong type has a regular non-forking extension, then it is itself regular.

REMARK 4.8.16 If G has a regular generic type p, then its generic types are foreign to all non-generic types. In particular, every subgroup of unbounded index in G is contained in $\mathrm{cl}_p(\emptyset)$.

PROOF: Consider a model $\mathfrak{M}$ and a type $q \in S_G(\mathfrak{M})$, and let $g \models p$ and $h \models q$, with $g \perp_{\mathfrak{M}} h$. Then $\mathrm{tp}(g/\mathfrak{M}, h)$ is regular, as are $\mathrm{tp}(gh/\mathfrak{M}, h)$ and $\mathrm{tp}(gh/\mathfrak{M})$. Moreover, $\mathrm{tp}(gh/\mathfrak{M})$ is generic. So if q is not generic, then $gh \not\perp_{\mathfrak{M}} g$ and p is foreign to $\mathrm{tp}(gh/\mathfrak{M}, g)$, whence to $\mathrm{tp}(h/\mathfrak{M}, g)$ and to q. ■

REMARK 4.8.17 If G has a regular generic type p, then a non-zero p-endomorphism of G is invertible.

PROOF: As any hyperdefinable subgroup of unbounded index in G is contained in $\mathrm{cl}_p(\emptyset)$, if r is a $\mathrm{cl}_p(\emptyset)$-definable p-endomorphism of G, then r is non-zero if and only if $\mathrm{im}(r)$ has bounded index in G. ■

THEOREM 4.8.18 *Let G be a hyperdefinable connected group in a simple theory, and suppose that a generic type p of G is locally modular and regular. If R denotes the ring of p-endogenies of G, then a tuple $\bar{g} = (g_0, g_1, \ldots, g_n)$ in G is dependent if and only if there are $r_i \in R$, not all zero, with $\sum_{i=0}^{n} r_i(g_i) \subset \mathrm{cl}_p(\emptyset)$.*

PROOF: Since G is connected (over $\emptyset$) and obviously p-internal and p-connected, $G = \tilde{Z}(G)$ by Theorem 4.8.4, so G is bounded-by-abelian. Clearly, if $\sum r_i g_i \subset \mathrm{cl}_p(\emptyset)$ for some non-trivial endogenies r_i, then $\bar{g}$ is dependent.

For the other direction, suppose $\bar{g}$ is dependent; we may assume that every proper subtuple of $\bar{g}$ is independent. Consider $\mathrm{tp}(\bar{g}/\mathrm{cl}_p(\emptyset))$. This type is p-minimal; by Proposition 4.8.10 there is a locally connected subgroup H of G^{n+1} which is hyperdefinable over $\mathrm{cl}_p(\emptyset)$, such that $\mathrm{tp}(\bar{g})$ is the generic type of a coset $H + \bar{h}$ with canonical parameter v, for some $\bar{h} \downarrow_{\mathrm{cl}_p(\emptyset)} \bar{g}$. For $i < n$ define a subgroup r_i by

$$r_i = \{(x_i, x_n) \in G \times G : (0, \ldots, 0, x_i, 0, \ldots, 0, x_n) \in H\}.$$

CLAIM. r_i induces a p-endomorphism of G for all $i < n$.

PROOF OF CLAIM: As $(g_0, \ldots, g_{n-1})$ is independent, its type over $\emptyset$ is p-minimal and cannot fork with $\mathrm{cl}_p(\emptyset)$. So $(g_0, \ldots, g_{n-1})$ is independent of v. If

$$\mathrm{tp}(g, g'/g_1, \ldots, g_{n-1}, v) = \mathrm{tp}(g_0, g_n/g_1, \ldots, g_{n-1}, v),$$

then $(g, g_1, \ldots, g_{n-1}, g') \in H + \bar{h}$, whence $(g_0 - g, 0, \ldots, 0, g_n - g') \in H$. We may choose g_0 and g to be independent generic elements of G over $g_1, \ldots, g_{n-1}, v$; then $g_0 - g$ is generic and the projection of r_0 to the first co-ordinate is a subgroup of bounded index in G. Similarly, the projection of r_0 to the second co-ordinate is a subgroup of bounded index in G, as are the two projections of r_i, for all $i < n$.

Next, consider $r_i(0) = \{g \in G : (0, g) \in r_i\}$. This is $\mathrm{cl}_p(\emptyset)$-definable and thus either contained in $\mathrm{cl}_p(\emptyset)$, or has bounded index in G. But in the latter case we could choose $g \in r_i(0)$ generic for G over $g_0, \ldots, g_{n-1}, g_n, \bar{h}$; then $(g_0, \ldots, g_{n-1}, g_n + g) \in H + \bar{h}$. But this would be a tuple of independent generic elements of G over $\bar{h}$ and thus generic for G^{n+1}, contradicting the fact that H and hence $H + \bar{h}$ is not generic for G^{n+1}. Therefore r_i induces a p-endomorphism for all $i < n$. ■

Put $S = \{g \in G : (\bar{0}, g) \in H\}$; this must be a subgroup of unbounded index in G, and hence be contained in $\mathrm{cl}_p(\emptyset)$. Let $r_n(x) = -x$, and put $h = \sum_{i=0}^{n} r_i(h_i)$. Then $(g_0 - h_0, \ldots, g_{n-1} - h_{n-1}, \sum_{i<n} r_i(g_i - h_i)) \in H$; as $\bar{g} \in H + \bar{h}$ implies $(g_0 - h_0, \ldots, g_n - h_n) \in H$, we get

$$[\sum_{i=0}^{n} r_i(g_i)] - h = [\sum_{i<n} r_i(g_i - h_i)] - (g_n - h_n) \in S \subseteq \mathrm{cl}_p(\emptyset);$$

as $h \mathop{\smile\!\!\!\!\!\!\;|}_{\mathrm{cl}_p(\emptyset)} \bar{g}$, this yields $\sum_{i\leq n} r_i(g_i) \in \mathrm{cl}_p(\emptyset)$. ■

4.9 BIBLIOGRAPHICAL REMARKS

The theory of generic types and stabilizers in a type-definable group in a simple theory (Section 4.1) was adapted from the stable case by Pillay [108]; Moosa [99] worked out the details for definable transitive group actions on sets. The rest follows [160] for the type-definable case (Section 4.2) and [163] for hyperdefinable groups, and incorporates ideas of Chatzidakis, Cherlin, Hrushovski and Pillay [30] for groups in a smoothly approximable structure (Example 4.1.14 is the least complicated such structure), a pseudo-finite field [57, 58], or an existentially closed field with automorphism [25]. For the locally modular groups of Section 4.8, Pillay has obtained similar results using local stability theory.

All results were previously known for type-definable groups in a stable theory [126, 161]: generic types were defined by Poizat [119, 121]; one-based groups were studied by Hrushovski and Pillay [56], locally modular ones by Hrushovski [47]; and the reconstruction of a generically given group is Hrushovski's version [46] of a theorem of Weil [165] for algebraic groups.

The group-theoretic Theorem 4.2.4 was proved by Schlichting [136] and rediscovered by Bergman and Lenstra [13]. Proposition 4.6.1 is due to Pourmahdian and myself.

Chapter 5

SUPERSIMPLE THEORIES

5.1 RANKS

Recall from Definition 2.8.12 that a simple theory is *supersimple* if for all finite tuples $\bar{a}$ and all A there is a finite subset $A_0 \subseteq A$ with $a \downarrow_{A_0} A$. The importance of supersimplicity stems from the fact that it allows a global, ordinal-valued rank, invariant under definable bijections, which orders definable sets and types and is compatible with independence. In fact, there are two (main) ranks; one suitable for complete types and one suitable for partial types. In this chapter, we shall again assume that we work in a simple theory (which need not be supersimple).

Let On^+ denote the class of ordinals together with an extra symbol ∞ (where $\alpha < \infty$ for every ordinal α, and $\alpha + \infty = \infty + \alpha = \infty + \infty = \infty$).

DEFINITION 5.1.1 The *SU-rank*, or *Lascar rank SU* is the least function from the collection of all types (over parameters in the monster model) to On^+ satisfying for every ordinal α:

$SU(p) \geq \alpha + 1$ if there is a forking extension q of p with $SU(q) \geq \alpha$.

Note that we do allow hyperimaginary types as well in the domain of SU. Clearly, SU is automorphism invariant, $SU(p) \geq SU(q)$ if $q \vdash p$, and $SU(p) = 0$ if and only if p is bounded. Furthermore, if q extends p and $SU(q) = SU(p) < \infty$, then q is a non-forking extension of p. Finally, if $SU(p) < \infty$ and $\alpha \leq SU(p)$, then p has an extension q with $SU(q) = \alpha$, by minimality of SU-rank.

REMARK 5.1.2 Every non-algebraic real type p in a simple theory has an extension of SU-rank 1.

PROOF: Enumerate all pairs (φ, k) of a formula φ and $k < \omega$ as $(\varphi_i, k_i : i < \alpha)$, and choose inductively non-algebraic extensions p_i of p such that p_i

extends $\bigcup_{j<i} p_j$ (with $\bigcup_{j<0} p_j = p$) and $D(p_i, \varphi_i, k_i)$ is minimal possible. By compactness this can be done at the limit stages, as a union of non-algebraic types is non-algebraic; $\bigcup_{i<\alpha} p_i$ will be a type whose only forking extensions are algebraic. Hence it has SU-rank 1. ■

REMARK 5.1.3 If a is a hyperimaginary element and $(A_i : i < \alpha)$ is an increasing sequence of sets with $SU(a/A_i) > 0$, it does not follow that $a \notin \mathrm{bdd}(A_i : i < \alpha)$. For instance, if $E = \bigwedge_{i<\omega} E_i$ for some $\emptyset$-definable equivalence relations E_i (with $E_{i+1} \vdash E_i$) and $a = \bar{a}_E$, it may well be that $\bar{a}_{E_{i+1}} \in \mathrm{acl}(A_{i+1}) - \mathrm{acl}(A_i)$ for all $i < \omega$, whence $SU(\bar{a}_E/A_i) > 0$ for all $i < \omega$, but $\bar{a}_E \in \mathrm{acl}(A_i : i < \omega)$. This does not happen if $SU(a/A_i) < \infty$ for some $i < \alpha$, as then $SU(a/A_i)$ must be constant from some i_0 onwards, and $a \mathop{\smile\!\!\!\!\!\mid}_{A_{i_0}} \bigcup_{i<\alpha} A_i$.

We shall now study the interaction between SU-rank and forking.

LEMMA 5.1.4 *If q is a non-forking extension of p, then $SU(q) = SU(p)$.*

PROOF: We use induction on α to show that if $SU(p) \geq \alpha$ then $SU(q) \geq \alpha$, the other inequality being obvious. This is clear for $\alpha = 0$ or limits. So suppose the assertion holds for rank α, and assume $p \in S(A_0)$ has a non-forking extension $q \in S(A_1)$ and a forking extension $q' \in S(A_2)$ with $SU(q') \geq \alpha$. By automorphism invariance, we may assume that there is $a \models q \cup q'$, and $A_2 \mathop{\smile\!\!\!\!\!\mid}_{A_0 a} A_1$. Since $A_1 \mathop{\smile\!\!\!\!\!\mid}_{A_0} a$, we get $A_1 \mathop{\smile\!\!\!\!\!\mid}_{A_0} A_2 a$, whence $a \mathop{\smile\!\!\!\!\!\mid}_{A_2} A_1$. By inductive hypothesis $SU(a/A_2) = SU(q') \geq \alpha$ implies $SU(a/A_2A_1) \geq \alpha$; since $a \not\mathop{\smile\!\!\!\!\!\mid}_{A_1} A_2$, we have $SU(q) = SU(a/A_1) \geq \alpha + 1$. ■

It follows that if q extends p and $SU(q) < \infty$, then q is a non-forking extension of p if and only if $SU(q) = SU(p)$.

THEOREM 5.1.5 *T is supersimple if and only if $SU(p) < \infty$ for every (real) type p.*

PROOF: Suppose $SU(\bar{a}/A) < \infty$. Let $A_0 \subseteq A$ be a finite subset of A such that $SU(\bar{a}/A_0)$ is minimal possible. Then for any finite subset $A_1 \subseteq A$ we have $SU(\bar{a}/A_0A_1) = SU(\bar{a}/A_0)$, so $\bar{a} \mathop{\smile\!\!\!\!\!\mid}_{A_0} A_1$ by the definition of SU-rank. By the finite character of forking, $\bar{a} \mathop{\smile\!\!\!\!\!\mid}_{A_0} A$.

For the other direction, suppose there is a type p with $SU(p) = \infty$. As every type does not fork over a small subset of its domain, there is some ordinal α such that $SU(q) = \infty$ if and only if $SU(q) \geq \alpha$. It follows that there is a sequence $p = p_0, p_1, p_2, \ldots$ of extensions of p, such that p_{i+1} is a forking extension of p_i with $SU(p_i) = \infty$, for all $i < \omega$. Let $q = \bigcup_{i<\omega} p_i$, and suppose $q = \mathrm{tp}(a/A)$. If A_0 is a finite subset of A, then A_0 is contained in the domain of p_i for some $i < \omega$. But $\mathrm{tp}(a/A_0) \subseteq p_i \subset p_{i+1} \subset q$ and p_{i+1} is a forking extension of p_i, so $\mathrm{tp}(a/A)$ is a forking extension of $\mathrm{tp}(a/A_0)$. Hence $a \not\perp_{A_0} A$ for all finite subsets A_0 of A. ∎

Recall that every ordinal α can be written as a finite sum $\sum_{i=1}^{k} \omega^{\alpha_i} n_i$ for ordinals $\alpha_1 > \cdots > \alpha_k$ and natural numbers $n_1, \ldots, n_k$; this sum is unique if we require all summands to be non-zero. (This is called the *Cantor normal form* of α.) If $\beta = \sum_{i=1}^{k} \omega^{\alpha_i} m_i$, then the *commutative sum* $\alpha \oplus \beta$ is defined to be $\sum_{i=1}^{k} \omega^{\alpha_i}(n_i + m_i)$; note that $\oplus$ is the smallest symmetric increasing function f from pairs of ordinals to ordinals which satisfies $f(\alpha+1, \beta) = f(\alpha, \beta) + 1$. Clearly, if α and β are finite, then $\alpha + \beta = \alpha \oplus \beta$.

From now on, whenever we write an expression of the form $\sum_{i \leq k} \omega^{\alpha_i} n_i$, this implies that $\alpha_0 > \alpha_1 > \cdots > \alpha_k$ and $n_0, \ldots, n_k < \omega$.

THEOREM 5.1.6 LASCAR INEQUALITIES

1. $SU(a/bA) + SU(b/A) \leq SU(ab/A) \leq SU(a/bA) \oplus SU(b/A)$.
2. *Suppose* $SU(a/Ab) < \infty$ *and* $SU(a/A) \geq SU(a/Ab) \oplus \alpha$. *Then* $SU(b/A) \geq SU(b/Aa) + \alpha$.
3. *Suppose* $SU(a/Ab) < \infty$ *and* $SU(a/A) \geq SU(a/Ab) + \omega^\alpha n$. *Then* $SU(b/A) \geq SU(b/Aa) + \omega^\alpha n$.
4. *If* $a \perp_A b$, *then* $SU(ab/A) = SU(a/A) \oplus SU(b/A)$.

PROOF:

1. We prove inductively on α that $SU(b/A) \geq \alpha$ implies $SU(ab/A) \geq SU(a/bA) + \alpha$. For $\alpha = 0$ this amounts to $SU(ab/A) \geq SU(a/bA)$, which is obvious. The limit case being trivial, we assume the inequality to hold for α and $SU(b/A) \geq \alpha + 1$. So there is $B \supseteq A$ with $b \not\perp_A B$ and $SU(b/B) \geq \alpha$. We may choose $B \perp_{Ab} a$, whence $SU(a/bA) = SU(a/bB)$. By the inductive hypothesis, $SU(ab/B) \geq SU(a/bB) + \alpha = SU(a/bA) + \alpha$. On the other hand, clearly $B \not\perp_A ab$, so $SU(ab/A) \geq SU(ab/B) + 1 \geq SU(a/bA) + \alpha + 1$.

   For the second inequality we prove inductively on α that $SU(ab/A) \geq \alpha$ implies $SU(a/bA) \oplus SU(b/A) \geq \alpha$. Again this is clear for $\alpha = 0$ and limit ordinals. So suppose $SU(ab/A) \geq \alpha + 1$, i.e. there is $B \supseteq A$ with $ab \not\perp_A B$ and $SU(ab/B) \geq \alpha$. By the inductive hypothesis, $SU(a/bB) \oplus SU(b/B) \geq \alpha$. If $b \not\perp_A B$, then $SU(b/A) \geq SU(b/B) + 1$ and $SU(a/bA) \oplus SU(b/A) \geq \alpha + 1$. Otherwise $b \perp_A B$ and $a \not\perp_{bA} B$, whence $SU(a/bA) \geq SU(a/bB) + 1$ and $SU(a/bA) \oplus SU(b/A) \geq \alpha + 1$.

2. By induction on $SU(a/Ab) \oplus \alpha$, the cases $\alpha = 0$ and α a limit ordinal being trivial. So suppose $SU(a/A) \geq SU(a/Ab) \oplus \alpha + 1$. There is some

$B \supseteq A$ such that $a \not\downarrow_A B$ and $SU(a/B) \geq SU(a/Ab) \oplus \alpha$. We may choose $B \downarrow_{Aa} b$, so $SU(b/Ba) = SU(b/Aa)$. Now if $b \not\downarrow_A B$, then since $SU(a/B) \geq SU(a/Bb) \oplus \alpha$, we get by the inductive hypothesis

$$SU(b/A) \geq SU(b/B) + 1 \geq SU(b/Ba) + \alpha + 1 = SU(b/Aa) + \alpha + 1.$$

Otherwise $a \not\downarrow_{bA} B$, so $SU(a/Ab) \geq SU(a/Bb) + 1$ and

$$SU(a/B) \geq SU(a/Ab) \oplus \alpha \geq SU(a/Bb) \oplus (\alpha + 1),$$

whence again by the inductive hypothesis

$$SU(b/A) = SU(b/B) \geq SU(b/Ba) + \alpha + 1 = SU(b/Aa) + \alpha + 1.$$

3. For $\alpha = 0$ ordinal and symmetric sums coincide. For $\alpha > 0$ and $\beta < \omega^\alpha n$ we have $SU(a/Ab) + \omega^\alpha n > SU(a/Ab) \oplus \beta$. The corollary now follows by continuity from part 2.

4. If, say, $SU(a/A) \geq \alpha + 1$, then there is a superset B of A such that $a \not\downarrow_A B$ and $SU(a/B) \geq \alpha$. We may choose $B \downarrow_{Aa} b$, whence $Ba \downarrow_A b$. Therefore $SU(b/B) = SU(b/A)$, $a \downarrow_B b$, and $B \not\downarrow_A ab$. By the inductive assumption

$$\begin{aligned} SU(ab/A) &\geq SU(ab/B) + 1 \geq (SU(a/B) \oplus SU(b/B)) + 1 \\ &\geq (\alpha + 1) \oplus SU(b/A). \end{aligned}$$

The zero and limit cases being trivial, the proposition now follows by induction and symmetry. ■

As every finitary hyperimaginary is definable over a finite tuple, it follows from the first Lascar inequality that a theory is supersimple if and only if every finitary hyperimaginary type has ordinal SU-rank. Note that the second and third inequalities can be seen as quantitative versions of forking symmetry.

LEMMA 5.1.7 *Let $SU(a/A) = \omega^\alpha n + \beta$ with $\beta < \omega^\alpha$. Then there is $a' \in \text{bdd}(Aa)$ such that $SU(a/a'A) = \omega^\alpha n$ and $SU(a'/A) = \beta$.*

PROOF: Let B be a set such that $SU(a/BA) = \omega^\alpha n$; we may assume that $B = \text{Cb}(a/BA)$. Choose a Morley sequence $(a_i : i \leq \omega)$ in $\text{lstp}(a/B)$. Then $SU(a_\omega/A, a_i : i < \omega) = \omega^\alpha n$ by Theorem 3.3.12; since $SU(a_\omega/A) = SU(a/A) < \infty$, there is $i_0 < \omega$ such that $SU(a_\omega/A, a_i : i < i_0) = \omega^\alpha n$, whence $B \in \text{bdd}(A, a_i : i < i_0)$ and $SU(B/A) < \infty$. As

$$\begin{aligned} SU(a_i/A, a_j : j < i) &\leq SU(a_i/A) = \omega^\alpha \cdot n + \beta < \omega^\alpha \cdot (n+1) \\ &\leq SU(a_i/BA) + \omega^\alpha = SU(a_i/BA, a_j : j < i) + \omega^\alpha \end{aligned}$$

for all $i < i_0$, we have $SU(B/A, a_j : j < i) < SU(B/A, a_j : j \leq i) + \omega^\alpha$ for all $i < i_0$ by Theorem 5.1.6, whence $SU(B/A) < \omega^\alpha$.

Now consider $a' \in \mathrm{Cb}(B/Aa)$ such that $SU(B/Aa') = SU(B/Aa)$. So $SU(a/Aa') = SU(a/ABa')$. But a' is definable over A in a Morley sequence in $\mathrm{lstp}(B/Aa)$ by Corollary 3.3.13, and hence is bounded over a finite initial segment of the Morley sequence, together with A. Therefore $SU(a'/A) < \omega^\alpha$, so Theorem 5.1.6 implies that $\omega^\alpha n = SU(a/BA) = SU(a/BAa') = SU(a/Aa')$.

Since $SU(a/A) = SU(aa'/A)$, Theorem 5.1.6 yields $SU(a'/A) = \beta$. ■

COROLLARY 5.1.8 *Suppose* $SU(a/A) = \sum_{i\leq k} \omega^{\alpha_i} n_i$. *Then there are* $a_0, \ldots, a_k \in \mathrm{bdd}(Aa)$ *with* $a_{i+1} \in \mathrm{bdd}(a_i)$ *for all* $i < k$, *such that* $SU(a/Aa_i) = \sum_{j<i} \omega^{\alpha_j} n_j$ *and* $SU(a_i/A) = \sum_{j=i}^{k} \omega^{\alpha_j} n_j$.

PROOF: We can take $a_0 = a$; Lemma 5.1.7 yields a_1. If we have found a_i, we apply Lemma 5.1.7 to type $\mathrm{tp}(a_i/A)$ and obtain $a_{i+1} \in \mathrm{bdd}(Aa_i) \subseteq \mathrm{bdd}(Aa)$ with $SU(a_i/Aa_{i+1}) = \omega^{\alpha_i} n_i$ and $SU(a_{i+1}/A) = \sum_{j=i+1}^{k} \omega^{\alpha_j} n_j$. Finally, Theorem 5.1.6 implies

$$\begin{aligned} SU(a/Aa_{i+1}) &= SU(aa_i/Aa_{i+1}) \\ &= SU(a/Aa_i) + SU(a_i/Aa_{i+1}) = \sum_{j=0}^{i} \omega^{\alpha_j} n_j, \end{aligned}$$

as $\sum_{j<i} \omega^{\alpha_j} n_j + \omega^{\alpha_i} n_i = \sum_{j<i} \omega^{\alpha_j} n_j \oplus \omega^{\alpha_i} n_i$. ■

Recall that a type is *regular* if it is unbounded and foreign to all its forking extensions.

REMARK 5.1.9 A type p of SU-rank of the form ω^α is regular.

PROOF: If q is a forking extension of p, then $SU(q) < \omega^\alpha$, so a realization of a forking extension of p cannot fork with a realization of (a non-forking extension of) p by Theorem 5.1.6. ■

DEFINITION 5.1.10 Two types p and q are *orthogonal over* a set A containing their domains, written $p \perp_A q$, if $a \mathop{\smile\!\!\!\!\!\!\;|}_A b$ for any realizations a and b of non-forking extensions of p and q, respectively, to A.

If $p, q \in S(A)$ and $p \perp_A q$, we say that p and q are *almost* orthogonal, denoted $p \perp^a q$.

Two types p and q are *orthogonal* if they are orthogonal over every set which contains their domains; we write $p \perp q$ for this.

In other words, two types are orthogonal if any two realizations (of non-forking extensions) are independent.

PROPOSITION 5.1.11 *In a supersimple theory, every unbounded type p is non-orthogonal to a regular type, even to a regular type of a real element.*

PROOF: Let q be a non-algebraic type of minimal SU-rank such that p is non-orthogonal to q. Such a type exists; in fact, since a hyperimaginary element cannot be independent of all real elements, we may even require q to be the type of a real singleton.

Now let q' be a forking extension of q, and suppose that q is non-orthogonal to q'. So there are a set A and realizations a, b, b' of non-forking extensions of p, q, q', respectively, to A, with $a \not\perp_A b$ and $b \not\perp_A b'$. Since $SU(b'/A) = SU(q') < SU(q)$, we have $a \perp_A b'$. So $a \not\perp_{Ab'} b$. But $SU(b/Ab') < SU(b/A) = SU(q)$; as $\text{tp}(a/Ab')$ is a non-forking extension of p, this contradicts our minimal choice of $SU(q)$. ■

If we want more information about q, we must allow hyperimaginary types.

PROPOSITION 5.1.12 *Suppose $SU(a/A) = \beta + \omega^\alpha n$, with $n > 0$ and $\infty > \beta \geq \omega^{\alpha+1}$ or $\beta = 0$. Then $\text{tp}(a/A)$ is non-orthogonal to a type of SU-rank ω^α.*

PROOF: Let b be such that $SU(a/Ab) = \beta + \omega^\alpha(n-1)$; as in the proof of Lemma 5.1.7 we may assume that $b = \text{Cb}(a/Ab)$, and $SU(b/A) < \infty$. As $a \not\perp_A b$, Theorem 5.1.6 implies that $SU(b/A) \geq \omega^\alpha$, and there is some $B \supseteq A$ with $SU(b/B) = \omega^\alpha$. We may assume that $B \perp_{Ab} a$. Then $SU(a/Bb) = SU(a/Ab) = \beta + \omega^\alpha(n-1)$. As b is contained in $\text{Cb}(a/Ab) = \text{Cb}(a/Bb)$, but is not bounded over B, a must fork with b over B; this must affect the SU-rank. Theorem 5.1.6 implies $SU(a/B) \geq \beta + \omega^\alpha n = SU(a/A)$, and equality holds. So $a \perp_A B$. ■

We shall now introduce the second rank.

DEFINITION 5.1.13 The *D-rank*, or *Shelah rank* D is the least function from the class of all consistent formulas to On^+ satisfying

> $D(\varphi(\bar{x})) \geq \alpha + 1$ if there is a formula $\varphi'(\bar{x})$ which forks over the parameters of $\varphi(\bar{x})$, such that $\varphi'(\bar{x}) \vdash \varphi(\bar{x})$ and $D(\varphi'(\bar{x})) \geq \alpha$.

For a partial type π we put $D(\pi) = \min\{D(\varphi) : \pi \vdash \varphi\}$.

In contrast to SU-rank, where we view a hyperimaginary type as a *complete* type (of some hyperimaginary sort), in the context of Shelah rank we treat hyperimaginary types as partial types of real (or imaginary) elements.

LEMMA 5.1.14 *$D(\varphi) = max\{D(p) : \varphi \in p\}$.*

PROOF: If $\varphi \in p$, then $D(\varphi) \geq D(p)$ by definition of $D(p)$, so $D(\varphi) \geq \sup\{D(p) : \varphi \in p\}$. For the other direction, and the attainment of the supremum, it is sufficient by Zorn's Lemma to show that for every partial type π and every formula φ, either $D(\pi \cup \{\varphi\}) = D(\pi)$ or $D(\pi \cup \{\neg\varphi\}) = D(\pi)$, since we can then complete any formula to a type of the same D-rank. We show by induction on α that if $D(\pi) \geq \alpha$, then $D(\pi \cup \{\varphi\}) \geq \alpha$ or $D(\pi \cup \{\neg\varphi\}) \geq \alpha$. This is clear for $\alpha = 0$ (as for consistent π either $\pi \cup \{\varphi\}$ or $\pi \cup \{\neg\varphi\}$ must be consistent), or limit ordinals.

Hence suppose $D(\pi) \geq \alpha + 1$, but $D(\pi \cup \{\varphi\}) \not\geq \alpha + 1$ and $D(\pi \cup \{\neg\varphi\}) \not\geq \alpha + 1$. By definition there are formulas ψ' and ψ'' implied by π with $D(\psi' \wedge \varphi) \not\geq \alpha + 1$ and $D(\psi'' \wedge \neg\varphi) \not\geq \alpha + 1$; put $\psi = \psi' \wedge \psi''$. As $D(\psi) \geq D(\pi) \geq \alpha + 1$, there is a formula $\vartheta \vdash \psi$ such that ϑ forks over the parameters in ψ, and $D(\vartheta) \geq \alpha$; we may choose the parameters of ϑ independent of the parameters of φ over the parameters of ψ. By inductive hypothesis, either $D(\vartheta \wedge \varphi) \geq \alpha$ or $D(\vartheta \wedge \neg\varphi) \geq \alpha$. But ϑ forks over the parameters of $\psi \wedge \varphi$, which are the same as the parameters of $\varphi \wedge \neg\varphi$. In the first case $\vartheta \wedge \varphi$ is consistent and forks over the parameters of $\psi \wedge \varphi$; since $\vartheta \wedge \varphi \vdash \psi \wedge \varphi$, we get $D(\psi \wedge \varphi) \geq \alpha + 1$, a contradiction. The second case is analogous. ■

Clearly, if $a \not\perp_A c$, then $D(a/A) > D(a/Ac)$. In particular, if q is a forking extension of p, then $D(q) < D(p)$.

EXAMPLE 5.1.15 Here is a simple structure $\mathfrak{M}$ (even supersimple, of SU-rank 2) with a type p and a non-forking extension q of p, such that $D(q) < D(p)$. The language consists of a unary predicate P and countably many unary predicates $(Q_i(x) : i < \omega)$, countably many equivalence relations $(E_i : i < \omega)$, and a binary relation R. The unary predicates are disjoint, and E_i is an equivalence relation on Q_i with infinitely many infinite classes for all $i < \omega$. Moreover, P is the disjoint union of sets P_I for all $I \subset \omega$ such that R is the random bipartite graph between P_I and $\neg Q \wedge \neg \bigvee_{i \in I} Q_i$. Hence P_I is type-defined by

$$P(x) \wedge \bigwedge_{i \in I} \forall y\, [Q_i(y) \rightarrow \neg R(x, y)] \wedge \bigwedge_{i \notin I} \exists y\, [Q_i(y) \wedge R(x, y)].$$

This structure is supersimple of SU-rank 2 (without the relation R it would be superstable of Lascar rank 2). The type $p = \{\neg Q_i(x) : i < \omega\}$ is complete and has D-rank 2 and SU-rank 1. However, if $a \in P$ and $b \models p$ are R-related and a is not related to any point in $\bigcup_{i<\omega} Q_i$, then $a \perp b$, but $1 = D(b/a) < D(b) = 2$, as $D(R(a, y)) = 1$.

PROPOSITION 5.1.16 *$SU(p) \leq D(p)$, for any type p.*

PROOF: We prove by induction on α that if $SU(p) \geq \alpha$, then $D(p) \geq \alpha$. This is trivial for $\alpha = 0$ or limit ordinals. So suppose $SU(p) \geq \alpha + 1$. Hence there is a forking extension q of p with $SU(q) \geq \alpha$. By inductive hypothesis,

$D(q) \geq \alpha$. But q contains a formula φ which forks over the domain of p. So if $\psi \in p$ is such that $D(\psi) = D(p)$, then φ forks over the parameters of ψ, and

$$D(p) = D(\psi) \geq D(\psi \wedge \varphi) + 1 \geq D(q) + 1 \geq \alpha + 1. \quad \blacksquare$$

PROPOSITION 5.1.17 *If T is supersimple, then $D(\varphi) < \infty$ for all formulas φ.*

PROOF: Suppose $D(\varphi) = \infty$ for some formula φ. As there are only boundedly many automorphism types of formulas, there is some ordinal α_0 such that $D(\psi) = \infty$ if and only if $D(\psi) \geq \alpha_0$. Hence we can inductively find a sequence $(\varphi_i : i < \omega)$ of formulas with $\varphi_0 = \varphi$, such that $D(\varphi_i) = \infty$, $\varphi_{i+1} \vdash \varphi_i$, and φ_{i+1} forks over the parameters of φ_i, for all $i < \omega$. Choosing the parameters of φ_{i+1} independent of the parameters for $(\varphi_j : j < i)$ over the parameters of φ_i, we may assume that φ_{i+1} forks over the parameters of $(\varphi_j : j \leq i)$; call this parameter set A_i. Then $\bigwedge_{i<\omega} \varphi_i$ is consistent, and can be completed to a type $p \in S(\bigcup_{i<\omega} A_i)$. But if A' is a finite subset of $\bigcup_{i<\omega} A_i$, then $A' \subseteq A_j$ for some $j < \omega$; since φ_{j+1} forks over A_j, it forks over A', and so does p. Hence p forks over every finite subset of its domain, and T is not supersimple. ■

So a supersimple theory allows two ordinal valued ranks, SU-rank and D-rank. One should be careful not to confuse the two: while SU-rank measures forking and has nice additivity properties, D-rank is continuous: every type contains a formula of the same D-rank. We may also define SU-rank for a partial type π as $SU(\pi) = \sup\{SU(p) : \pi \subseteq p\}$. However, even in a superstable theory it is not always the case that a partial type (or even a formula) contains a complete type of the same SU-rank.

EXAMPLE 5.1.18 Let $\mathfrak{M}$ be the structure which consists of disjoint predicates Q and $(P_i : i < \omega)$, and bijections $f_i : Q^i \to P_i$. Then $\mathfrak{M}$ is superstable, $D(Q) = SU(Q) = SU(\neg Q \wedge \bigwedge_{i<\omega} \neg P_i) = 1$, and $SU(P_i) = D(P_i) = i$ for all $i < \omega$. So $D(x = x) = SU(x = x) = \omega$, but all types have finite SU-rank.

REMARK 5.1.19 If P is an $\emptyset$-invariant family of types closed under nonforking extensions, one can localize both SU-rank and D-rank to P, the crucial conditions being

- $SU_P(a/A) \geq \alpha+1$ if there are some $B \supseteq A$ and some b with $\mathrm{tp}(b/B) \in P$, such that $a \not\perp_B b$ and $SU_P(a/Bb) \geq \alpha$.
- $D_P(\varphi(x)) \geq \alpha + 1$ if there are some set B containing the parameters of φ and some b with $\mathrm{tp}(b/B) \in P$, such that there is a formula $\varphi'(x, b, B)$ which forks over B, with $\varphi'(x, b, B) \vdash \varphi(x)$ and $D(\varphi'(x, b, B)) \geq \alpha$.

Then $a \in \mathrm{cl}_P(A)$ if and only if $\mathrm{tp}(a/A)$ is P-analysable and $SU_P(a/A) = 0$.

EXERCISE 5.1.20 Prove Theorem 5.1.6 for SU_P-rank.

EXERCISE 5.1.21 Prove Propositions 5.1.16 and 5.1.17 for SU_P- and D_P-rank.

EXERCISE 5.1.22 Show that a type p with $SU_P(p) < \infty$ is P-minimal if and only if every forking extension of p has smaller SU_P-rank.

EXERCISE 5.1.23 Prove Corollary 5.1.8, Proposition 5.1.11 and Proposition 5.1.12 for P-minimal types and SU_P-rank.

5.2 WEIGHT AND DOMINATION

DEFINITION 5.2.1 Let p be a type, and λ a cardinal. The *weight* of p, denoted $w(p)$, is at least λ if there is a non-forking extension $\text{tp}(a/A)$ of p and a sequence $(a_i : i < \lambda)$ independent over A, such that $a \not\perp_A a_i$ for all $i < \lambda$. The weight of p is λ if $w(p) \geq \lambda$ but $w(p) \not\geq \lambda^+$.

As usual, we write $w(a/A)$ for $w(\text{tp}(a/A))$.

LEMMA 5.2.2 *If $a \perp_A B$, then $w(a/A) = w(a/B)$.*

PROOF: By definition $w(a/A) \geq w(a/B)$. Conversely, if $a \perp_A C$ and I is independent over C with $c \not\perp_C a$ for all $c \in I$, we may choose $IC \perp_{Aa} B$. So $B \perp_A ICa$, whence $a \perp_A BC$ and I is independent over BC; we thus get $a \not\perp_{BC} c$ for all $c \in I$ and $w(a/B) \geq w(a/A)$. ■

LEMMA 5.2.3 *$w(a/A) < \lambda$ if and only if for any $B \perp_A a$ and any independent sequence I over B there is $J \subseteq I$ with $|J| < \lambda$ such that $I - J \perp_B a$.*

PROOF: Suppose $w(a/A) < \lambda$, and let I be an independent sequence over B for some $B \perp_A a$. If J is a maximal subsequence of I with $J \perp_B a$, then $a \not\perp_{BJ} c$ for all $c \in I - J$; as $I - J$ is independent over BJ, it has cardinality less than λ.

Conversely, if $B \perp_A a$ and I is an independent sequence over B such that $c \not\perp_B a$ for all $c \in I$, then the only subsequence J of I with $J \perp_B a$ is empty. ■

LEMMA 5.2.4 *If $a \perp_A b$, then $w(ab/A) = w(a/A) + w(b/A)$.*

PROOF: Let $B \perp_A ab$ and I be an independent sequence over B. There is a subsequence $J \subseteq I$ of cardinality at most $w(a/A)$ such that $I - J \perp_B a$, and a subsequence $J' \subseteq I - J$ of cardinality at most $w(b/A)$ such that $I - J - J' \perp_{Ba} b$, whence $I - J - J' \perp_B ab$ and $w(ab/A) \leq w(a/A) + w(b/A)$ by Lemma 5.2.3.

Conversely, if $B \perp_A ab$, and I is independent over B with $c \not\perp_B a$ for all $c \in I$, and J is independent over B with $c \not\perp_B b$ for all $c \in J$, we may choose

$I \mathop{\smile\!\!\!\!\!\!\mid}_B J$. Then $I \cup J$ is independent over B and $c \not\mathop{\smile\!\!\!\!\!\!\mid}_B ab$ for all $c \in I \cup J$, whence $w(ab/A) \geq w(a/A) + w(b/A)$. ■

THEOREM 5.2.5 *In a supersimple theory, a type p of SU-rank $\sum_{i<k} \omega^{\alpha_i} n_i$ has weight at most $\sum_{i<k} n_i$.*

PROOF: Suppose $I = (b_i : i < m)$ is a finite independent sequence over some parameter set A, and a realizes a non-forking extension of p to A such that $a \not\mathop{\smile\!\!\!\!\!\!\mid}_A b_i$ for all $i < m$. Put $\alpha_{-1} = \alpha_0 + 1$, and $A_i = A \cup \{b_j : j < i\}$ for $i \leq m$. As $SU(a/A) < \omega^{\alpha_{-1}}$, Theorem 5.1.6 yields a maximal $i < k$ is such that $SU(b_j/A_j) < SU(b_j/A_j a) + \omega^{\alpha_{i-1}}$ for all $j < m$. By Theorem 5.1.6 we have $SU(a/A_j) < SU(a/A_{j+1}) + \omega^{\alpha_{i-1}}$, whence $SU(a/A) < SU(a/A_j) + \omega^{\alpha_{i-1}}$ and $SU(a/A_j) \geq \sum_{s<i} \omega^{\alpha_s}$, for all $j \leq m$. On the other hand, if

$$I_i = \{b_j \in I : SU(b_j/A_j) \geq SU(b_j/A_j a) + \omega^{\alpha_i}\},$$

then $SU(a/A_j) \geq SU(a/A_{j+1}) + \omega^{\alpha_i}$ for all $b_j \in I_i$ again by Theorem 5.1.6; it follows that $|I_i| \leq n_i$, as the coefficient of ω^{α_i} in the SU-rank of a, the first coefficient which is affected by forking with I, can only decrease n_i times.

Let $I' = I - I_i$, and $A'_j = A_j \cap (A \cup I')$. Then by definition of I_i and independence of I we get

$$\begin{aligned} SU(b_j/A'_j) &= SU(b_j/A) = SU(b_j/A_j) \\ &< SU(b_j/A_j a) + \omega^{\alpha_i} \leq SU(b_j/A'_j a) + \omega^{\alpha_i} \end{aligned}$$

for all $b_j \in I'$. So we can repeat the above argument for I'; by the Lascar inequalities 5.1.6 and induction we obtain $|I| = |I_0| + \cdots + |I_{k-1}| \leq n_0 + \cdots + n_{k-1}$. Thus $w(p) \leq \sum_{i<k} n_i$. ■

Note that this result requires not only $SU(p) < \infty$, but also $SU(b_i/A) < \infty$ for all $b_i \in I$ (and hence supersimplicity of the ambient theory).

DEFINITION 5.2.6 An element a *dominates* an element b *over* a set A, written $a \vartriangleright_A b$, if $b \mathop{\smile\!\!\!\!\!\!\mid}_A c$ for all $c \mathop{\smile\!\!\!\!\!\!\mid}_A a$. Two elements a and b are *equidominant over* a set A, denoted $a \doteq_A b$, if $a \vartriangleright_A b$ and $b \vartriangleright_A a$.
A type p is *more dominant* than a type q *over* a set A containing their domains if there are realizations a and b of non-forking extensions of p and q to A, such that a dominates b over A; we also write $p \vartriangleright_A q$ for this relation. We say that p is *more dominant* than q, written $p \vartriangleright_0 q$, if p is more dominant than q over some set A.
Two types p and q are *equidominant over* a set A containing their domains, in symbols $p \doteq_A q$, if they have realizations of non-forking extensions to A which are equidominant over A; they are *equidominant*, witten $p \doteq_0 q$, if they are equidominant over some set A.

It is easy to see that if a dominates b over A and $B \mathop{\smile\!\!\!\!\!\!\,|}_A a$, then a dominates b over AB. Furthermore, domination over a set A is transitive on elements, and equidominance over A is an equivalence relation on elements.

REMARK 5.2.7 If $a \rhd_A b$ and I is independent over some $B \mathop{\smile\!\!\!\!\!\!\,|}_A b$ with $c \not\mathop{\smile\!\!\!\!\!\!\,|}_B b$ for all $c \in I$, we may choose $IB \mathop{\smile\!\!\!\!\!\!\,|}_{Ab} a$. Then $B \mathop{\smile\!\!\!\!\!\!\,|}_A ab$, whence $a \rhd_B b$ and $c \not\mathop{\smile\!\!\!\!\!\!\,|}_B a$ for all $c \in I$. Hence $w(a/A) \geq w(b/A)$. In particular, $p \rhd_0 q$ implies $w(p) \geq w(q)$.

LEMMA 5.2.8 *If $a \rhd_A b$ and $c \mathop{\smile\!\!\!\!\!\!\,|}_A ab$, then $ac \rhd_A bc$.*

PROOF: Suppose $d \mathop{\smile\!\!\!\!\!\!\,|}_A ac$. Then $d \mathop{\smile\!\!\!\!\!\!\,|}_{Ac} a$, whence $a \mathop{\smile\!\!\!\!\!\!\,|}_A dc$. So $b \mathop{\smile\!\!\!\!\!\!\,|}_A dc$, whence $b \mathop{\smile\!\!\!\!\!\!\,|}_{Ac} d$, and $d \mathop{\smile\!\!\!\!\!\!\,|}_A bc$. ■

Equidominance of two types p and q does not imply that *every* pair of non-forking extensions of p and q to the same parameter set has equidominant realizations. For instance, it may happen that p has two orthogonal non-forking extensions q and q'. Clearly $p \doteq_0 q$ and $p \doteq_0 q'$, but two orthogonal types can never be equidominant.

If $p \rhd_0 q$ and $q \rhd_0 p$, this does not imply that p and q are equidominant:

EXAMPLE 5.2.9 Let X be an infinite set and σ a bijection between X and $\{(x, y, z) \in X^3 : x \neq y \neq z \neq x\}$, without any cycles. Then $\mathrm{Th}(X, \sigma)$ is stable, with a unique stationary 1-type p over $\emptyset$. If a and b are two independent realizations of p, then (a, b) dominates a over $\emptyset$; if $c \models p$ independently of a, b, then $\sigma^{-1}(a, b, c)$ realizes p and dominates (a, b) over $\emptyset$. However, p and $\mathrm{tp}(a, b)$ are not equidominant. In fact, there are two equidominance classes among types of independent n-tuples of realizations of p: those with n odd and those with n even.

LEMMA 5.2.10 *For any A-invariant family P of types, $\mathrm{cl}_P(aA) \doteq_{\mathrm{cl}_P(A)} a$. In particular, $\mathrm{tp}(a/\mathrm{cl}_P(A))$ is regular if and only if $\mathrm{tp}(\mathrm{cl}_P(aA)/\mathrm{cl}_P(A))$ is.*

PROOF: The first assertion is immediate from Lemma 3.5.5; the second follows. ■

LEMMA 5.2.11 *1. A regular type p has weight* 1.

2. *If $w(p) = 1$ and $q \not\mathop{\smile\!\!\!\!\!\!\,|}_A p$, then $q \rhd_A p$.*

3. *If $w(p) = w(q) = 1$ and $p \not\mathop{\smile\!\!\!\!\!\!\,|}_A q$, then p and q are equidominant over A.*

PROOF:

1. Suppose a realizes a non-forking extension of p over some set A, and consider two elements b and c which are independent over A. Suppose

$a \not\perp_A b$. Replacing b by an initial segment $(b_i : i \leq n)$ of a Morley sequence in $\mathrm{lstp}(a/Ab)$ independent of c over bA, we may choose n such that $a \perp_A (b_i : i < n)$ but $a \not\perp_{A\cup(b_i:i<n)} b_n$. By regularity of $\mathrm{tp}(a/A) = \mathrm{tp}(b_n/A)$ this implies $b_n \perp_A (b_i : i < n)$; since $c \perp_A (b_i : i \leq n)$, we get $b_n \perp_A (c, b_i : i < n)$. Suppose $a \not\perp_A c$. By regularity of $\mathrm{tp}(b_n/A) = \mathrm{tp}(a/A)$ we obtain $b_n \perp_{A(c,b_i:i<n)} a$; this yields $b_n \perp_A (a, c, b_i : i < n)$, a contradiction. Hence $w(p) = 1$.

2. Let a and b realize non-forking extensions of p and q over A with $a \not\perp_A b$. If $c \perp_A b$, then $w(a/A) = 1$ and $a \not\perp_A b$ together imply $a \perp_A c$, so b dominates a over A.

3. This follows immediately from part 2. ■

In particular, two non-orthogonal types of weight 1 are equidominant.

COROLLARY 5.2.12 *Non-orthogonality over a set A is an equivalence relation on (unbounded) types over A of weight 1 (and in particular on regular types).*

PROOF: Clearly $\not\perp_A$ is reflexive and symmetric. If $p, q, r \in S(A)$ have weight 1 with $p \not\perp_A q$ and $q \not\perp_A r$, then for any $b \models q$ we can find $a \models p$ and $c \models r$ with $a \not\perp_A b$ and $b \not\perp_A c$. Since $w(b/A) = 1$, we must have $a \not\perp_A c$, and $p \not\perp_A r$. ■

LEMMA 5.2.13 *Suppose $p \in S(A)$ is a Lascar strong type which has a non-forking regular extension $q \in S(B)$. Then p is regular.*

PROOF: Suppose $a, a' \models p$, $C \perp_A a$, $C \not\perp_A a'$ and $a \not\perp_C a'$. Since $w(q) = 1$, we have $w(a'/A) = 1$ and $a' \perp_A a$. Clearly we may assume that $a \models q$. Since $\mathrm{lstp}(a/A) = \mathrm{lstp}(a'/A)$, there is a strong A-automorphism σ mapping a to a'; if $B' = \sigma(B)$, then $\mathrm{lstp}(B'/A) = \mathrm{lstp}(B/A)$ and by the Independence Theorem 3.2.15 we find $B'' \perp_A aa'$ realizing $\mathrm{tp}(B/Aa) \cup \mathrm{tp}(B'/Aa')$. Using another A-automorphism, we may assume $B'' = B$ and $a, a' \models q$. Finally, we may suppose $C \perp_{Aaa'} B$, whence $B \perp_A Caa'$; as then $a \perp_B C$, $a' \not\perp_B C$ and $a \not\perp_{BC} a'$, we contradict regularity of q. ■

We have already seen that transitivity of domination, or of equidominance, may fail for types (even for Lascar strong types). However, an easy application of the Independence Theorem yields the following partial transitivity.

LEMMA 5.2.14 *Suppose $A_1 \perp_{A_0} A_2$. If $p_i \in S(A_i)$ are Lascar strong types for $i = 0, 1, 2$ with $p_1 \rhd_0 p_0 \rhd_0 p_2$, then $p_1 \rhd_0 p_2$.*

PROOF: By assumption there are $B_1 \supseteq A_1 \cup A_0$ and $B_2 \supseteq A_0 \cup A_2$, as well as $a_1 \models p_1$, $a_0, a_0' \models p_0$ and $a_2 \models p_2$ with $a_1 \perp_{A_1} B_1$, $a_0 \perp_{A_0} B_1$,

$a_0' \mathop{\smile\!\!\!\!\!\!|}_{A_0} B_2$ and $a_2 \mathop{\smile\!\!\!\!\!\!|}_{A_2} B_2$, such that $a_1 \trianglerighteq_{B_1} a_0$ and $a_0' \trianglerighteq_{B_2} a_2$. Since $A_1 \mathop{\smile\!\!\!\!\!\!|}_{A_0} A_2$, we may assume that $B_1 \mathop{\smile\!\!\!\!\!\!|}_{A_0} B_2$; by the Independence Theorem 3.2.15 we may assume that $a_0 = a_0'$ and $a_0 \mathop{\smile\!\!\!\!\!\!|}_{A_0} B_1B_2$. We may also assume that $a_1 \mathop{\smile\!\!\!\!\!\!|}_{B_1a_0} B_2$ and $a_2 \mathop{\smile\!\!\!\!\!\!|}_{B_2a_0} B_1$; since $a_0 \mathop{\smile\!\!\!\!\!\!|}_{B_1} B_2$ and $a_0 \mathop{\smile\!\!\!\!\!\!|}_{B_2} B_1$, this yields $B_2 \mathop{\smile\!\!\!\!\!\!|}_{B_1} a_0a_1$ and $B_1 \mathop{\smile\!\!\!\!\!\!|}_{B_2} a_0a_2$. Put $B = B_1B_2$. As domination is preserved under the addition of independent parameters, we get $a_1 \trianglerighteq_B a_0$ and $a_0 \trianglerighteq_B a_2$, whence $a_1 \trianglerighteq_B a_2$ and $p_1 \trianglerighteq_0 p_2$. ■

DEFINITION 5.2.15 Let $\trianglerighteq$ be the transitive closure of $\trianglerighteq_0$ on the class of Lascar strong types, and put $p \doteq q$ if $p \trianglerighteq q$ and $q \trianglerighteq p$. If $p \trianglerighteq q$, we say that p *dominates* q; if $p \doteq q$ we say that p and q are *domination-equivalent.*

So domination is transitive for Lascar strong types, and domination-equivalence is an equivalence relation on Lascar strong types. Note that domination and domination-equivalence are preserved under non-forking extensions, since $p \doteq_0 q$ whenever q is a non-forking extension of p. Furthermore, example 5.2.9 shows that $\doteq$ is not the transitive closure of $\doteq_0$, not even for stable theories.

PROBLEM 5.2.16 Is $\trianglerighteq$ the two-step iteration of $\trianglerighteq_0$? In other words, if p and q are Lascar strong types with $p \trianglerighteq q$, is there some Lascar strong type r with $p \trianglerighteq_0 r \trianglerighteq_0 q$?

If we knew that domination is type-definable (for the parameter sets of the types?), then Lemma 3.3.1 together with Lemma 5.2.14 would yield a positive answer to question 5.2.16.

LEMMA 5.2.17 *Suppose $w(a/A) < \omega$, and $(b_0, \ldots, b_k, b_k')$ is independent over A, with $a \not\mathop{\smile\!\!\!\!\!\!|}_A b_k'$ and $a \not\mathop{\smile\!\!\!\!\!\!|}_A b_i$ for all $i \leq k$. Then there is $B \mathop{\smile\!\!\!\!\!\!|}_A a$ and b with $Bb \mathop{\smile\!\!\!\!\!\!|}_A b_0 \ldots b_k$, $b \not\mathop{\smile\!\!\!\!\!\!|}_B a$ and $w(b/B) = 1$.*

PROOF: Suppose otherwise. We define inductively sets $A = B_k \subseteq B_{k+1} \subseteq \cdots$ and elements $(b_n, b_n' : n > k)$ such that for all $n > k$

1. $B_n \mathop{\smile\!\!\!\!\!\!|}_A a$,
2. $a \not\mathop{\smile\!\!\!\!\!\!|}_{B_n} b_n$ and $a \not\mathop{\smile\!\!\!\!\!\!|}_{B_n} b_n'$, and
3. $b_0 \ldots b_{n-1} \mathop{\smile\!\!\!\!\!\!|}_{B_{n-1}} B_nb_nb_n'$ and $b_n \mathop{\smile\!\!\!\!\!\!|}_{B_n} b_n'$.

Note that condition 3. implies inductively that $(b_0, \ldots, b_n, b_n')$ is independent over B_n. Suppose for some $n \geq k$ we have found $(B_i, b_i, b_i' : k \leq i \leq n)$, and let $B \supseteq B_n$ be such that $B \mathop{\smile\!\!\!\!\!\!|}_{B_n} a$ and $Bb_n' \mathop{\smile\!\!\!\!\!\!|}_{B_n} b_0 \ldots b_n$, but whenever $B' \supseteq B$ with $B' \not\mathop{\smile\!\!\!\!\!\!|}_B b_n'$ and $B'b_n' \mathop{\smile\!\!\!\!\!\!|}_{B_n} b_0 \ldots b_n$, then $a \not\mathop{\smile\!\!\!\!\!\!|}_{B_n} B'$. Such a B exists by the local and finite character of forking.

Now $(b_0, \ldots, b_n, b'_n)$ is independent over B, and every element in the sequence forks with a over B. Hence $w(b'_n/B) \geq 2$ by assumption, and there are $B_{n+1} \mathop{\smile\!\!\!\!\!\!\mid}_B b'_n$ with $B_{n+1} \supseteq B$ and b_{n+1}, b'_{n+1} independent over B_{n+1}, both of which fork with b'_n over B_{n+1}. We may choose $B_{n+1}b_{n+1}b'_{n+1} \mathop{\smile\!\!\!\!\!\!\mid}_{Bb'_n} ab_0 \ldots b_n$, whence $b_0 \ldots b_n \mathop{\smile\!\!\!\!\!\!\mid}_{B_n} B_{n+1}b'_n b_{n+1}b'_{n+1}$. Since $b'_n \not\mathop{\smile\!\!\!\!\!\!\mid}_B B_{n+1}b_{n+1}$ and $b'_n \not\mathop{\smile\!\!\!\!\!\!\mid}_B B_{n+1}b'_{n+1}$, our choice of B implies $a \not\mathop{\smile\!\!\!\!\!\!\mid}_{B_n} B_{n+1}b_{n+1}$ and $a \not\mathop{\smile\!\!\!\!\!\!\mid}_{B_n} B_{n+1}b'_{n+1}$. As $B_{n+1} \mathop{\smile\!\!\!\!\!\!\mid}_{B_n} a$, we are done.

Put $B = \bigcup_{n<\omega} B_n$. Then $a \mathop{\smile\!\!\!\!\!\!\mid}_A B$, and $(b_n : n < \omega)$ is an infinite independent family over B such that a forks over B with every element of the sequence, contradicting $w(a/A) < \omega$. ■

THEOREM 5.2.18 *Suppose $w(p) < \omega$. Then p is equidominant with a finite product (i.e. an independent tuple) of types of weight 1; in particular, p has finite weight. Moreover, if types of finite weight are non-orthogonal to regular types, then p is equidominant with a finite product of regular types.*

PROOF: If $w(p) = 1$, we are done. Otherwise there is a realization a of a non-forking extension of p to some set A_0 and b_0, b'_0 independent over A_0, each of which fork with a over A_0. So p is non-orthogonal to a type of weight 1 by Lemma 5.2.17. As p does not have infinite weight, there is a realization a of some non-forking extension of p to some set A and a maximal finite tuple $\bar{b}$ of independent realizations of types over A of weight 1, such that a dominates $\bar{b}$ over A. Possibly after adding parameters A' with $A' \mathop{\smile\!\!\!\!\!\!\mid}_A a$ (whence $A' \mathop{\smile\!\!\!\!\!\!\mid}_A \bar{b}$), we may assume that $a \rhd_A a\bar{b}$.

Now suppose that $\bar{b}$ does not dominate a over A. So there is some $b \mathop{\smile\!\!\!\!\!\!\mid}_A \bar{b}$ with $b \not\mathop{\smile\!\!\!\!\!\!\mid}_A a$. Possibly adding some parameters A' with $A' \mathop{\smile\!\!\!\!\!\!\mid}_A a$, whence $A' \mathop{\smile\!\!\!\!\!\!\mid}_A a\bar{b}$, we may assume by Lemma 5.2.17 that $w(b/A) = 1$. Hence a dominates b over A by Lemma 5.2.11. If $x \mathop{\smile\!\!\!\!\!\!\mid}_A a$ for some element x, we get $x \mathop{\smile\!\!\!\!\!\!\mid}_A a\bar{b}$, whence $x \mathop{\smile\!\!\!\!\!\!\mid}_{A\bar{b}} a$. But $b \mathop{\smile\!\!\!\!\!\!\mid}_A \bar{b}$ and $b \not\mathop{\smile\!\!\!\!\!\!\mid}_{A\bar{b}} a$, so $w(b/A) = 1$ implies $b \mathop{\smile\!\!\!\!\!\!\mid}_{A\bar{b}} x$ and finally $x \mathop{\smile\!\!\!\!\!\!\mid}_A b\bar{b}$: it follows that a dominates $b\bar{b}$ over A. This contradicts the maximality of $\bar{b}$.

By Lemma 5.2.4 a tuple of n types of weight 1 has weight n, and so does $\text{tp}(a/A)$ by Remark 5.2.7. Finally, if types of finite weight are non-orthogonal to regular types, we may replace every type of weight 1 in the decomposition by an equidominant regular type. ■

COROLLARY 5.2.19 *A finitary type in a supersimple theory is equidominant with a finite product of regular types.*

PROOF: By Theorem 5.2.5 and Proposition 5.1.11, the assumptions of Theorem 5.2.18 hold in a supersimple theory. ■

REMARK 5.2.20 It is not obvious from the definition that the weight of a type is a cardinal: it might happen that $w(p) < \lambda$, but $w(p) \geq \lambda'$ for all $\lambda' < \lambda$. Theorem 5.2.18 shows that this cannot happen for $\lambda = \omega$; a type has either finite or infinite weight.

PROBLEM 5.2.21 If $w(p) \geq \aleph_n$ for all $n < \omega$, is $w(p) \geq \aleph_\omega$?

As $w(p) \leq |T|$, this assumes an uncountable language.

5.3 ELIMINATION OF HYPERIMAGINARIES

Recall from Section 3.6 that a theory *eliminates hyperimaginaries* if for every hyperimaginary element a there is a set A of imaginary elements, such that a is interdefinable with A. The main result of this section is:

THEOREM 5.3.1 *A supersimple theory eliminates hyperimaginaries.*

The proof of Theorem 5.3.1 will use induction on yet another rank.

DEFINITION 5.3.2 The rank $R(a/A)$ is the least ordinal α with the following property: there are $A = a_0, a_1, \ldots, a_n$ in $\mathrm{dcl}(Aa)$ such that $a \in \mathrm{bdd}(a_n)$, $a_i \in \mathrm{dcl}(a_{i+1})$, $\mathrm{tp}(a_{i+1}/a_i)$ is internal in formulas of D-rank α_i for some $\alpha_i \leq \alpha$, and $\mathrm{tp}(a/a_i)$ is orthogonal to all types of D-rank less than α_i, for all $i < n$.

Note that $R(a/A) = 0$ if and only if $a \in \mathrm{bdd}(A)$.

LEMMA 5.3.3 *In a supersimple theory, every (hyperimaginary) type of ordinal SU-rank has ordinal R-rank.*

PROOF: Consider the hyperimaginary type $\mathrm{tp}(a_E/A)$, where a is a possibly infinite tuple. If $SU(a_E/A) = 0$, then $a_E \in \mathrm{bdd}(A)$ and we are done. Otherwise $a_E \not\perp_A a$, so there is a finite tuple $a' \subseteq a$ with $a_E \not\perp_A a'$; by supersimplicity $D(a'/A) < \infty$. Let α_0 be minimal such that $\mathrm{tp}(a_E/A)$ is non-orthogonal to a type q of D-rank α_0. By Proposition 3.4.12 there is $a'' \in \mathrm{dcl}(a_E A) - \mathrm{bdd}(A)$, such that $\mathrm{tp}(a''/A)$ is internal in the family of A-conjugates of q, and hence in formulas of D-rank α_0. By minimality of α_0 the type $\mathrm{tp}(a_E/A)$ is orthogonal to all formulas of D-rank less than α_0. Furthermore $a'' \in \mathrm{dcl}(Aa_E) - \mathrm{bdd}(A)$ implies $a_E \not\perp_A a''$, whence $SU(a_E/Aa'') < SU(a_E/A)$. We put $a_0 = A$ and $a_1 = Aa''$; the Lemma now follows by induction on $SU(a_E/A)$. ■

Note that if we had allowed a sequence $(a_0, a_1, \ldots, a_n, \ldots)$ of infinite length in Definition 5.3.2, then every type in a supersimple theory would have ordinal R-rank.

LEMMA 5.3.4 *Suppose $SU(A/B) < \infty$ and $R(a/B) < \alpha$ for all elements $a \in A$. Then $R(A/B) < \alpha$.*

PROOF: As $SU(A/B) < \infty$, there is a finite tuple $\bar{a}$ of elements of A such that $SU(A/B\bar{a}) = 0$. Hence $A \in \mathrm{bdd}(B\bar{a})$, and it is enough to show the lemma for $A = \bar{a}$, where it is obvious from Lemma 3.4.5 and the proof of Lemma 5.3.3. ■

LEMMA 5.3.5 *Suppose $SU(a/A) < \infty$, and put $A' = \{a' \in \mathrm{dcl}(Aa) : R(a'/A) < \alpha\}$. Then $R(A'/A) < \alpha$, and $\mathrm{tp}(a/A')$ is orthogonal to all types of Shelah rank less than α.*

PROOF: $R(A'/A) < \alpha$ follows from Lemma 5.3.4, since $SU(A'/A) \leq SU(a/A) < \infty$. Now suppose $\mathrm{tp}(a/A')$ is non-orthogonal to some formula of Shelah rank less than α. By Proposition 3.4.12 there is some $a' \in \mathrm{dcl}(A'a) - \mathrm{bdd}(A')$ such that $\mathrm{tp}(a'/A')$ is internal in formulas of Shelah rank less than α. As $R(A'/A) < \alpha$, there are $n < \omega$ and $A = A_0, A_1, \ldots, A_n \in \mathrm{dcl}(AA')$ such that $A' \in \mathrm{bdd}(A_n)$, $A_i \in \mathrm{dcl}(A_{i+1})$, and $\mathrm{tp}(A_{i+1}/A_i)$ is internal in formulas of Shelah rank less than α, for all $i < n$. Then $\mathrm{bdd}(A') = \mathrm{bdd}(A_n)$, so $\mathrm{tp}(a'/A_n)$ is internal in formulas of Shelah rank less than α. Hence $R(a'A_n/A) < \alpha$; as $A' \in \mathrm{dcl}(Aa)$ and thus $a' \in \mathrm{dcl}(Aa)$, we get $a'A_n \in A'$, contradicting $a' \notin \mathrm{bdd}(A')$. ■

The following Proposition lies at the heart of our proof of Theorem 5.3.1.

PROPOSITION 5.3.6 *Let a_0 be a hyperimaginary element in a supersimple theory. If there is a (possibly infinite) tuple a of imaginary elements with $SU(a) < \infty$ and $a_0 \in \mathrm{dcl}(a)$, then a_0 is interdefinable with a sequence of imaginary elements.*

PROOF: We shall use induction on $R(a/a_0)$ to prove the theorem. If $R(a/a_0) = 0$, then $a_0 \in \mathrm{dcl}(a)$ and $a \subseteq \mathrm{acl}(a_0)$; the assertion now follows from Lemma 3.6.3. So assume that whenever a' is a sequence of imaginaries with $SU(a') < \infty$ and $a_0' \in \mathrm{dcl}(a')$ is a hyperimaginary with $R(a'/a_0') < \alpha$, then a_0' is interdefinable with a sequence of imaginaries.

Consider a hyperimaginary element $a_0 \in \mathrm{dcl}(a)$ for some sequence a of imaginary elements with $SU(a) < \infty$ and $R(a/a_0) = \alpha$. By definition there is a sequence $a_0, a_1, \ldots, a_n$ of hyperimaginaries in $\mathrm{dcl}(a)$ with $a \in \mathrm{bdd}(a_n)$, such that $a_i \in \mathrm{dcl}(a_{i+1})$, $\mathrm{tp}(a_{i+1}/a_i)$ is internal in formulas of Shelah rank α_i for some $\alpha_i \leq \alpha$ and $\mathrm{tp}(a/a_i)$ is orthogonal to all types of Shelah rank less than α_i, for all $i < n$. We shall show that a_{n-1} is interdefinable with a sequence a' of imaginary elements; clearly $SU(a') = SU(a_{n-1}) \leq SU(a) < \infty$. Replacing a by a' and repeating the argument then yields that $a_{n-2}, a_{n-3}, \ldots, a_0$ are interdefinable with sequences of imaginary elements. Note that by the case $R(a/a_0) = 0$ already treated, a_n is interdefinable with a sequence of imaginary elements, and we may indeed assume $a = a_n$. Furthermore, if $\alpha_{n-1} < \alpha$, we are done by inductive hypothesis. Thus we may assume $\alpha_{n-1} = \alpha$, and $\mathrm{tp}(a/a_{n-1})$ is orthogonal to all types of Shelah rank less than α.

Consider the canonical base c of $\mathrm{lstp}(a/a_{n-1})$. Since $a_{n-1} \in \mathrm{dcl}(a)$, we obtain $a_{n-1} \in \mathrm{dcl}(c)$ and $c \in \mathrm{bdd}(a_{n-1})$. By the case $R(a/a_0) = 0$, it is sufficient to show that c is interdefinable with a sequence of imaginary elements. Note that $\mathrm{lstp}(a/a_{n-1})$ is determined by the class of a modulo every bounded equivalence relation which is type-definable over a_{n-1}. But these classes are all definable over a, whence $c = \mathrm{bdd}(a_{n-1}) \cap \mathrm{dcl}(a)$. By Corollary 3.1.17 there is a type-definable equivalence relation E such that c is interdefinable with the class a_E of a modulo E. Put $p_a(x) = \mathrm{tp}(a/a_E) = E(x, a)$. Then p_a is essentially the same as $\mathrm{lstp}(a/a_{n-1})$.

It is easy to see that for any $a' \models \mathrm{tp}(a)$ the following conditions are equivalent:

1. $E(a, a')$.

2. $p_a \cup p_{a'}$ is consistent.

3. $p_a \cup p_{a'}$ is consistent and forks neither over a_E nor over a'_E.

4. For every formula $\psi(x, a) \in p_a$ the conjunction $\psi(x, a) \wedge \psi(x, a')$ is consistent and forks neither over a_E nor over a'_E.

However, while p_a is a_E-invariant (i.e. invariant under all automorphisms fixing a_E), it is a partial type with parameters in a. Thus any particular formula in p_a need *not* be invariant under all automorphisms fixing a_E, and 2. cannot be used for a local description of the condition $E(a, a')$. For this reason, we shall use 4.

Since $SU(a) < \infty$, there is a finite subtuple $\bar{a} \subseteq a$ such that $a \in \mathrm{acl}(\bar{a})$. Let $p'_a(\bar{x}) = \mathrm{tp}(\bar{a}/a_E)$ be the corresponding restriction of p_a.

CLAIM. There is a set A containing a with $A - a$ finite, a non-forking extension p'_A of p'_a to A, some $k < \omega$, and a formula $\psi_0(\bar{x}) \in p'_A$, such that every realization of ψ_0 can fork over A with at most k independent elements b_i whose type over A is orthogonal to types of Shelah rank less than α.

PROOF OF CLAIM: By construction, $\mathrm{tp}(a_n/a_{n-1})$ is internal in formulas of Shelah rank α; as $a = a_n$ and $\mathrm{bdd}(a_{n-1}) = \mathrm{bdd}(a_E)$, this is also true of $\mathrm{tp}(a/a_E)$. Therefore there are a set A' with $A' \mathop{\smile\!\!\!\!\!\!|}_{a_E} a$, formulas $(\varphi_i : i \in I)$ over A' with $D(\varphi_i) = \alpha$ for all $i \in I$, and realizations $c_i \models \varphi_i$ for $i \in I$, such that $a \in \mathrm{dcl}(A', c_i : i \in I)$. Since $\bar{a}$ is an imaginary element in $\mathrm{dcl}(a)$, we may choose the index set such that $\bar{a} \in \mathrm{dcl}(A', c_i : i < k)$ for some $k < \omega$, say $\bar{a} = f(c_0, \ldots, c_{k-1})$ for some A'-definable function f. Put

$$\psi_0(\bar{x}) = \exists\, y_0, \ldots, y_{k-1}\, [\bar{x} = f(y_0, \ldots, y_{k-1}) \wedge \bigwedge_{i<k} \varphi_i(y_i)].$$

Clearly $\psi_0 \in \mathrm{tp}(\bar{a}/A'a_E)$, and we may choose A' to be finite. Put $A = A' \cup a$, and let $p'_A(\bar{x})$ be a non-forking extension of $\mathrm{tp}(\bar{a}/A'a_E)$ to A; as $\bar{a} \perp_{a_E} A'$, this is a non-forking extension of p'_a.

Consider any $\bar{a}' \models \psi_0$, and a sequence $(b_i : i < n)$ independent over A, such that $\bar{a}' \not\perp_A b_i$ and $\mathrm{tp}(b_i/A)$ is orthogonal to all types of Shelah rank less than α, for all $i < n$. Since $\bar{a}' \models \psi_0$ there are $c'_i \models \varphi_i$ for $i < k$, such that $\bar{a}' = f(c'_0, \ldots, c'_{k-1})$. Then $(c'_i : i < k) \not\perp_A b_i$ for all $i < n$. Now $D(c'_i/A) \leq D(\varphi_i) = \alpha$ for all $i < k$. As $\mathrm{tp}(b_i/A)$ is orthogonal to types of Shelah rank less than α, and in particular to forking extensions of $\mathrm{tp}(c'_i/A)$, independence of $(b_i : i < n)$ over A implies $n \leq k$. ■

As $A - a$ is finite and $SU(a) < \infty$, supersimplicity implies $SU(A) < \infty$. Let p_A be a non-forking extension of p_a extending p'_A. Put $q = \mathrm{tp}(a)$, $q' = \mathrm{tp}(A)$, and whenever $C \models q'$, let c denote the subtuple of C corresponding to a in A. Let $A_0 = \{a' \in \mathrm{dcl}(A) : R(a'/a_E) < \alpha\}$. By Corollary 3.1.17 there is some type-definable equivalence relation F such that A_0 is interdefinable with A_F; as $\mathrm{lstp}(A/A_F)$ is given by $\mathrm{tp}(A/A_F)$ together with the classes of A modulo every A_F-definable bounded equivalence relation, and since such a class is in $\mathrm{bdd}(A_F) \cap \mathrm{dcl}(A) \subseteq A_0$, we see that $\mathrm{tp}(A/A_F) \vdash \mathrm{lstp}(A/A_F)$. Hence $F(A, A')$ implies $\mathrm{lstp}(A/A_F) = \mathrm{lstp}(A'/A_F)$. Furthermore $a_E \in \mathrm{dcl}(A_F)$, $SU(A_F) \leq SU(A) < \infty$, $R(A_F/a_E) < \alpha$, and $\mathrm{tp}(A/A_F)$ is orthogonal to all types of Shelah rank less than α by Lemma 5.3.5.

Next, consider $\mathrm{tp}(a/a_E) = E(x, a)$. Since $a \in \mathrm{acl}(\bar{a})$, for any finite $\bar{a}' \subseteq a$ containing $\bar{a}$ we have $D(\bar{a}'/a_E) = D(\bar{a}/a_E)$. By supersimplicity, there is a formula $\vartheta_0(x, a) \in E(x, a)$ of minimal D-rank which contains at least the free variables $\bar{x}$; since E is an equivalence relation, by compactness there is a symmetric formula $\vartheta(x, y) \in E$ such that $\exists z, z' [\vartheta(x, z) \wedge \vartheta(z, z') \wedge \vartheta(z, y)]$ proves $\vartheta_0(x, y)$.

CLAIM. If a and b realize q and $\vartheta(a, b)$ holds, then a_E and b_E are interbounded.

PROOF OF CLAIM: Let $(b_i : i < \omega)$ be an indiscernible sequence of realizations of q satisfying $\exists z [E(a, z) \wedge \vartheta(z, x)]$. Let $\psi(x, y) \in E$ be any formula, and consider a formula $\psi' \in E$ implying ϑ, such that $\exists z [\psi'(x, z) \wedge \psi'(y, z)]$ implies $\psi(x, y)$. Then $\models \vartheta(b_i, c)$ for any $c \models \psi'(b_i, y)$, so $\psi'(b_i, y) \vdash \vartheta_0(a, y)$. By minimality of D-rank, $D(\psi'(b_i, y)) = D(\psi'(a, y)) = D(\vartheta_0(a, y))$, so the family $\{\psi'(b_i, y) : i < \omega\}$ must be consistent. But then for any $i, j < \omega$ there is c_{ij} such that $\psi'(b_i, c_{ij}) \wedge \psi'(b_j, c_{ij})$ holds, whence $\models \psi(b_i, b_j)$. As $\psi \in E$ was arbitrary, $\models E(b_i, b_j)$ for all i, j and $(b_i)_E$ is bounded over a_E. Hence $b_E \in \mathrm{bdd}(a_E)$; by symmetry $a_E \in \mathrm{bdd}(b_E)$. ■

Let (*) denote the following condition: A and C realize q', b realizes q, $F(A, C)$ holds and $C \perp_{A_F} b$. This is type-definable, since the independence may be expressed as $D(C/A_F b, \varphi, k) \geq D(A/A_F, \varphi, k)$ for all formulas φ and all $k < \omega$. Note that $F(A, C)$ implies $A_F = C_F$ and thus $a_E = c_E$.

CLAIM. There is $\psi_1(x,a) \in p_a$ such that whenever $b, c, c' \models q$ with $E(c',c)$ and $\psi_1(x,c) \wedge \psi_1(x,b)$ is consistent, then $\models \vartheta(c',b)$.

PROOF OF CLAIM: Suppose $b, c, c' \models q$ with $E(c,c')$ and $p(x,c) \cup p(x,b)$ consistent. Then $E(x,c) \wedge E(x,b)$ is consistent, whence $E(c,b)$ and $E(c',b)$. Thus $\models \vartheta(c',b)$; the existence of ψ_1 follows by compactness. ■

Let us remark that even though we view a formula $\psi(x,a) \in p_a$ as being from $\mathrm{tp}(a/a_E)$, it does not follow that for c_E and b_E interbounded (in the notation from the claim above) $\psi(x,c) \wedge \psi(x,b)$ does not fork over c_E if and only if it is consistent. This would be true if $\psi(x,a)$ were a formula with parameter a_E; the problems arise from the fact that the parameter really occuring in the formula is a.

Let Ψ be the set of formulas $\psi(x,y)$ such that $\psi(x,a) \in p_a$ and $\psi(x,a)$ implies $\psi_1(x,a)$ and $x \in \mathrm{acl}(\bar{x})$ (i.e. ψ implies that the subset of variables from x occuring in ψ is algebraic over $\bar{x}$, where $\bar{x}$ are the variables in p'_a and ψ_0). For $\psi \in \Psi$ define $S_\psi(A,b)$ to hold if there is C such that (A,b,C) satisfies (*), $\vartheta(c,b)$ holds, and $\psi_0(\bar{x},C) \wedge \psi(x,c) \wedge \psi(x,b)$ is consistent and does not fork over A_F. Note that whenever (A,b,C) satisfy (*) and $\vartheta(c,b)$ holds, then $\mathrm{bdd}(b_E) = \mathrm{bdd}(c_E) = \mathrm{bdd}(a_E) \subseteq \mathrm{bdd}(A_F)$. But $\mathrm{tp}(b/b_E)$ is orthogonal to $\mathrm{tp}(A_F/a_E)$ and hence to $\mathrm{tp}(A_F/b_E)$, whence $b \mathop{\smile\!\!\!\!\!\!\:|}_{A_F} C$ implies $b \mathop{\smile\!\!\!\!\!\!\:|}_{b_E} C$. We may therefore type-define the condition "$\psi_0(\bar{x},C) \wedge \psi(x,c) \wedge \psi(x,b)$ does not fork over A_F" by saying that there exists a sequence $(C_i b_i : i < \omega)$ indiscernible over A_F, with $\mathrm{tp}(C_0 b_0/A_F) = \mathrm{tp}(Cb/A_F)$, such that $D(\mathrm{tp}(C_i/A_F, C_j b_j : j < i), \varphi, n) \geq D(\mathrm{tp}(A/A_F), \varphi, n)$ and $D(\mathrm{tp}(b_i/A_F, C_i, C_j b_j : j < i), \varphi, n) \geq D(\mathrm{tp}(a/a_E), \varphi, n)$ for all formulas φ and all $n < \omega$, and such that $\bigwedge_{j<i} \psi_0(\bar{x}, C_j) \wedge \psi(x,c_j) \wedge \psi(x,b_j)$ is consistent, for all $i < \omega$. Thus S_ψ is a type-definable condition.

CLAIM. If $F(A,A')$, then $S_\psi(A,b)$ if and only if $S_\psi(A',b)$. Moreover, if $S_\psi(A,b)$ holds for all $\psi \in \Psi$, then $a_E = b_E$.

PROOF OF CLAIM: The first assertion is is obvious from the definition of S_ψ. For the moreover part, suppose $S_\psi(A,b)$ holds for all $\psi \in \Psi$. Then there is C such that A, b, C satisfies (*), $\vartheta(c,b)$ holds, and $\{\psi_0(\bar{x},C)\} \cup p_c \cup p_b$ is consistent and does not fork over A_F. But then $\models E(c,b)$, and $a_E = c_E = b_E$. ■

CLAIM. $\neg S_\psi(A,b)$ holds if and only if (†) there is C such that (*) holds of (A,b,C), and either $\not\models \vartheta(c,b)$, or there is the beginning of a Morley-sequence $(C_i, b_i : i < k+2)$ in $\mathrm{lstp}(C, b/A_F)$ such that $\bigwedge_{i<k+2} \psi_0(\bar{x}, C_i) \wedge \psi(x,c_i) \wedge \psi(x,b_i)$ is inconsistent.

Note that (†) is type-definable for the same reason as S_ψ.

PROOF OF CLAIM: First we show that S_ψ and (†) are disjoint on $q' \times q$. So suppose not, and let $A \models q'$ and $b \models q$ be such that there is C witnessing

$S_\psi(A, b)$, and C' witnessing (†). Then $c_E = a_E = c'_E$; since $\psi \vdash \psi_1$, in addition $\psi_1(x, c) \wedge \psi_1(x, b)$ is consistent, whence $\models \vartheta(c', b)$ by the choice of ψ_1. By the Independence Theorem 3.2.15, since $\mathrm{lstp}(C/A_F) = \mathrm{lstp}(C'/A_F)$ and both C and C' are independent of b over A_F, the fact that $\psi_0(\bar{x}, C) \wedge \psi(x, c) \wedge \psi(x, b)$ is consistent and does not fork over A_F implies the same for $\psi_0(\bar{x}, C') \wedge \psi(x, c') \wedge \psi(x, b)$, contradicting our choice of C' as a witness for (†). Hence S_ψ and (†) are disjoint on $q' \times q$.

Next we show that S_ψ and (†) cover $q' \times q$. Note that there always is some C such that (A, b, C) satisfies (*). If $\not\models \vartheta(c, b)$, or if $\psi_0(\bar{x}, C) \wedge \psi(x, c) \wedge \psi(x, b)$ is consistent and does not fork over A_F, either (A, b) satisfies (†) or S_ψ. Otherwise, consider any initial segment $(C_i, b_i : i < k + 2)$ of a Morley sequence in $\mathrm{lstp}(C, b/A_F)$ and suppose that there is $e \models \bigwedge_{i<k+2} \psi_0(\bar{x}, C_i) \wedge \psi(x, c_i) \wedge \psi(x, b_i)$. Then $e \not\perp_{A_F} C_i b_i$ for all $i < k + 2$; since $C_0 \perp_{A_F} (C_i b_i : 0 < i < k + 2)$ and $A_F = (C_0)_F \in \mathrm{dcl}(C_0)$, we obtain that $e \not\perp_{C_0} C_i b_i$ for all $0 < i < k + 2$ and $\{C_i b_i : 0 < i < k + 2\}$ is a family of independent tuples over C_0, whose types over C_0 are orthogonal to all types of R-rank less than α. But by definition of Ψ, the tuple e is algebraic over its subtuple $\bar{e} \models \psi_0(\bar{x}, C_0)$, whence $\bar{e} \not\perp_{C_0} C_i b_i$ for all $0 < i < k + 2$. This contradicts the choice of ψ_0 and k. Therefore $\bigwedge_{i<k+2} \psi_0(x, C_i) \wedge \psi(x, c_i) \wedge \psi(x, b_i)$ is inconsistent, and we are in case (†) again. ■

Hence for any $\psi \in \Psi$ both S_ψ and $\neg S_\psi$ are type-definable on $q' \times q$; by compactness S_ψ is definable on $q' \times q$ by some formula $\rho(X, y)$ (which depends on ψ, of course). Since $q'(X) \wedge q'(Y) \wedge q(y) \wedge F(X, Y) \models S_\psi(X, y) \leftrightarrow S_\psi(Y, y)$, compactness yields a formula $\pi(y) \in q$ such that

$$q'(X) \wedge q'(Y) \wedge \pi(y) \wedge F(X, Y) \vdash \rho(X, y) \leftrightarrow \rho(Y, y).$$

Thus the equivalence relation $E_\psi(X, Y)$ defined by $\forall y \{\pi(y) \rightarrow [\rho(X, y) \leftrightarrow \rho(Y, y)]\}$ is coarser than F on q'. Hence $A_{E_\psi} \in \mathrm{dcl}(A_F)$. On the other hand, if $E_\psi(A, A')$ for all $\psi \in \Psi$, then in particular $S_\psi(A, a')$ holds for all $\psi \in \Psi$, whence $a_E = a'_E$. So if $\tilde{a}$ denotes the sequence $(A_{E_\psi} : \psi \in \Psi)$ of imaginary elements, then $a_E \in \mathrm{dcl}(\tilde{a})$, $SU(\tilde{a}) \leq SU(A_F) \leq SU(A) < \infty$, and $R(\tilde{a}/a_E) \leq R(A_F/a_E) < \alpha$. By inductive hypothesis a_E is interdefinable with a sequence of imaginary elements. ■

COROLLARY 5.3.7 *If p is a type in a supersimple theory with $SU(p) < \infty$, then the canonical base of p is interdefinable with a set of imaginary elements.*

PROOF: Since a type of ordinal SU-rank does not fork over a finite subset of its domain, it is sufficient to show that the canonical base of $\mathrm{lstp}(a/A)$ consists of imaginary elements, for any finite tuples a and A (of real or imaginary elements). But $\mathrm{Cb}(a/A)$ is a hyperimaginary element in $\mathrm{dcl}(aA) \cap \mathrm{bdd}(A)$. So the assertion follows from Proposition 5.3.6. ■

We can now prove Theorem 5.3.1: a finitary type has ordinal SU-rank, and its canonical base is therefore interdefinable with a set of imaginary elements by Corollary 5.3.7; elimination of hyperimaginaries follows by Lemma 3.6.4. ∎

Thus, canonical bases in a supersimple theory live in $\mathfrak{C}^{eq}$, and Lascar strong type is the same as strong type (Proposition 3.6.5). This provides a characterization of canonical bases in supersimple theories

THEOREM 5.3.8 *Let p be a Lascar strong type in a supersimple theory. Then* $\mathrm{Cb}(p)$ *is the definable closure of the canonical parameters for p-normal formulas in p. More precisely, it is the definable closure of the canonical parameters for formulas $\varphi(x, a)$ which are (ψ, k)-normal for some formula ψ and some $k < \omega$, with $D(\varphi(x, a), \psi, k) = D(p, \psi, k)$.*

Note that Lemma 3.6.12.3 then implies that $\mathrm{Cb}(p)$ is the definable closure of the φ-definitions of p, where φ runs through the set of p-stable formulas.

PROOF: By Lemma 3.6.12.1 it is sufficient to prove the second assertion. By Lemma 3.6.12.1 and .2 all the canonical parameters of (ψ, k)-normal formulas $\varphi(x, a)$ with $D(\varphi(x, a), \psi, k) = D(p, \psi, k)$ are in $\mathrm{Cb}(p)$; it remains to show the reverse inclusion.

By Theorem 5.3.1, the canonical base of p is a set C of imaginary elements; let $p_0 = p\restriction_C$. By supersimplicity, there is a finite subtuple $c \in C$ such that p (and p_0) does not fork over c, and $C \in \mathrm{acl}(c)$. Write $p_0(x) = p(x, C)$. As C is the canonical base of p, for any $C' \models \mathrm{tp}(C)$ with $C' \neq C$ the join $p(x, C) \cup p(x, C')$ is either inconsistent or forks over C. Now let c_0 be any subtuple of C containing c. If $(C', c_0') \models \mathrm{tp}(C, c_0)$, then $c_0' \neq c_0$ implies that $p(x, C) \cup p(x, C')$ is inconsistent or forks over C. By compactness, there is some formula $\varphi(x, a) \in p(x, C)$, some formula ψ and some $k < \omega$, such that for any $(C', c_0', a') \models \mathrm{tp}(C, c_0, a)$ with $c_0' \neq c_0$ we have $D(\varphi(x, a) \wedge \varphi(x, a'), \psi, k) < D(p, \psi, k)$; by strengthening $\varphi(x, a)$ we may assume $D(\varphi(x, a), \psi, k) = D(p, \psi, k)$.

Suppose $\vartheta(y, c_0)$ isolates the algebraic type $\mathrm{tp}(a/c_0)$, and put $\varphi'(x; c_0) = \exists y\, [\vartheta(y, c_0) \wedge \varphi(x, y)]$. Then $\varphi'(x, c_0)$ is the union of the finitely many c_0-conjugates of $\varphi(x, a)$. As $D(\varphi \vee \varphi', \psi, k) = \max\{D(\varphi, \psi, k), D(\varphi', \psi, k)\}$ (see Lemma 2.3.6), it follows that $c_0' \models \mathrm{tp}(c_0)$ and $c_0' \neq c_0$ imply

$$D(\varphi'(x, c_0') \wedge \varphi'(x, c_0), \psi, k) = D(\varphi(x, a) \wedge \varphi(x, a'), \psi, k) < D(p, \psi, k),$$

where a' is such that $\mathrm{tp}(a'c_0') = \mathrm{tp}(ac_0)$. Thus $\varphi'(x, c_0)$ is (ψ, k)-normal, with canonical parameter c_0. As c_0 was arbitrary (containing c), we finish. ∎

5.4 SUPERSIMPLE GROUPS

In this section we shall analyse the structure of groups with ordinal SU-rank in a simple theory; most of our considerations will hold in general for hyperdefinable groups. Throughout, G will denote a hyperdefinable group over $\emptyset$ in a simple theory, unless stated otherwise.

Consider a generic element g over a set A of parameters and some $h \in G$ independent of g over A. Then $gh \perp_A h$, and

$$\begin{aligned} SU(g/A) &= SU(g/A,h) = SU(gh/A,h) = SU(gh/A) \\ &\geq SU(gh/A,g) = SU(h/A,g) = SU(h/A); \end{aligned}$$

it follows that all generic types have maximal SU-rank possible. So we can define $SU(G) = SU(p)$, where p is any generic type for G. Conversely, if $p \in S_G(A)$ has maximal SU-rank possible, then $SU(gp) = SU(G)$ for any $g \in G$, so gp does not fork over $\emptyset$ and p is generic by Proposition 4.3.10. In particular, if H is a hyperdefinable subgroup of G over A, then $SU(H) = SU(G)$ if and only if H has bounded index. More generally, we can consider the coset space G/H; it follows by a similar argument as above that its generic types (which are the images of the generic types of G, see Remark 4.3.11) are exactly its types of maximal rank. More precisely, if E denotes the equivalence relation $x^{-1}y \in H$ and g is generic for G over A, then $SU(G/H) = SU(g_E/A)$. On the other hand, by Lemma 4.3.12 the class g_E is the canonical parameter for the coset gH over A, and g is generic for that coset over A, g_E. Since $SU(H) = SU(gH) = SU(g/A, g_E)$, Theorem 5.1.6 implies the following Lascar inequality for groups:

$$SU(H) + SU(G/H) \leq SU(G) \leq SU(H) \oplus SU(G/H). \tag{5.1}$$

We can slightly generalize this: if K is another hyperdefinable subgroup of G over A, put $KH = \{gh : g \in K, h \in H\}$. If g, h are independent generic elements over A for K and H, respectively, then gh has maximal rank in KH. Furthermore, with E as above, $(gh)_E$ is generic for KH/H and gh is generic for gH over the canonical parameter $(gh)_E$ (and A). Thus:

$$SU(H) + SU(KH/H) \leq SU(KH) \leq SU(H) \oplus SU(KH/H).$$

The situation with respect to Shelah rank is less straightforward.

PROBLEM 5.4.1 Is a type $p \in S_G(A)$ generic if and only if $D(p) = D(G)$?

Clearly, a type $p \in S_G$ with $D(p) = D(G)$ is generic (and there always are such types).

LEMMA 5.4.2 *Let $\mathfrak{M}$ be a model and $p \in S_G(\mathfrak{M})$ have SU-rank $\sum_{i\leq k} \omega^{\alpha_i} n_i$. Suppose $SU(ab^{-1}/\mathfrak{M}) < SU(p) + \omega^{\alpha_k}$ for any two independent realizations a and b of p. Then (some non-forking extension of) p is a right translate of a generic type for the stabilizer* $\mathrm{stab}(p)$. *In particular,* $SU(\mathrm{stab}(p)) = SU(p)$.

PROOF: Choose two independent realizations a and b of p. Since

$$SU(ab^{-1}/\mathfrak{M}) \geq SU(ab^{-1}/\mathfrak{M},b) = SU(a/\mathfrak{M},b) = SU(a/\mathfrak{M}),$$

there is some $A \supseteq \mathfrak{M}$ with $SU(ab^{-1}/A) = SU(p)$. Clearly we may assume $A \mathop{\smile\!\!\!\!\!\vert}_{\mathfrak{M},ab^{-1}} a,b$. As $\mathrm{tp}(ab^{-1}/A)$ is also based on a Morley sequence $(c_i : i < \omega)$ in $\mathrm{lstp}(ab^{-1}/A)$; replacing A by such a Morley sequence, we may further assume that every tuple in A has ordinal SU-rank over $\mathfrak{M}$. Now $SU(\bar{a}/\mathfrak{M},a,b) = SU(\bar{a}/\mathfrak{M},ab^{-1})$ for all $\bar{a} \in A$; since $SU(ab^{-1}/\mathfrak{M}) < SU(ab^{-1}/\mathfrak{M},\bar{a}) + \omega^{\alpha_k}$ by assumption, the Lascar inequalities 5.1.6 yield first $SU(\bar{a}/\mathfrak{M}) < SU(\bar{a}/\mathfrak{M},ab^{-1}) + \omega^{\alpha_k} = SU(\bar{a}/\mathfrak{M},a,b) + \omega^{\alpha_k}$, and then $SU(a,b/\mathfrak{M}) < SU(a,b/\mathfrak{M},\bar{a}) + \omega^{\alpha_k}$. But $SU(a,b/\mathfrak{M}) = SU(p) \oplus SU(p)$ by Theorem 5.1.6, and the smallest monomial in $SU(a,b/\mathfrak{M})$ is ω^{α_k}. Therefore $SU(a,b/\mathfrak{M}) = SU(a,b/\mathfrak{M},\bar{a})$ by Theorem 5.1.6 again, and $a,b \mathop{\smile\!\!\!\!\!\vert}_{\mathfrak{M}} A$. Hence $ab^{-1} \mathop{\smile\!\!\!\!\!\vert}_{\mathfrak{M}} A$ and $SU(ab^{-1}/\mathfrak{M}) = SU(p)$. So

$$SU(ab^{-1}/\mathfrak{M},a) = SU(b^{-1}/\mathfrak{M},a) = SU(p) = SU(ab^{-1}/\mathfrak{M}),$$

whence $ab^{-1} \mathop{\smile\!\!\!\!\!\vert}_{\mathfrak{M}} a$. By Remark 4.5.12 a right translate of a non-forking extension of p is generic for $\mathrm{stab}(p)$. ■

PROPOSITION 5.4.3 *Suppose* $SU(G) = \sum_{i<k} \omega^{\alpha_i} n_i$, *and put* $\beta_i = \sum_{j<i} \omega^{\alpha_j} n_j$ *for* $i \leq k$. *Then* G *has a hyperdefinable normal subgroup* G_i *of* SU*-rank* β_i*; it is unique up to commensurativity.*

PROOF: Choose a model $\mathfrak{M}$ and some $p \in S_G(\mathfrak{M})$ of SU-rank β_i. Then p satisfies the requirements of Lemma 5.4.2, as for independent realizations a and b of p we get

$$SU(ab^{-1}/\mathfrak{M}) \leq SU(G) < \beta_i + \omega^{\alpha_i}.$$

Put $H = \mathrm{stab}(p)$; this is a hyperdefinable subgroup of SU-rank β_i.
CLAIM. H and H^g are commensurate for any $g \in G$.
PROOF OF CLAIM: If $H \cap H^g$ has unbounded index in H, then $\beta_i = SU(H) > SU(H \cap H^g)$; since the smallest monomial in β_i is ω^{α_i}, we get $SU(H) \geq SU(H \cap H^g) + \omega^{\alpha_i}$. The Lascar inequality (5.1) now implies $SU(H/(H \cap H^g)) \geq \omega^{\alpha_i}$. As HH^g/H^g is isomorphic to $H/(H \cap H^g)$ and must have the same SU-rank, we get

$$\begin{aligned} SU(G) &\geq SU(HH^g) \geq SU(H^g) + SU(HH^g/H^g) \\ &\geq SU(H^g) + SU(H/(H \cap H^g)) \geq \beta_i + \omega^{\alpha_i} > SU(G), \end{aligned}$$

a contradiction. Hence $H \cap H^g$ has bounded index in H, and by symmetry in H^g. ■

By Corollary 4.5.14 there is a hyperdefinable normal subgroup G_i of G commensurate with a bounded intersection of G-conjugates of H; as these are all commensurate with H, so is G_i. Hence $SU(G_i) = \beta_i$. Finally, if N is another normal subgroup of G with $SU(N) = \beta_i$, commensurativity of G_i and N is shown in the same way as commensurativity of H and H^g. ∎

REMARK 5.4.4 As G_i is commensurate with all its automorphic images, Theorem 4.5.13 implies that we can choose G_i to be hyperdefinable over $\emptyset$.

We shall now formulate the appropriate version of Zil'ber's Indecomposability Theorem.

THEOREM 5.4.5 *Let G be a hyperdefinable group in a simple theory with $SU(G) < \omega^{\alpha+1}$, and $\mathfrak{X}$ a family of hyperdefinable subsets of G. Then there is a hyperdefinable subgroup H of G with $H \subseteq X_1^{\pm 1} \cdots X_m^{\pm 1}$ for some $X_1, \ldots, X_m \in \mathfrak{X}$, such that $SU(XH) < SU(H) + \omega^\alpha$ for all hyperdefinable $X \subseteq \langle \mathfrak{X} \rangle$ (and in particular all $X \in \mathfrak{X}$). Moreover, H is unique up to commensurativity.*

REMARK 5.4.6 A hyperdefinable subset X of G is called *α-indecomposable* if for all hyperdefinable subgroups H of G either X is contained in a single coset of H, or $SU(XH) \geq SU(H) + \omega^\alpha$ (or equivalently, $SU(XH/H) \geq \omega^\alpha$). So if $X \in \mathfrak{X}$ is α-indecomposable and contains 1, then X is contained in H; if this holds for all $X \in \mathfrak{X}$, then $\langle \mathfrak{X} \rangle = H$ is hyperdefinable.

PROOF: As $SU(G) < \omega^{\alpha+1}$, there must be some maximal $k < \omega$ such that for some $X_1, \ldots, X_{m'} \in \mathfrak{X}$ the set $X_1^{\pm 1} \cdots X_{m'}^{\pm 1} =: Y$ contains a type of SU-rank at least $\omega^\alpha k$. Choose a model $\mathfrak{M}$ and $p \in S_G(\mathfrak{M})$ extending the partial type "$x \in Y$" with $SU(p) = \omega^\alpha k$. Put $H = \text{stab}(p)$; then $S(p) \subseteq YY^{-1}$ and $H = \text{stab}(p) \subseteq (YY^{-1})^2$. By maximality of k, the assumptions of Lemma 5.4.2 are satisfied, and $SU(H) = \omega^\alpha k$. Since k was chosen maximal and $H \subseteq (YY^{-1})^2$, we have $SU(ZH) < SU(H) + \omega^\alpha$ for any finite product Z of sets in $\mathfrak{X}$ or their inverses. Note that by compactness any hyperdefinable $X \subseteq \langle \mathfrak{X} \rangle$ is contained in such a product.

If K is another group contained in a finite product of sets in $\mathfrak{X}$ and their inverses (e.g. if K is an image of H under an automorphism stabilizing $\mathfrak{X}$, or merely $\{X_1, \ldots, X_{m'}\}$), then either $SU(H \cap K) < \omega^\alpha k$, or H and K are commensurate. But the first case cannot happen as it would imply $SU(HK) \geq \omega^\alpha(k+1)$ by the Lascar inequality (5.1), contradicting maximality of k. So H is unique up to commensurativity. ∎

REMARK 5.4.7 By Theorem 4.5.13 we may even find such an H which is invariant under all automorphisms stabilizing $\mathfrak{X}$, or $X_1, \ldots, X_{m'}$. If all

$X \in \mathfrak{X}$ are G-invariant subsets, or if $\mathfrak{X}$ is G-invariant, we can also get a normal subgroup H.

Recall that a group is *simple* if it has no normal subgroups. Simplicity of a group has nothing to do with simplicity of a theory, and is an unfortunate clash of terminology.

DEFINITION 5.4.8 A group G is *definably simple* if it has no definable normal subgroups; it is *hyperdefinably simple* if it has no hyperdefinable normal subgroups.

Here are two related results:

PROPOSITION 5.4.9 *Suppose G is a hyperdefinable, hyperdefinably simple non-Abelian group in a simple theory, with $SU(G) < \infty$. Then G is simple, and $SU(G) = \omega^\alpha n$ for some ordinal α and some $n < \omega$.*

PROOF: If $SU(G) = \omega^\alpha n + \beta$ for some ordinal α, some non-zero $n < \omega$, and some $\beta < \omega^\alpha$, then Proposition 5.4.3 implies that there is a hyperdefinable normal subgroup H of G with $SU(H) = \omega^\alpha n$; by hyperdefinable simplicity we have $H = G$ and $SU(G) = \omega^\alpha n$. Furthermore, $G = G^0_A$ for any set A.

CLAIM. If $g \in G$ is non-trivial, then $SU(g^G) \geq \omega^\alpha$.

PROOF OF CLAIM: Suppose $SU(g^G) < \omega^\alpha$. Since g^G is in definable bijection with $G/C_G(g)$, the Lascar inequality (5.1) implies $SU(C_G(g)) = \omega^\alpha n$, and $C_G(g)$ is commensurate with G. Therefore $\tilde{Z}(G)$ is non-trivial (it contains g) normal, and must be equal to G. So G' is bounded by Proposition 4.4.10.3, and centralized by G (but not necessarily hyperdefinable). Therefore $[g, G]$ is a normal central hyperdefinable subgroup of G for all $g \in G$, and non-trivial for $g \notin Z(G)$; by hyperdefinable simplicity $G = Z(G)$. ■

Now if N is a normal subgroup of G and $1 \neq n \in N$, then by Theorem 5.4.5 there is a normal (as n^G is normal) hyperdefinable subgroup $H \leq \langle n^G \rangle \leq N$ of G with $SU(n^G H/H) < \omega^\alpha$. As $SU(n^G) \geq \omega^\alpha$, H must be non-trivial; since $H \leq N$, hyperdefinable simplicity implies $H = N = G$ and G is simple. ■

PROPOSITION 5.4.10 *Suppose G is a definable, definably simple non-Abelian group in a simple theory, with $SU(G) < \infty$. Then G is simple, and $SU(G) = \omega^\alpha n$ for some ordinal α and some $n < \omega$.*

PROOF: Suppose $SU(G) = \omega^\alpha n + \beta$ with $\beta < \omega^\alpha$.

CLAIM. $SU(g^G) \geq \omega^\alpha$ for any $g \neq 1$ in G.

PROOF OF CLAIM: Suppose $SU(g^G) < \omega^\alpha$. As g^G and $G/C_G(g)$ are in definable bijection, we get $SU(G/C_G(g)) < \omega^\alpha$. By Proposition 4.2.7 applied to the family of G-conjugates of $C_G(g)$ there is a definable normal subgroup N commensurable with some intersection of the form $C_G(g)^{g_0} \cap \cdots \cap C_G(g)^{g_n}$. Now $SU(G/C_G(g)^{g_i}) = SU(G/C_G(g))$ for all $i \leq n$; as any coset $h \cdot \bigcap_{i \leq n} C_G(g)^{g_i}$ is definable over the tuple $(hC_G(g)^{g_0}, \ldots, hC_G(g)^{g_n})$,

$$SU(G/N) = SU(G/\bigcap_{i \leq n} C_G(g)^{g_i}) \leq \bigoplus_{i \leq n} SU(G/C_G(g)^{g_i}) < \omega^\alpha.$$

So N is non-trivial and must equal G by definable simplicity. Hence $C_G(g)$ has finite index in G, and the intersection of the G-conjugates of $C_G(g)$ is a definable normal subgroup of finite index in G. Again by definable simplicity, $C_G(g) = G$. So $g \in Z(G)$, which is a definable normal subgroup and thus trivial. ■

CLAIM. G has no type-definable normal subgroup.

PROOF OF CLAIM: Choose a (non-trivial) type-definable normal subgroup N of minimal SU-rank possible. For nontrivial $n \in N$ and $k < \omega$ put $X_k = (n^G \cup n^{-G})^k$. Then X_k contains a type-definable normal subgroup N_1 for big $k < \omega$ by Theorem 5.4.5; by minimality the index of N_1 in N must be bounded. By compactness finitely many translates of X_k must cover N, and the sub-union of those translates by elements in N must be equal to N, contradicting definable simplicity. ■

Finally, if N is a non-trivial normal subgroup of G and $1 \neq n \in N$, then $(n^G \cup n^{-G})^k$ contains a non-trivial type-definable normal subgroup of G for big $k < \omega$ by Theorem 5.4.5, which must be proper since it is contained in N.

It follows that G is simple, and $SU(G) = \omega^\alpha n$ by Proposition 5.4.9. ■

REMARK 5.4.11 A type-definable *stable* group G of SU-rank $\sum_{i \leq k} \omega^{\alpha_i} n_i$ has an Abelian subgroup of SU-rank at least ω^{α_0}.

PROOF: Let H be a type-definable connected subgroup of G of minimal SU-rank $\geq \omega^{\alpha_0}$; recall that connectivity does not depend on the parameter set for a stable group. Then $SU(H) = \omega^{\alpha_0} k$ for some non-zero $k \leq n_0$ by Proposition 5.4.3, and $SU(C_H(h)) < \omega^{\alpha_0}$ for all non-central $h \in H$. It follows from the Lascar inequality (5.1) that

$$SU(h^H) = \omega^{\alpha_0} k = SU(H),$$

so all non-central conjugacy classes are generic. But a stable connected group has a unique generic type by Remark 4.1.24, so two generic conjugacy classes must intersect, and hence be equal: all non-central elements of H are conjugate.

If H is not Abelian, consider $\bar{H} = H/Z(H)$. This has a unique non-trivial conjugacy class; in particular it cannot be Abelian either. So it does not have exponent 2. If $h \in \bar{H}$ is non-trivial and $g \in \bar{H}$ is such that $h^g = h^{-1}$, then $h \notin C_{\bar{H}}(g) < C_{\bar{H}}(g^2) \ni h$. But g and g^2 are conjugate, so $C_{\bar{H}}(g)$

and $C_{\tilde{H}}(g^2)$ are conjugate; we thus obtain an infinite increasing chain of centralizers, contradicting stability. ∎

The above argument, easy as it is, heavily depends on stability; it is puzzling that a corresponding result has not yet been shown for a supersimple group. The easiest case ought to be the following:

PROBLEM 5.4.12 If G is a hyperdefinable group in a simple theory of SU-rank 1, does G have a hyperdefinable bounded-by-Abelian subgroup A of bounded index?

REMARK 5.4.13 If P is an $\emptyset$-invariant family of types, all the results of this section generalize to groups G with $0 < SU_P(G) < \infty$.

5.5 TYPE-DEFINABLE SUPERSIMPLE GROUPS

Given a definable group G in a supersimple theory, a type-definable subgroup H induces a type-definable equivalence relation E via $xEy \Leftrightarrow x^{-1}y \in H$. According to Theorem 5.3.1, on every complete type $p \in S_G$ this equivalence relation is the intersection of definable equivalence relations; however it does not follow that there is a sequence $(E_i : i < \alpha)$ of equivalence relations such that E is equivalent to $\bigwedge_{i<\alpha} E_i$ on the whole group G. If this were indeed the case, we could put $H_i = \{g \in G : \forall x \in G\; xgE_ix\}$; it is easy to check that H_i is a definable subgroup of G, and $\bigcap_{i<\alpha} H_i = H$. So H would be the intersection of definable subgroups of G. More generally, the question arises whether a type-definable group (not necessarily a subgroup of a definable group) in a supersimple theory is the intersection of definable groups. This is indeed the case, and constitutes the main result of this section.

REMARK 5.5.1 A type-definable group in a *stable* theory is the intersection of definable groups.

PROOF: Suppose G is given as $\bigwedge_{i<\alpha} X_i$ for a family $(X_i : i < \alpha)$ of definable sets closed under finite conjunctions; we may assume by compactness that products are well-defined and associative on X_0 (but may go outside X_0), and for every $x \in X_0$ there is a unique two-sided inverse $x^{-1} \in X_0$. For every $i < \alpha$ put

$$Y_i = \{x \in X_0 : \bigwedge_j d_{p_j}y\; yx \in X_i\},$$

where $d_{p_j}y$ runs through the finitely many generic definitions for the formula $yx \in X_i$ given by Proposition 4.1.25. Then Y_i is definable for all $i < \alpha$; if $x \in \bigwedge_{i<\alpha} Y_i$, then $yx \in \bigwedge_{i<\alpha} X_i = G$ for all generic $y \in G$, whence $x \in G$. Conversely, it follows immediately from the definition that $Y_i \supseteq G$. So $\bigwedge_{i<\alpha} Y_i$ also type-defines G; by compactness there is $i_0 < \alpha$ such that $Y_{i_0}^2 \subseteq X_0$. Note that $(Y_i : i < \alpha)$ is also closed under finite conjunctions; we may thus assume that $Y_i^2 \subseteq X_0$ for all $i < \alpha$. Put

$$Z_i = \{x \in Y_i : \forall\, y \in Y_i\ xy \in Y_i\}.$$

If $x \in G, y \in Y_i$ and $g \in G$ is generic over x, y, then gx is generic for G over y, so $(gx)y \in X_i$. By associativity $g(xy) = (gx)y \in X_i$ and $xy \in Y_i$, so $x \in Z_i$ and $Z_i \supseteq G$. On the other hand, if $x, x' \in Z_i$, then $xx' \in Y_i$, so Z_i is easily seen to be closed under multiplication. It follows that the set G_i of invertible elements of Z_i forms a definable super-group of G, with $G = \bigcap_{i<\alpha} G_i$. ∎

Recall that a group is *radicable* if every element has an n-th root for all $n < \omega$.

COROLLARY 5.5.2 *A type-definable stable radicable group is connected.*

PROOF: If H is a relatively definable subgroup of finite index in a radicable group G, then G/H is a finite radicable group, and must be trivial. As a type-definable group of bounded index is the intersection of relatively definable groups of finite index by Remark 5.5.1, the assertion follows. ∎

Let us return to simple theories.

LEMMA 5.5.3 *Suppose G is a type-definable group over $\emptyset$ in a simple theory, with $SU(G) = \omega^\alpha n$ and $D(G) < \infty$. Then there are a definable super-group G_0 of G and definable subgroups G_i of G_0, with $G = \bigcap_i G_i$.*

PROOF: Clearly we may assume that we work over a model $\mathfrak{M}$ of the ambient theory. By compactness there is a definable superset X_0 of G with $D(X_0) = D(G)$, such that multiplication is defined and associative on X_0 (but may go outside) and containing, for every $x \in X_0$, a unique two-sided inverse x^{-1}. Again by compactness, there is some definable set $X_1 \supseteq G$ with $X_1^2 \subseteq X_0$.

CLAIM. $SU(X_1) = \omega^\alpha n$.

PROOF OF CLAIM: Suppose X_1 contains a type p with $SU(p) > \omega^\alpha n$. Let $(x_i : i \in I)$ be a Morley sequence in p and consider the sets x_iG, for $i \in I$. Since $X_1^2 \subseteq X_0$, they are all contained in X_0. On the other hand, since $SU(x_i/x_j) > SU(G)$ for $i \neq j$, we get $x_i \notin x_jG$; by compactness and indiscernibility there is a definable superset X' of G contained in X_1 such that $x_iX' \cap x_jX' = \emptyset$ for all $i \neq j$. It follows that $D(X') < D(X_0)$, contradicting $D(G) \leq D(X') < D(X_0) = D(G)$. ∎

Let X_2 be a definable superset of G closed under inverse, with $X_2^2 \subseteq X_1$.

CLAIM. Given a formula $\varphi(x, y)$ with $\varphi(x, y) \vdash x \in X_2$, there is a formula $\vartheta(y)$ over $\mathfrak{M}$ such that $\models \vartheta(m)$ if and only if $\varphi(x, m)$ contains a type of SU-rank $\omega^\alpha n$.

PROOF OF CLAIM: We show that the condition "$\varphi(x, m)$ contains a type of SU-rank $\omega^\alpha n$" is type-definable over $\mathfrak{M}$, and that its negation is type-definable over $\mathfrak{M}$ as well.

Let $p \in S_G(\mathfrak{M})$ be a generic type for G. If $\varphi(x,m)$ contains a type q with $SU(q) = \omega^\alpha n$, consider a realization $x_0 \models q$ and $g \models p$ with $g \mathop{\smile\!\!\!\!\!\mid}_{\mathfrak{M}} x_0, m$. Then $g^{-1}x_0 \models \varphi(gx,m)$, and

$$\begin{aligned}\omega^\alpha n = SU(X_1) &\geq SU(g^{-1}x_0/\mathfrak{M},m) \geq SU(g^{-1}x_0/\mathfrak{M},m,g)\\ &= SU(x_0/\mathfrak{M},m,g) = SU(x_0/\mathfrak{M},m) = SU(q) = \omega^\alpha n,\end{aligned}$$

whence $g^{-1}x_0 \mathop{\smile\!\!\!\!\!\mid}_{\mathfrak{M},m} g$ and $\varphi(gx,m)$ does not fork over $\mathfrak{M}\cup\{m\}$. Conversely, suppose $\varphi(gx,m)$ does not fork over $\mathfrak{M}\cup\{m\}$ for some $g \models p$ with $g \mathop{\smile\!\!\!\!\!\mid}_{\mathfrak{M}} m$. Then there is $x_0 \models \varphi(gx,m)$ with $x_0 \mathop{\smile\!\!\!\!\!\mid}_{\mathfrak{M},m} g$, and

$$\begin{aligned}SU(gx_0/\mathfrak{M},m) &\geq SU(gx_0/\mathfrak{M},m,x_0) = SU(g/\mathfrak{M},m,x_0)\\ &= SU(g/\mathfrak{M},m) = SU(g/\mathfrak{M}) = \omega^\alpha n.\end{aligned}$$

Since $gx_0 \models \varphi(x,m)$, the formula $\varphi(x,m)$ contains a type of SU-rank $\omega^\alpha n$. But the condition "$\exists\, g \models p\,[g \mathop{\smile\!\!\!\!\!\mid}_{\mathfrak{M}} m \wedge \varphi(gx,m)$ does not fork over $\mathfrak{M}\cup\{m\}]$" is type-definable over $\mathfrak{M}$, as we only have to express that there is an independent sequence $(g_i : i<\omega)$ in p with $(g_i : i<\omega) \mathop{\smile\!\!\!\!\!\mid}_{\mathfrak{M}} m$ and indiscernible over $\mathfrak{M}\cup\{m\}$, such that $\bigwedge_{i<\omega}\varphi(g_i x,m)$ is consistent, which we can do by Lemma 2.3.15. Hence there is a partial type $\pi(y)$ such that $\models \pi(m)$ if and only if $\varphi(x,m)$ contains a type of SU-rank $\omega^\alpha n$.

Now consider the following condition: "there is an independent sequence $(g_i : i<\omega)$ in p with $(g_i : i<\omega) \mathop{\smile\!\!\!\!\!\mid}_{\mathfrak{M}} m$ and indiscernible over $\mathfrak{M}\cup\{m\}$, such that $\bigwedge_{i<n}\varphi(g_i x,m)$ is inconsistent". This is again type-definable by Lemma 2.3.15. Suppose it is true of m, as witnessed by some sequence $(g_i : i<\omega)$. Then $\varphi(g_0x,m)$ forks over $\mathfrak{M}\cup\{m\}$, and the formula $\varphi(x,m)$ cannot contain a type of SU-rank $\omega^\alpha n$.

Conversely, suppose $\varphi(x,m)$ does not contain a type of SU-rank $\omega^\alpha n$, and consider an independent sequence $(g_i : i<\omega)$ in p with $(g_i : i<\omega) \mathop{\smile\!\!\!\!\!\mid}_{\mathfrak{M}} m$ and indiscernible over $\mathfrak{M}\cup\{m\}$. Then $\varphi(g_0x,m)$ forks over $\mathfrak{M}\cup\{m\}$, so $\bigwedge_{i<\omega}\varphi(g_ix,m)$ is inconsistent. If $\bigwedge_{i<n}\varphi(g_ix,m)$ is consistent and contains an element x_0, then $x_0 \not\mathop{\smile\!\!\!\!\!\mid}_{\mathfrak{M},m} g_i$ for all $i \leq n$. Since $g_i \mathop{\smile\!\!\!\!\!\mid}_{\mathfrak{M}} (m, g_j : j<i)$ and hence

$$SU(g_i/\mathfrak{M},m,g_j : j<i) = SU(g_i/\mathfrak{M}) = \omega^\alpha n,$$

we get

$$SU(g_i/\mathfrak{M},m,g_j : j<i) \geq SU(g_i/\mathfrak{M},m,x_0,g_j : j<i) + \omega^\alpha,$$

whence by the Lascar inequality 5.1.6

$$SU(x_0/\mathfrak{M},m,g_j : j<i) \geq SU(x_0/\mathfrak{M},m,g_j : j\leq i) + \omega^\alpha$$

for all $i \leq n$, and $SU(x_0/\mathfrak{M}, m) \geq \omega^\alpha(n+1)$, contradicting $SU(x_0/\mathfrak{M}, m) \leq SU(X_1) = \omega^\alpha n$. It follows that $\bigwedge_{i<n} \varphi(g_i x, m)$ is inconsistent.

Hence there is a partial type $\pi'(y)$ such that $\models \pi'(m)$ if and only if $\varphi(x, m)$ does not contain a type of SU-rank $\omega^\alpha n$. By compactness, we find the required formula ϑ. ∎

Now consider a definable superset X of G closed under inverse, with $X^4 \subseteq X_2$. By the previous claim there is a formula $\vartheta(x, y)$ over $\mathfrak{M}$ such that $\models \vartheta(x, y)$ if and only if $x \in X \wedge y \in X^3$ and $xX \cap yX$ contains a type of SU-rank $\omega^\alpha n$.

CLAIM. ϑ is a stable formula.

PROOF OF CLAIM: Let $(a_i, b_i : i < \omega)$ be an $\mathfrak{M}$-indiscernible sequence such that $\models \vartheta(a_i, b_j)$ if and only if $i \leq j$. Then $a_0 X \cap b_0 X$ contains a type $p(x, a_0, b_0)$ of SU-rank $\omega^\alpha n$, and clearly $p(x, a_0, b_0)$ does not fork over $\mathfrak{M}$. But then $\bigwedge_{i<\omega} p(x, a_i, b_i)$ does not fork over $\mathfrak{M}$ by Theorem 2.5.4 and must have SU-rank $\omega^\alpha n$, so $\models \vartheta(a_i, b_j)$ for all $i, j < \omega$, a contradiction. ∎

Put $Y = \{y \in X^3 : \bigwedge_i d_{p_i} x\, \vartheta(x, y)\}$, where d_{p_i} runs through the finitely many generic ϑ-definitions for G given by Proposition 4.1.25. For all $x, y \in G$ we have $xX \cap yX \supseteq G$, so $\vartheta(x, y)$ holds and $Y \supseteq G$. On the other hand, if $y \in Y$, $g \in G$, and h is generic for G over $\mathfrak{M} \cup \{y, g\}$, then $g^{-1}h$ is generic for G over $\mathfrak{M} \cup \{y, g\}$; moreover $hX \cap gyX$ contains a type of SU-rank $\omega^\alpha n$ if and only if $g^{-1}hX \cap yX$ does, and $hX \cap gyX \neq \emptyset$ implies $gy \in X^3$. Therefore $\models \vartheta(h, gy)$ for all generic h, and $gy \in Y$. So

$$Z = \{x \in Y : \forall y \in Y\ xy \in Y\}$$

is a definable superset of G closed under multiplication, and the set G_X of invertible elements of Z forms a definable super-group G_X of G. Finally, as $G_X \subseteq Z \subseteq Y \subseteq X^3$, we have that $\bigcap_{X \supseteq G} G_X = G$. ∎

THEOREM 5.5.4 *A type-definable group G in a supersimple theory is the intersection of definable groups.*

PROOF: Clearly, we may assume that the ambient model is sufficiently saturated. By compactness, there is a definable superset X of G on which multiplication is defined (but may go outside) and is associative (whenever all partial products are in X), and such that any $x \in X$ has a two-sided inverse in X. We use induction on $SU(G)$. Suppose $SU(G) = \omega^\alpha n + \beta$ with $\beta < \omega^\alpha$. If $\beta = 0$, we are done by Lemma 5.5.3. If $\beta > 0$, Proposition 5.4.3 yields a type-definable normal subgroup N of G with $SU(N) = \omega^\alpha n$, and N is the intersection of definable groups by Lemma 5.5.3, say $N = \bigcap_{i \in I} H_i$ for some definable groups H_i. As $\bigcap_{i \in I} H_i = N \trianglelefteq G$, we may fix a definable group $H_0 \geq N$ such that $GH_0 \subseteq X$, and assume $H_i^g \subseteq H_0$ for all $i > 0$ by compactness. By supersimplicity we may further assume $SU(H_i) = SU(N)$

for all $i > 0$, so

$$SU(H_i^g) = SU(N) \leq SU(H_i^g \cap H_i^{g'})$$

for all $g, g' \in G$. Hence for all $i > 0$ any two G-conjugates of H_i are commensurate subgroups of H_0, and thus commensurable by definability; by saturation and compactness this commensurability must be uniform.

Pick any $i > 0$. By Theorem 4.2.4 applied to the family of G-conjugates of H_i in $\langle H_i^g : g \in G \rangle \leq H_0$ there is a G-invariant definable group H commensurable with H_i and contained in H_0. Then $N_X(H)/H$ is a definable (imaginary) set containing the type-definable (imaginary) group GH/H. But $GH/H \cong G/(G \cap H)$; since $SU(G \cap H) = SU(N) = \omega^\alpha n$, we get

$$SU(G/(G \cap H)) = SU(G/N) = \beta < SU(G)$$

by the Lascar inequality (5.1). By inductive hypothesis there are definable groups $\bar{G}_j \subseteq N_X(H)/H$ for $j \in J$ with $\bigcap_{j \in J} \bar{G}_j = GH/H$. Clearly the pre-image G_0 of $\bar{G}_0$ in X is a definable group with $G \leq G_0 \subseteq X$. Therefore every definable superset of G contains a definable super-group of G; it follows that G is the intersection of definable groups. ■

COROLLARY 5.5.5 *A type-definable radicable supersimple group or a supersimple division ring of characteristic zero is (additively) connected.*

PROOF: As Corollary 5.5.2, noting that the additive group of a field of characteristic zero is divisible. ■

COROLLARY 5.5.6 *A type-definable division ring F in a supersimple theory is definable.*

PROOF: If F is finite, the assertion is obvious. We may hence assume that F is infinite. By Theorem 5.5.4 the additive and the multiplicative group of F are both intersections of definable groups. By supersimplicity and compactness there is a definable multiplicative group M containing $F^\times$ and a definable additive group $A \subseteq M \cup \{0\}$ containing F^+ with $SU(A) = SU(F)$, such that the distributive laws hold on A. By compactness again, there is a definable additive group A' containing F^+ such that $kA' \subseteq A$ for all $k \in F^\times$. Then for every $k \in F^\times$ the group kA' is an isomorphic image of A' inside A; since

$$SU(A) \geq SU(A' + kA') \geq SU(A') = SU(kA') \geq SU(F) = SU(A),$$

this means that $A' + kA'$ is commensurate with A' and kA', and hence commensurable by definability of A'. By compactness A' is uniformly commensurable with its $F^\times$-translates; by Theorem 4.2.4 (applied to the family of $F^\times$-translates of A' inside $\langle kA' : k \in F^\times\rangle \leq A$) there is a definable $F^\times$-invariant subgroup A_0 of A commensurable with A' containing $\bigcap_{k\in F^\times} kA'$ and hence F. Let $F_0 = \{m \in A_0 : mA_0 \leq A_0\}$, a definable subgroup of A_0 containing F. Then F_0 is a ring; since $F_0^\times \leq M$, it has no zero divisors. But then for any non-zero $a \in F_0$ the sequence $(a^i F_0 : i < \omega)$ cannot descend infinitely, as otherwise the formula $xA_0 < yA_0$ would define a partial order with infinite chains, contradicting simplicity. Hence $a^i F_0 = a^{i+j} F_0$ for some $i \geq 0, j > 0$, and $a^i(1 - a^j k) = 0$ for some $k \in F_0$. It follows that a is invertible, and F_0 is a division ring.

Starting with a smaller A, we see that F is the intersection of definable division rings F_i; as the additive index $|F_i^+ : F_j^+|$ is infinite for any $F_i > F_j$ and therefore $SU(F_i) > SU(F_j)$, supersimplicity yields $F_i = F_j$, whence $F = F_0$. ■

However, the precise relation between Theorem 5.3.1 and Theorem 5.5.4 remains mysterious, as elimination of hyperimaginaries only implies that on every complete type a type-definable equivalence relation is the intersection of definable equivalence relations. If E is given as $x^{-1}y \in H$ for some type-definable subgroup H of a definable group G in a simple theory with elimination of hyperimaginaries, varying over all the types in G it seems difficult to glue the resulting equivalence relations together definably.

PROBLEM 5.5.7 If H is a type-definable subgroup of a definable group in a simple theory with elimination of hyperimaginaries, is it the intersection of definable groups?

Even in a supersimple theory there remain open questions:

PROBLEM 5.5.8 If G is a hyperdefinable group in a supersimple theory, are there definable groups $(G_i : i \in I)$ and $(H_j : j \in J)$ with $G = \bigcap_{i\in I} G_i / \bigcap_{j\in J} H_j$?

This is true for a hyperdefinable group in a stable theory.

5.6 SUPERSIMPLE DIVISION RINGS

In this section, we shall study the structure of a supersimple division ring. In particular, we shall prove that a supersimple division ring is commutative of monomial SU-rank, and has only boundedly many extensions of degree n for all $n < \infty$.

PROPOSITION 5.6.1 *A hyperdefinable division ring of ordinal SU-rank in a simple theory has SU-rank of the form $\omega^\alpha n$ for some ordinal α and some $n < \omega$.*

PROOF: If D is a hyperdefinable division ring in a simple theory with $SU(D) = \omega^\alpha n + \beta$ for some $n > 0$ and $0 < \beta < \omega^\alpha$, Proposition 5.4.3

yields a type-definable additive subgroup H of D^+ with $SU(H) = \omega^\alpha n$. If $d \in D^\times$ and H and dH are not commensurate, then $SU(H \cap dH) < \omega^\alpha n$, so $SU(H) \geq SU(H \cap dH) + \omega^\alpha$. Therefore $SU(H/(H \cap dH)) \geq \omega^\alpha$ by the Lascar inequality (5.1), whence

$$\begin{aligned} SU(H + dH) &\geq SU(H) + SU((H + dH)/dH) \\ &= SU(H) + SU(H/(H \cap dH)) \geq \omega^\alpha(n+1) > SU(D), \end{aligned}$$

a contradiction. Hence H is commensurate with all its $D^\times$-translates. As $D^\times$ acts on D^+ by multiplication, Theorem 4.5.13 applied to the family of $D^\times$-translates of H implies that we may replace H by a commensurate group invariant under multiplication by elements in $D^\times$. Then H is an ideal in D of SU-rank $\omega^\alpha n > 0$, thus non-trivial, and must be the whole of D. ■

Recall that a field of characteristic $p > 0$ is *perfect* if every element has a p-th root. If K is perfect, so is every finite extension of K.

LEMMA 5.6.2 *Let K be an unbounded hyperdefinable field over $\emptyset$ in a simple theory, with $SU(K) < \infty$. Then for each $n < \omega$ the multiplicative group $(K^\times)^n$ of n-th powers has bounded index in $K^\times$. In particular, K is perfect. Moreover, if char$(K) = p > 0$, then the additive subgroup $\{x^p - x : x \in K\}$ has bounded index in K^+.*

PROOF: Note first that both $(K^\times)^n$ and $\{x^p - x : x \in K\}$ are hyperdefinable over $\emptyset$. Let $a \in K$ be generic over $\emptyset$, and put $b = a^n$. Then $a \in \mathrm{acl}(b)$, so $SU(a/b) = 0$. Thus $SU(b) = SU(a)$ by the Lascar inequalities 5.1.6, so $SU((K^\times)^n) \geq SU(b) = SU(a) = SU(K)$. It follows that $(K^\times)^n$ has bounded index in $K^\times$. If char$(K) = p > 0$, then K^p is a hyperdefinable subfield of K. So K is a vector space over K^p; as $SU(K) = SU(K^p)$, the additive index $|K : K^p|$ must be bounded. It follows that $K = K^p$ and K is perfect.

Next consider the additive endomorphism $\sigma : x \mapsto x^p - x$ (in characteristic $p > 0$). This has finite kernel, so for generic $a \in K$ and $b = a^p - a$ we have $a \in \mathrm{acl}(b)$ and $SU(a/b) = 0$, whence $SU(b) = SU(a)$. Therefore the image of σ is an additive subgroup of bounded index. ■

Let us look first at the superstable case.

FACT 5.6.3 [77, Theorem 15.8] *Let D be a division ring, and K a commutative subfield of D such that D is finite-dimensional over K. Then D is finite-dimensional over its centre $Z(D)$; in fact $[D : Z(D)] \leq [D : K]^2$.*

THEOREM 5.6.4 *An infinite type-definable division ring D in a stable theory with $SU(D) < \infty$ is an algebraically closed field.*

PROOF: We prove first that an infinite type-definable commutative field K of ordinal SU-rank is algebraically closed. So suppose not. Since K is perfect, there is a finite extension K' of K which has a finite Galois extension L; we may assume that the degree $[L : K']$ is minimal possible. As K' is interpretable in K and hence also type-definable superstable, we may assume $K = K'$.

The Galois group $G(L/K)$ is finite and has an element σ of order q for some prime number q. If $L^\sigma \geq K$ is the fixed field of σ, then L is also a finite Galois extension of L^σ; minimality yields $L^\sigma = K$ and the extension is cyclic. If ξ is a primitive q-th root of unity (if q is different from the characteristic p of K), then $K(\xi)$ is a Galois extension of degree at most $q-1$; minimality again implies $K = K(\xi)$. But then the extension is *radical*: there is some $a \in L$ with $L = K(a)$ and either $a^q \in K$ (if $q \neq p$), or $a^p - a \in K$ (if $q = p$).

K is connected both additively and multiplicatively by Remark 4.5.8; so $(K^\times)^q = K^\times$ and $\{x^p - x : x \in K\} = K^+$, as these are subgroups of bounded index by Lemma 5.6.2. So if L is a Kummer extension and $a^q \in K$, then a^q has a q-th root in K; since $\xi \in K$, all q-th roots of a^q are in K, whence $a \in K$ and $L = K$, a contradiction. Similarly, if $q = p$ and L is an Artin-Schreier extension, there is $b \in K$ with $b^p - b = a^p - a$; as $b - a$ lies in the prime field, $a \in K$ and $L = K$. Hence K must be algebraically closed.

Next consider an infinite stable division ring D with $SU(D) = \omega^\alpha n$. By Remark 5.4.11 it has an Abelian multiplicative subgroup A with $SU(A) \geq \omega^\alpha$. Put $K = Z(C_D(A))$. Then K is a commutative subfield of D containing A, and $SU(K) \geq \omega^\alpha$. If $d_0, \ldots, d_n \in D$ are linearly independent over K, then

$$SU(d_0K \oplus \cdots \oplus d_nK) = SU(K) \cdot (n+1) \geq \omega^\alpha (n+1) > SU(D),$$

a contradiction. Hence D is finite-dimensional over a commutative subfield; by Fact 5.6.3 it is finite-dimensional over $Z(D)$. So $Z(D)$ is an infinite superstable, whence algebraically closed, field. But $Z(D)(d)$ is a finite-dimensional extension of $Z(D)$ for any $d \in D$, so $d \in Z(D)$ and D is commutative. ■

Even though a supersimple field need not be algebraically closed, it does have a small absolute Galois group (i.e. with only boundedly many closed subgroups of every finite index).

THEOREM 5.6.5 *Let K be a hyperdefinable field of ordinal SU-rank over $\emptyset$ in a supersimple theory. Then for every $n < \omega$ there are only boundedly many extensions of K of degree n (up to K-isomorphism).*

PROOF: Fix $n < \omega$, and suppose $f(x) \in K[x]$ and $g(y) \in K[y]$ are irreducible monic polynomials over K of degree n, such that $K[x]/(f(x))$ and $K[y]/(g(y))$ are isomorphic over K. Then there are polynomials $h(x)$ and $h'(y)$ such that $g(h(x)) \in (f(x))$ and $f(h'(y)) \in (g(y))$; we may reduce $h(x)$ modulo $f(x)$ and $h'(y)$ modulo $g(y)$ and assume that they are both of degree at most $n-1$. But existence of such h and h' is a hyperdefinable condition, which induces an equivalence relation E on the set X of n-tuples

$(a_0, \ldots, a_{n-1}) \in K^n$ such that $\sum_{i<n} a_i x^i + x^n$ is irreducible (for a properly hyperdefinable field, X need not *a priori* be given by a partial type, but is a union of n-types over $\emptyset$). Note that E is hyperdefinable on every complete type in X over $\emptyset$; by abuse of notation we shall identify a monic polynomial of degree n with the tuple of its first n coefficients and consider the relation E on such polynomials.

Given an irreducible monic polynomial $f(x) \in K[x]$ of degree n, consider the hyperdefinable field $L = K[x]/(f(x))$. If $SU(K) = \omega^\alpha m$, then $SU(L) = \omega^\alpha mn$, as L is an n-dimensional vector space over K. If $\alpha \in L$ is generic, there is a monic polynomial $g(y) \in K[y]$ of minimal degree (at most n) such that $g(\alpha) = 0$; since α is algebraic over the coefficients of g, each of which has rank at most $\omega^\alpha m$, it follows that g has degree n and its coefficients are independent generic elements of K. Clearly g is irreducible, whence gEf. However, since f and thus the E-class g_E of g was already fixed, $SU(g/g_E) = \omega^\alpha mn$. As

$$SU(g/g_E) + SU(g_E) \leq SU(g) = \omega^\alpha mn$$

by the Lascar inequalities 5.1.6, we get $SU(g_E) = 0$, and there are only boundedly many equivalence classes. ∎

Next, we shall prove that a supersimple division ring is commutative. In order to do this, we shall prove first that a supersimple field has trivial Brauer group, and then reduce the general case to the case of a division ring which is finite-dimensional over its centre. So we shall start with an explanation of the relevant field-theoretic facts concerning the Brauer group and Galois cohomology.

Let K be a perfect field. A *central simple algebra* over K is a finite dimensional K-algebra A whose centre is K and which has no nontrivial two-sided ideals. If A and B are two central simple K-algebras, so is their tensor product $A \otimes_K B$. We call A and B *equivalent* if for some $m, n < \omega$ the matrix algebras $M_m(A)$ and $M_n(B)$ are isomorphic as K-algebras; this equivalence is respected by the tensor product. The set of equivalence classes forms an Abelian group under the tensor product operation, called the *Brauer group* $\mathrm{Br}(K)$ of K. For any central simple K-algebra A there is a division ring D of finite dimension over its centre $Z(D) = K$ such that A is isomorphic to a matrix algebra over D; moreover D is uniquely determined by the equivalence class of A. Conversely of course, every division ring D of finite dimension over its centre is a central simple $Z(D)$-algebra, so there is a one-one correspondence between division rings of finite dimension over their centre K and $\mathrm{Br}(K)$; the trivial element of $\mathrm{Br}(K)$ corresponds to K itself.

If L is a finite extension of K, then tensoring with L induces a homomorphism from $\mathrm{Br}(K)$ into $\mathrm{Br}(L)$, whose kernel is denoted $\mathrm{Br}(L/K)$. Since for any central simple algebra A over K there is some finite extension L of K (called a *splitting field* for A) such that $A \otimes_K L$ is isomorphic to a matrix algebra over L and thus represents the trivial element of $\mathrm{Br}(L)$,

$$\mathrm{Br}(K) = \bigcup\{\mathrm{Br}(L/K) : \ L \text{ a finite Galois extension of } K\}.$$

If L is a finite Galois extension of K with Galois group G, there is a classical isomorphism between $\mathrm{Br}(L/K)$ and the Galois cohomology group $H^2(G, L^\times)$: given a 2-cocycle $f : G \times G \to L^\times$, define a K-algebra structure on the L-vector space A with basis $\{u_g : g \in G\}$ by: $u_g \cdot u_{g'} = f(g, g')u_{gg'}$, and $u_g x = g(x)u_g$ for $g \in G$ and $x \in L$. This turns A into a central simple K-algebra A_f; the map $f \to A_f$ is the required isomorphism.

FACT 5.6.6 [138, X.6] *If $K < L < L'$ is a sequence of finite Galois extensions, then there is an exact sequence*

$$0 \to H^2(Gal(L/K), L^\times) \to H^2(Gal(L'/K), L'^\times) \to H^2(Gal(L'/L), L'^\times).$$

FACT 5.6.7 [138, IX] *Let G be a finite group and A a G-module. Fix $n \geq 1$, and suppose that $H^n(G_p, A) = 0$ for all primes p, where G_p is a Sylow p-subgroup of G. Then $H^n(G, A) = 0$.*

Recall that if L is a finite Galois extension of K, the *norm* map $N_{L/K} : L^\times \to K^\times$ maps any $a \in L^\times$ to the product of its $Gal(L/K)$-conjugates.

FACT 5.6.8 [61, Theorem 8.14] *If L is a cyclic extension of K, then $H^2(Gal(L/K), L^\times)$ is isomorphic to $K^\times/N_{L/K}(L^\times)$.*

LEMMA 5.6.9 *The Brauer group of every finite extension of a perfect field K is trivial if and only if for any finite extension L of K and Kummer extension L' of L the norm map $N_{L'/L}$ is surjective.*

PROOF: If $\mathrm{Br}(L)$ is trivial for every finite extension L of K, so is $\mathrm{Br}(L'/L)$ for all Kummer extensions L' of L; surjectivity of the norm map $N_{L'/L}$ follows from Fact 5.6.8.

Conversely, suppose there is a finite extension L of K and a Galois extension L' of L such that $\mathrm{Br}(L'/L)$ is non-trivial; we choose it such that the degree $[L' : L]$ is minimal. By Facts 5.6.6 and 5.6.7 we may assume that $Gal(L'/L)$ is a simple p-group for some prime p, and hence cyclic of order p. If p is the characteristic of K, then the extension is an Artin-Schreier extension and cyclic; since L is perfect, even $N_{L'/L}(L^\times) = L^\times$ and $H^2(Gal(L'/L), L'^\times)$ is trivial by Fact 5.6.8.

If p is different from the characteristic of K, let ξ be a primitive p-th root of unity. As $[L(\xi) : L] < p = [L' : L]$, minimality yields $\mathrm{Br}(L(\xi)/L) = \{0\}$. But $L'(\xi)$ is a Kummer extension of $L(\xi)$, so $\mathrm{Br}(L'(\xi)/L(\xi)) = \{0\}$ by surjectivity of the norm map $N_{L'(\xi)/L(\xi)}$, whence $\mathrm{Br}(L'(\xi)/L) = \{0\}$ by Fact 5.6.6, and finally $\mathrm{Br}(L'/L) = \{0\}$. ■

PROPOSITION 5.6.10 *Let K be an unbounded hyperdefinable field in a simple theory, with $SU(K) < \infty$. Then* $\mathrm{Br}(K)$ *is trivial.*

PROOF: K is perfect by Lemma 5.6.2; since a finite extension of K is again hyperdefinable unbounded of ordinal SU-rank, Lemma 5.6.9 implies that it is enough to show that for any Kummer extension L of K the norm map $N_{L/K}$ is surjective. Suppose $L = K(a)$ for some a with $a^p \in K$. Then $\{1, a, a^2, \ldots, a^{p-1}\}$ is a basis for L as a vector space over K. An easy computation shows that for any $x_1, x_2 \in K$ we have $N_{L/K}(x_1 + x_2 a) = x_1^p + a x_2^p$ (if $p \neq 2$), or $N_{L/K}(x_1 + a x_2) = x_1^p - a x_2^p$ (if $p = 2$). Let T be the multiplicative subgroup of the p-th powers of the elements in $K^\times$. Then T is hyperdefinable and has bounded index in K by Lemma 5.6.2, so $T \pm aT = K^\times$ by Corollary 4.5.7, and the norm map is surjective. ∎

COROLLARY 5.6.11 *A finite dimensional division algebra over an unbounded hyperdefinable field K of ordinal SU-rank in a simple theory is commutative.*

PROOF: K is a finite extension of the centre Z of the division algebra by Fact 5.6.3. So Z is an unbounded hyperdefinable field and has trivial Brauer group by Proposition 5.6.10. ∎

We now have everything to show that a type-definable supersimple division ring is commutative.

THEOREM 5.6.12 *A type-definable division ring D of ordinal SU-rank is commutative.*

PROOF: As a finite division ring is a finite field, we may assume that D is infinite. Since centralizers are type-definable division subrings of D, for any $A \subseteq B$ either $C_D(A)$ is finite, $C_D(A) = C_D(B)$, or $C_D(B)$ has infinite index in $C_D(A)$ and $SU(C_D(A)) > SU(C_D(B))$. It follows that the descending (and thus the ascending as well by compactness) chain condition on centralizers holds and $Z(D)$ is relatively definable. Replacing D by a suitable centralizer, we may assume that $C_D(a)$ is either commutative or finite (and hence commutative) for every $a \in D - Z(D)$.

By Proposition 5.6.1 there is some ordinal α and non-zero $n < \omega$ with $SU(D) = \omega^\alpha n$; moreover $SU(C_D(a)) < \omega^\alpha$ for any $a \in D - Z(D)$ since D cannot be finite-dimensional over a commutative subfield by Corollary 5.6.11. As $SU(a^G) = SU(D/C_D(a))$ for any $a \in D - Z(D)$, the Lascar inequality (5.1) yields $SU(a^D) = SU(D)$: every non-central conjugacy class is generic. But belonging to the same conjugacy class is a type-definable equivalence relation, so there can only be boundedly many non-central conjugacy classes. Similarly the additive subgroup $\{x^a - x : x \in D\}$ is generic for $a \in D - Z(D)$ and has bounded index in D^+.

If every element of $D^\times$ has finite order modulo $Z(D)$, then D is commutative by Kaplanski's Theorem [77, Theorem 5.15], a contradiction. So there is an element $a \in D - Z(D)$ of infinite (multiplicative) order modulo $Z(D)$; since $C_D(a^n) \geq C_D(a)$ for any $n < \omega$, we may assume that $C_D(a)$ is maximal possible by the chain condition on centralizers.

CLAIM. D has characteristic 0.

PROOF OF CLAIM: Suppose $\mathrm{char}(D) = p > 0$. Since $C_D(a)$ is infinite and $\{x^a - x : x \in D\}$ has bounded index in D^+, there is non-zero $a' \in C_D(a)$ of the form $b^a - b$ for some $b \in D$. Clearly $b \notin C_D(a)$; since $b^{a^p} = b + pa' = b$, we get $C_D(a^p) > C_D(a)$, contradicting maximality of $C_D(a)$. ∎

CLAIM. If $c^2 \in Z(D)$ for some $c \in D$, then $c \in Z(D)$.

PROOF OF CLAIM: Suppose $c \notin Z(D)$ and consider the endomorphism $\sigma : x \mapsto x^c + x$ from D to $C_D(c)$. Since $SU(\mathrm{im}(\sigma)) \leq SU(C_D(c)) < \omega^\alpha$, the kernel of σ must be generic and of bounded index by the Lascar inequality (5.1). As $Z(D)$ is infinite, there is non-zero $z \in Z(D) \cap \ker(\sigma)$, whence $z^c + z = z + z = 0$, contradicting $\mathrm{char}(D) = 0$. ∎

CLAIM. There is some $c \in Z(D)$ without a square root in D.

PROOF OF CLAIM: Suppose otherwise. Using the previous claim, we obtain a sequence $-1 = \xi_0, \xi_1, \ldots$ of elements in $Z(D)$ with $\xi_{i+1}^2 = \xi_i$ for all $i < \omega$. By compactness there is a definable superset X of D such that X is closed under multiplicative inverse, the distributive laws hold, and products of six elements in X exist (not necessarily inside D) and are associative. As D has only boundedly many non-central conjugacy classes, there is some $n < \omega$ such that among any n elements in $D - Z(D)$ at least two are conjugate by some element in X.

Take any $b \in D - Z(D)$. By Ramsey's Theorem there must be an infinite sequence $I = (i_0, i_1, \ldots) \subseteq \omega$ such that all elements $\{\xi_i b : i \in I\}$ are conjugate under X. For any $j < \omega$ fix $x_j \in X$ with $(\xi_{i_0} b)^{x_j} = \xi_{i_j} b$. Then for any $k > i_0$

$$(b^{2^k})^{x_j} = ((\xi_{i_0} b)^{2^k})^{x_j} = ((\xi_{i_0} b)^{x_j})^{2^k} = (\xi_{i_j} b)^{2^k} = \xi_{i_j - k} b^{2^k}$$

(where $\xi_s = 1$ for $s < 0$). It follows that $b^{2^k} \in C_D(x_j)$ if and only if $k > i_j$, and we obtain a descending chain $C_D(x_0) > C_D(x_0, x_1) > C_D(x_0, x_1, x_2) > \cdots$. However, by distributivity on X these centralizers are division subrings of D, as the SU-rank must decrease every time, we obtain a contradiction. ∎

CLAIM. If $b \in D - Z(D)$, then $Z(D) \subseteq b^D - b$.

PROOF OF CLAIM: As $b + Z(D)$ is covered by boundedly many non-central conjugacy classes, one of them, say $d^D \cap (b + Z(D))$, is generic in $b + Z(D)$. We may assume that $d \in b + Z(D)$, so $d^D - d$ is generic in $Z(D)$. If

$d^x - d = z \in Z(D)$ and $d^y - d = z' \in Z(D)$ for some $x, y \in D$, then

$$d^{xy^{-1}} - d = (d^x - d)^{y^{-1}} + d^{y^{-1}} - d = z^{y^{-1}} - (d^y - d)^{y^{-1}} = z - z',$$

so $(d^D - d) \cap Z(D)$ is an additive generic subgroup of $Z(D)$, say A. For any $z \in Z(D)$ we have

$$zA = z[(d^D) - d] \cap Z(D) = [(zd)^D - zd] \cap Z(D).$$

So if zd and $z'd$ are conjugate, say $(zd)^x = z'd$ for some $x \in D$, then

$$zA = (zA)^x = [(zd)^{Dx} - (zd)^x] \cap Z(D) = [(z'd)^D - z'd] \cap Z(D) = z'A;$$

as there are only boundedly many conjugacy classes in $Z(D)d$, the intersection of all $Z(D)$-translates of A is a bounded intersection, and hence an ideal in $Z(D)$ of bounded index. But in characteristic zero $Z(D)$ is infinite, whence unbounded, and has no proper ideal of bounded index. Therefore $A = Z(D)$, so $b \in d + Z(D) \subseteq d^D$, whence $b^D = d^D$ and $b^D - b \supseteq Z(D)$. ■

As $a^2 \notin Z(D)$ and $c \in Z(D)$, there is some $x \in D$ with $(a^2)^x - a^2 = c$; note that

$$C_D(a^x) = C_D(a)^x = C_D(a^2)^x = C_D((a^2)^x) = C_D(a^2) = C_D(a),$$

so a and a^x commute. Put $a' = (a^x/a)^2$ (a square) and $b' = c/a^2$ (a non-square); then $a' - b' = 1$. Since $b' \notin Z(D)$, there is $y \in D$ with $b'^y - b' = 1$, whence $b'^y = a'$. So a square in D is conjugate to a non-square, which yields the final contradiction. ■

Note that in a supersimple theory we could have assumed that D is outright definable by Corollary 5.5.6, which would have slightly simplified the proof. However, the assumption that $SU(D) < \infty$ is weaker than supersimplicity; furthermore the results all generalize to a division ring with $0 < SU_P(D) < \infty$ for some $\emptyset$-invariant family P of types.

PROBLEM 5.6.13 Is an unbounded hyperdefinable division ring D of ordinal SU-rank in a simple theory commutative?

The problem in generalizing the above proof lies in the fact that an infinite hyperdefinable set could still be bounded.

REMARK 5.6.14 A field K is *pseudo-algebraically closed (PAC)* if every absolutely irreducible variety over K (i.e which remains irreducible over the algebraic closure of K) has a K-rational point. Hrushovski [49] has shown that a perfect PAC field with only boundedly many extensions of every degree is supersimple of SU-rank 1.

CONJECTURE 5.6.15 *A supersimple field is perfect PAC with only boundedly many extensions of every degree.*

Triviality of the Brauer group at least implies that every absolutely irreducible rational variety over a supersimple field K has a K-rational point. However, showing pseudo-algebraic closedness in general appears to be beyond current methods.

5.7 BIBLIOGRAPHICAL REMARKS

Shelah rank is of course due to Shelah [143, 147], Lascar rank (for stable theories, where it is called U-rank) and the Lascar inequalities 5.1.6 to Lascar [78, 81], and the stable analogue of Proposition 5.1.17 to Poizat [116]; the proofs transfer easily to the simple context. Kim and Pillay [70] develop the basic properties of Shelah and Lascar rank in a simple theory. Example 5.1.15 is essentially due to Kim.

Weight, regularity and orthogonality were defined by Shelah [147]; the bound for the weight in a superstable theory, as well as the decomposition of a superstable type into an equidominant finite product of regular types (Corollary 5.2.19) were shown by Lascar [81, 83, 84]. Our exposition again follows the stable case, but the matter is complicated by non-stationarity of types. Lemma 5.2.17 and Theorem 5.2.18 and are immediate localizations of Hyttinen [60], who proves that if every type has weight less than ω, then every type has finite weight (see also [106]). Problem 5.2.21 was posed by Poizat [125].

Elimination of hyperimaginaries was first shown by Buechler [17] for a low supersimple theory; the full result appeared in [19]. Theorem 5.3.8 is due to Kim [68] and Pillay [111].

Supersimple groups are studied in [160] (in particular Theorem 5.5.4) and [163]. Remark 5.5.1 is due to Hrushovski [46]. The original version of Zil'ber's Indecomposability Theorem 5.4.5 for uncountably categorical groups appeared in [167]; it was generalized to superstable groups by Berline and Lascar [14]; the version here was inspired by the treatment in Chatzidakis and Hrushovski [25] for difference fields, and the proof by Poizat [126]. The argument in Remark 5.4.11 is due to Reineke [133].

Macintyre [93], Cherlin and Shelah [31] classify superstable division rings as algebraically closed fields. Pillay and Poizat [113] prove that a supersimple field has small Galois group. Hrushovski [49] shows that a perfect PAC field with small Galois group is supersimple of SU-rank 1; this was extended by Chatzidakis and Pillay [27], who prove that PAC fields with small Galois group are simple. Commutativity of a supersimple division ring was shown by Pillay, Scanlon and myself [114]. More information about the Brauer group can be found in Jacobson [61] and Serre [138].

Chapter 6

MISCELLANEOUS

6.1 SMALL THEORIES

DEFINITION 6.1.1 A countable theory is *small* if $S(\emptyset)$ is countable.

Note that smallness is preserved under naming finitely many parameters, as n-types over a parameter set of size m induce $(m+n)$-types over $\emptyset$. Small theories are of particular interest in connection with Vaught's conjecture, which states that every countable theory should either have countably or continuum many countable models, up to isomorphism; this has so far been shown only for superstable theories of finite SU-rank [18]. Since a countable model can only realize countably many types and $S(\emptyset)$ is either countable or has size continuum, it follows that a theory with fewer than continuum many non-isomorphic countable models must be small. Further examples of small theories are ω-categorical theories, where the Ryll-Nardzewski Theorem states that $S_n(\emptyset)$ is finite for all $n < \omega$. These shall be studied later in section 6.2.

As in a small theory there are only countably many inequivalent formulas, we can assume in this section that the language is countable.

DEFINITION 6.1.2 Let A be a parameter set. The *Cantor-Bendixson rank* *CB* (over A) is the smallest function from the set of all consistent A-formulas to On^+ satisfying:

> $CB(\varphi(\bar{x})) \geq \alpha+1$ if there are pairwise contradictory A-formulas $\varphi_i(\bar{x})$ for $i < \omega$ with $\varphi_i(\bar{x}) \vdash \varphi(\bar{x})$ and $CB(\varphi_i(\bar{x})) \geq \alpha$ for all $i < \omega$.

For a partial type π over A, put $CB(\pi) = \min\{CB(\varphi) : \varphi \in \pi\}$.
The *Cantor-Bendixson degree* of a partial type $\pi(\bar{x})$ over A is the number of complete types in $[\pi(\bar{x})]$ of Cantor-Bendixson rank $CB(\pi)$.

Note that the Cantor-Bendixson rank depends on the parameter set A; increasing A will in general increase the rank. Furthermore, the Cantor-Bendixson degree is a finite number.

LEMMA 6.1.3 *In a small theory, Cantor-Bendixson rank is ordinal valued over finite parameter sets.*

PROOF: Suppose A is a finite parameter set, and $\varphi(x)$ is an A-formula with $CB(\varphi) = \infty$. As there are only countably many formulas over A, Cantor-Bendixson rank is ∞ if and only if it is at least uncountable. Since $CB(\varphi) \geq \omega_1 + 1$, there are two disjoint A-formulas $\varphi_0(x)$ and $\varphi_1(x)$ which imply $\varphi(x)$ and have Cantor-Bendixson rank at least ω_1. Hence $CB(\varphi_0) = CB(\varphi_1) = \infty$; we can now split up φ_0 and φ_1 into two formulas of rank ∞; continuing like this, we obtain continuum many types over A. As A was finite, $S(\emptyset)$ is uncountable. ■

COROLLARY 6.1.4 *In a small theory, for every finite set A every consistent formula $\varphi(\bar{x})$ over A extends to an isolated type in $S(A)$.*

PROOF: Choose a formula $\psi(\bar{x}) \vdash \varphi(\bar{x})$ of minimal Cantor-Bendixson rank and degree over A. If $\vartheta(\bar{x})$ were a formula over A such that both $\vartheta \wedge \psi$ and $\neg\vartheta \wedge \psi$ are consistent, then one of the two conjunctions would have smaller Cantor-Bendixson rank or degree, contradicting our minimal choice. Hence ψ isolates a type over A. ■

COROLLARY 6.1.5 *Let T be a small countable theory. Then over every finite parameter set A there is a countable atomic model $\mathfrak{M}(A)$. Any such model is prime, and if $\mathfrak{M}$ is any prime model over A, then $\mathfrak{M}$ and $\mathfrak{M}(A)$ are isomorphic over A.*

PROOF: Let $\mathfrak{N}$ be a countable model containing A, and enumerate all consistent $\mathfrak{N}$-formulas with free variable x as $\{\varphi_i(x) : i < \omega\}$. Put $A_0 = A$; given A_i, choose minimal $j < \omega$ such that φ_j is over A_i and $\models \neg\varphi_j(a)$ for all $a \in A_i$. By Corollary 6.1.4 there is $a_i \in \mathfrak{N}$ such that $\text{tp}(a_i/A_i)$ is isolated and $\models \varphi_j(a_i)$; put $A_{i+1} = A_i \cup \{a_i\}$. Put $\mathfrak{M}(A) = \bigcup_{i<\omega} A_i$; by the Tarski Test 1.2.14 it is easy to see that $\mathfrak{M}(A)$ is an elementary substructure of $\mathfrak{N}$ containing A. Note that $\text{tp}(\bar{a}/A)$ is isolated for every finite subset $\bar{a} \in \mathfrak{M}(A)$ by Remark 1.2.3, so $\mathfrak{M}(A)$ is atomic over A. If $\mathfrak{M}$ is any model containing A and $\sigma_i : A_i \to \mathfrak{M}$ is an elementary embedding extending id_A, then $\sigma_i(\text{tp}(a_i/A_i))$ must be realized in $\mathfrak{M}$ by isolation, so we can extend σ_i to A_{i+1}, and hence to the whole of $\mathfrak{M}(A)$. This shows that $\mathfrak{M}$ is prime.

Now let $\mathfrak{M}$ be a prime model over A. As $\mathfrak{M}$ embeds elementarily into $\mathfrak{M}(A)$ over A by primality, all types over A realized in $\mathfrak{M}$ must be isolated. Choose an enumeration $\{a'_i : i < \omega\}$ of $\mathfrak{M} - A$; we shall construct an increasing sequence of partial A-isomorphisms σ_i between $\mathfrak{M}(A)$ and $\mathfrak{M}$, such that $\mathrm{dom}(\sigma_i) - A$ is finite, $a_i \in \mathrm{dom}(\sigma_{i+1})$ and $a'_i \in \mathrm{range}(\sigma_{i+1})$ for all $i < \omega$.

Put $\sigma_0 = \mathrm{id}_A$, and suppose σ_i has been constructed. As $\mathrm{tp}(a_i \mathrm{dom}(\sigma_i)/A)$ is isolated, so are $\mathrm{tp}(a_i/\mathrm{dom}(\sigma_i))$ and its image under σ_i. Hence there is some $a' \in \mathfrak{M}$ with $\mathrm{tp}(a' \mathrm{range}(\sigma_i)/A) = \mathrm{tp}(a_i \mathrm{dom}(\sigma_i)/A)$. Similarly, there is $a'' \in \mathfrak{M}(A)$ with $\mathrm{tp}(a'' a_i \mathrm{dom}(\sigma_i)/A) = \mathrm{tp}(a'_i a' \mathrm{range}(\sigma_i)/A)$; we may put $\sigma_{i+1} = \sigma_i \cup \{(a_i, a'), (a'', a'_i)\}$.

Clearly, $\bigcup_{i<\omega} \sigma_i$ is the required isomorphism. ∎

We can now prove the Ryll-Nardzewski Theorem alluded to before:

COROLLARY 6.1.6 RYLL-NARDZEWSKI THEOREM *A countable complete theory T is ω-categorical if and only if $S_n(\emptyset)$ is finite for all $n < \omega$.*

PROOF: If $S_n(\emptyset)$ is finite for all $n < \omega$, then every type over a finite set is isolated. So every countable model is atomic, and isomorphic to the prime model over $\emptyset$ by Corollary 6.1.5. Hence T is ω-categorical.

For the converse, suppose $S_n(\emptyset)$ is uncountable for some $n < \omega$. Then there must be uncountably many non-isomorphic countable models, since each of them realizes only countably many types, and every type is realized in some countable model. Otherwise T is small, and there is a countable atomic model by Corollary 6.1.5. By ω-categoricity, all countable models must be atomic, all types over $\emptyset$ are isolated, and $S_n(\emptyset)$ is finite for all $n < \omega$. ∎

6.1.1 ELIMINATION OF HYPERIMAGINARIES

In this subsection, we shall prove that a small theory allows the best elimination of type-definable equivalence relations possible (Theorem 6.1.9).

LEMMA 6.1.7 *Let A be a finite parameter set in a small theory, and E a finitary type-definable equivalence relation over A which is coarser than equality of types (over A). Then there are A-definable equivalence relations E_i (for $i < \omega$) such that $E(\bar{x}, \bar{y}) \Leftrightarrow \bigwedge_{i<\omega} E_i(\bar{x}, \bar{y})$.*

PROOF: If $p(\bar{x}), q(\bar{y}) \in S(A)$ are any two types, then $p(\bar{x}) \wedge q(\bar{y})$ either proves $E(\bar{x}, \bar{y})$ (so p and q are E-related), or it proves $\neg E(\bar{x}, \bar{y})$ (so p and q are not E-related). Suppose $p, q \in S_1(A)$ are not E-related. By compactness there are formulas $\varphi(\bar{x}) \in p(\bar{x})$ and $\psi(\bar{y}) \in q(\bar{y})$ such that $\varphi(\bar{x}) \wedge \psi(\bar{y}) \vdash \neg E(\bar{x}, \bar{y})$; we choose φ and ψ such that $\neg[\varphi(\bar{x}) \vee \psi(\bar{x})]$ has minimal Cantor-Bendixson rank and degree possible.

CLAIM. $\neg[\varphi(\bar{x}) \vee \psi(\bar{x})]$ is inconsistent.

PROOF OF CLAIM: If not, it contains a type $p'(\bar{x})$ with $CB(p'(\bar{x})) = CB(\neg[\varphi(\bar{x}) \vee \psi(\bar{x})])$. Since $\varphi(\bar{x}) \wedge p'(\bar{y}) \wedge E(\bar{x}, \bar{y})$ and $p'(\bar{x}) \wedge \psi(\bar{y}) \wedge E(\bar{x}, \bar{y})$ cannot both be consistent, there is a formula $\vartheta(\bar{x}) \in p'(\bar{x})$ such that

$\varphi(\bar{x}) \wedge \vartheta(\bar{y}) \wedge E(\bar{x}, \bar{y})$ (say) is inconsistent. We put $\psi'(\bar{y}) = \psi(\bar{y}) \vee \vartheta(\bar{y}) \in q(\bar{y})$. Then $\varphi(\bar{x}) \wedge \psi'(\bar{y}) \vdash \neg E(\bar{x}, \bar{y})$, and $\neg[\varphi(\bar{x}) \vee \psi'(\bar{x})]$ either has smaller Cantor-Bendixson rank, or the same rank and smaller degree, contradicting our minimal choice for φ and ψ. ■

It follows that for any types $p, q \in S(A)$ which are not E-related we can find an E-invariant A-formula $\varphi_{p,q}(\bar{x}) \in p - q$. Then $E(\bar{x}, \bar{y})$ is equivalent to

$$\bigwedge\{\varphi_{p,q}(\bar{x}) \leftrightarrow \varphi_{p,q}(\bar{y}) : p(\bar{x}) \wedge q(\bar{y}) \vdash \neg E(\bar{x}, \bar{y})\},$$

and thus to an intersection of A-definable equivalence relations. ■

LEMMA 6.1.8 *Let $E(\bar{x}, \bar{y})$ be a finitary type-definable equivalence relation over a finite set A, and $p(\bar{x}) \in S(A)$ a complete type. If every type-definable equivalence relation which is coarser than equality of types is equivalent to an intersection of definable equivalence relations, then E is equivalent to an intersection of A-definable equivalence relations on p.*

PROOF: Let $\bar{a} \models p$, and let $R(\bar{x}, \bar{y})$ be the type-definable relation over $A\bar{a}$ given by

$$\exists \bar{z}\, [\text{tp}(\bar{a}\bar{x}/A) = \text{tp}(\bar{z}\bar{y}) \wedge E(\bar{a}, \bar{z})].$$

If $R(\bar{x}, \bar{y})$, as witnessed by some $\bar{z}$, let σ be the A-automorphism mapping $\bar{z}\bar{y}$ to $\bar{a}\bar{x}$. Then $\text{tp}(\bar{a}\bar{y}) = \text{tp}(\sigma(\bar{a}), \bar{x})$ and $\models E(\sigma(\bar{a}), \bar{a})$, so $R(\bar{y}, \bar{x})$ holds. Similarly $R(\bar{x}, \bar{y})$ and $R(\bar{y}, \bar{z})$ imply $R(\bar{x}, \bar{z})$, so R is an equivalence relation. If $\text{tp}(\bar{x}/A\bar{a}) = \text{tp}(\bar{y}/A\bar{a})$, clearly $R(\bar{x}, \bar{y})$ holds, so by assumption there are definable equivalence relations $(R_i(\bar{x}, \bar{y}, \bar{a}) : i < \omega)$ such that $R(\bar{x}, \bar{y}) \Leftrightarrow \bigwedge_{i<\omega} R_i(\bar{x}, \bar{y}, \bar{a})$. Note that R has only boundedly many classes, so every R_i has only boundedly many $A\bar{a}$-conjugates. Hence every R_i has only finitely many $A\bar{a}$-conjugates by compactness, and we may replace every R_i by the intersection of its $A\bar{a}$-conjugates (which is still coarser than R) and assume that it is definable over $A\bar{a}$.

For each $i < \omega$ put

$$E_i(\bar{x}, \bar{y}) = \forall \bar{z}\, [R_i(\bar{z}, \bar{x}, \bar{x}) \leftrightarrow R_i(\bar{z}, \bar{y}, \bar{y})].$$

Clearly E_i is an A-definable equivalence relation, for all $i < \omega$. We claim that is has the desired properties. So suppose $E(\bar{x}, \bar{y})$ holds for some $\bar{x}, \bar{y} \models p$, and choose any $\bar{z}$. We can then find some $\bar{z}'$ with $\text{tp}(\bar{x}\bar{z}') = \text{tp}(\bar{y}\bar{z})$, whence $\models R(\bar{z}', \bar{z}, \bar{x})$, and $\models R_i(\bar{z}', \bar{z}, \bar{x})$ for all $i < \omega$. Therefore

$$R_i(\bar{z}, \bar{x}, \bar{x}) \Leftrightarrow R_i(\bar{z}', \bar{x}, \bar{x}) \Leftrightarrow R_i(\bar{z}, \bar{y}, \bar{y}),$$

whence $\models E_i(\bar{x}, \bar{y})$ for all $i < \omega$.

Conversely, suppose $\models E_i(\bar{x}, \bar{y})$ for all $i < \omega$. Since $R_i(\bar{x}, \bar{x}, \bar{x})$ holds for all $i < \omega$, we get $\models R_i(\bar{x}, \bar{y}, \bar{y})$ and therefore $\models R(\bar{y}, \bar{x}, \bar{y})$, whence $\models E(\bar{x}, \bar{y})$. ■

THEOREM 6.1.9 *Let $E(\bar{x},\bar{y})$ be a finitary type-definable equivalence relation over a finite set A in a small theory. Then there are A-definable equivalence relations $(E_i : i < \omega)$ such that $E(\bar{x},\bar{y}) \Leftrightarrow \bigwedge_{i<\omega} E_i(\bar{x},\bar{y})$.*

PROOF: Let R be the relation $\exists\bar{z}\,[\mathrm{tp}(\bar{z}/A) = \mathrm{tp}(\bar{x}/A) \wedge E(\bar{z},\bar{y})]$. Then R is easily seen to be a type-definable equivalence relation over A which is coarser than equality of types, and must be equivalent to an intersection $\bigwedge_{i<\omega} R_i(\bar{x},\bar{y})$ of A-definable equivalence relations by Lemma 6.1.7; we clearly may assume that R_{i+1} refines R_i for all $i < \omega$.

By Lemma 6.1.8, for every type $p(\bar{x}) \in S(A)$ there are A-definable equivalence relations E_i^p for $i < \omega$ such that $p(\bar{x}) \wedge p(\bar{y}) \vdash E(\bar{x},\bar{y}) \leftrightarrow \bigwedge_{i<\omega} E_i^p(\bar{x},\bar{y})$. For every formula $\varphi(\bar{x},\bar{y}) \in E(\bar{x},\bar{y})$ and every type $p(\bar{x}) \in S(A)$ compactness yields an A-formula $\psi_\varphi^p(\bar{x}) \in p$ and some $i_\varphi^p < \omega$ such that

$$\psi_\varphi^p(\bar{x}) \wedge \psi_\varphi^p(\bar{y}) \wedge E_{i_\varphi^p}^p(\bar{x},\bar{y}) \vdash \varphi(\bar{x},\bar{y}),$$

restricting ψ_φ^p, we may also assume that there is some $\varphi'(\bar{x},\bar{y}) \in E(\bar{x},\bar{y})$ with

$$\psi_\varphi^p(\bar{x}) \wedge \psi_\varphi^p(\bar{y}) \wedge \varphi'(\bar{x},\bar{y}) \vdash E_{i_\varphi^p}^p(\bar{x},\bar{y}).$$

As E is an equivalence relation, compactness yields a symmetric formula $\varphi'' \in E$ such that $\exists\bar{x}',\bar{y}'\,[\varphi''(\bar{x},\bar{x}') \wedge \varphi''(\bar{x}',\bar{y}') \wedge \varphi''(\bar{y}',\bar{y})]$ implies $\varphi'(\bar{x},\bar{y})$. Then $\exists\bar{x}\,[\psi_\varphi^p(\bar{x}) \wedge \varphi''(\bar{x},\bar{y})]$ contains the R-class of any realization $\bar{a} \models p$ (which does not depend on the choice of $\bar{a}$); by compactness there is some $i < \omega$ such that it contains the whole R_i-class C_p of $\bar{a}$. Note that C_p is A-definable and contains p.

By compactness, there is a finite set P of types such that $\forall\bar{x}\,\bigvee_{p\in P} \bar{x} \in C_p$ holds. If i_0 is such that C_p is a finite union of R_{i_0}-classes for all $p \in P$, we may modify the C_p and assume that they are all finite unions of R_{i_0}-classes and disjoint. Define $E_\varphi(\bar{x},\bar{y})$ as

$$\bigvee_{p\in P} [\bar{x} \in C_p \wedge \bar{y} \in C_p \wedge \forall\,\bar{x}',\bar{y}' \models \psi_\varphi^p\,\{[\varphi''(\bar{x},\bar{x}') \wedge \varphi''(\bar{y},\bar{y}')] \to E_{i_\varphi^p}^p(\bar{x}',\bar{y}')\}.$$

It is clear that E_φ is an A-definable equivalence relation. Suppose $E(\bar{x},\bar{y})$ holds. Then there is a unique $p \in P$ such that $\bar{x} \in C_p$; since $R_{i_0}(\bar{x},\bar{y})$ holds, we also have $\bar{y} \in C_p$. If $\bar{x}',\bar{y}' \models \psi_\varphi^p$ and $\varphi''(\bar{x},\bar{x}')$ and $\varphi''(\bar{y},\bar{y}')$ hold, then $\models \varphi'(\bar{x}',\bar{y}')$, whence $\models E_{i_\varphi^p}^p(\bar{x}',\bar{y}')$, and $E_\varphi(\bar{x},\bar{y})$ holds.

Conversely, if $\models E_\varphi(\bar{x},\bar{y})$, then since C_p is contained in $\exists\bar{z}\,[\psi_\varphi^p(\bar{z}) \wedge \varphi''(\bar{z},\bar{x})]$ and as $\varphi'' \vdash \varphi$, there are $\bar{x}',\bar{y}'$ such that $\varphi(\bar{x},\bar{x}') \wedge \varphi(\bar{x}',\bar{y}') \wedge \varphi(\bar{y}',\bar{y})$ holds. Therefore $\bigwedge_{\varphi\in E} E_\varphi(\bar{x},\bar{y}) \vdash E(\bar{x},\bar{y})$. ∎

REMARK 6.1.10 The above proof shows in general that a type-definable equivalence relation is equivalent to an intersection of definable equivalence relations if its restriction to every complete type is equivalent to an intersection of definable equivalence relations, and if the type-definable equivalence relation $\exists \bar{z}\,[\mathrm{tp}(\bar{z}) = \mathrm{tp}(\bar{y}) \wedge E(\bar{x}, \bar{z})]$ is equivalent to an intersection of definable equivalence relations.

COROLLARY 6.1.11 *In a small simple theory, Lascar strong type is the same as strong type. In other words, for any A and any type $p \in S(A)$ there are A-definable equivalence relations $(E_i : i < \alpha)$ such that* $\mathrm{lstp}(x/A) = \mathrm{lstp}(y/A) \Leftrightarrow \bigwedge_{i<\alpha} E_i(x, y)$.

PROOF: By Corollary 2.7.9 we have $\mathrm{lstp}(x/A) = \mathrm{lstp}(y/A)$ if and only if $\mathrm{lstp}(x/\bar{a}) = \mathrm{lstp}(y/\bar{a})$ for all finite $\bar{a} \in A$, so we may restrict ourselves to types over finite sets A. But equality of Lascar strong type over A is a type-definable equivalence relation by Corollary 2.7.9, which for finite A must be intersection of definable equivalence relations by Theorem 6.1.9. The assertion follows. ■

COROLLARY 6.1.12 *Let G be a definable group in a small theory, and H a type-definable subgroup of G over finitely many parameters. Then H is the intersection of definable subgroups.*

PROOF: Let A be the finite set of parameters needed to define G and type-define H. Then $xy^{-1} \in H$ is a type-definable equivalence relation over A, and equivalent to an intersection of definable equivalence relations $(E_i : i < \alpha)$ by Theorem 6.1.9. Put

$$H_i = \{g \in G : \forall x \in G\; E_i(x, gx)\}.$$

If $h \in H$ and $x \in G$, then $x(hx)^{-1} \in H$, whence $E_i(x, hx)$ for all i, and $H \subseteq H_i$. On the other hand, if $h \in H_i$ for all i, then $E_i(1, h)$ for all E_i (put $x = 1$), whence $h \in H$. Furthermore, if $g, h \in H_i$, then $E_i(h^{-1}x, hh^{-1}x)$, whence $E_i(x, h^{-1}x)$, as well as $E_i(x, hx) \wedge E_i(hx, ghx)$, whence $E_i(x, ghx)$. It follows that H_i is a definable subgroup of G, and $H = \bigcap_i H_i$. ■

PROBLEM 6.1.13 Is a type-definable subgroup of a type-definable group G in a small theory (even over finitely many parameters) an intersection of definable groups?

PROBLEM 6.1.14 Is a type-definable group (over finitely many parameters) in a small theory the intersection of definable groups?

One might also wonder whether one could use the methods of this section to give a positive answer to Problem 5.5.7. As far as rings are concerned, it is known [162] that a small field is algebraically closed.

PROBLEM 6.1.15 Is a small division ring commutative? Is a small simple division ring commutative?

6.1.2 LOCALLY MODULAR THEORIES

In this subsection we shall study local modularity in small theories. More precisely, we shall establish that every locally modular real type p (over a finite set) in a small theory has finite weight, and that the weight is witnessed by a decomposition into regular components inside the p-closure of any realization of the type. This is in particular applicable to small one-based theories.

LEMMA 6.1.16 *Let P be an $\emptyset$-invariant family of types in a small theory. For a finite tuple $\bar{a}$ let $\mathcal{L}$ be the lattice of P-closed subclasses of* $\mathrm{cl}_P(\bar{a})$. *Then no dense order can be embedded into $\mathcal{L}$.*

PROOF: Suppose $\{L_i : i \in \mathbb{Q}\}$ is a collection of P-closed subclasses of $\mathrm{cl}_P(\bar{a})$ with $L_i \subset L_j$ for $i < j$. For $r \in \mathbb{R}$ put $L_r = \bigcup_{i<r} L_i$, and $p_r = \mathrm{tp}(\bar{a}/L_r)$. Then p_r has a non-forking extension to $\mathrm{cl}_P(\bar{a})$, realized by some element $\bar{a}_r$. Let $A_r = \mathrm{cl}_P(\bar{a}) \cap \mathrm{dcl}(\bar{a}\bar{a}_r)$. Then $\bar{a}_r \perp_{A_r} \mathrm{cl}_P(\bar{a})$ by Lemma 3.5.3, so $\mathrm{tp}(\bar{a}_r/\mathrm{cl}_P(\bar{a}))$ does not fork over A_r. But for $s > r$ there are rational i and j such that $r < i < j < s$ and $\bar{a} \not\perp_{L_i} L_j$. Hence there are a formula φ and some $k < \omega$ such that $D(\bar{a}/L_i, \varphi, k) > D(\bar{a}/L_j, \varphi, k)$, whence

$$\begin{aligned} D(\bar{a}_r/A_r, \varphi, k) &= D(\bar{a}/L_r, \varphi, k) \geq D(\bar{a}/L_i, \varphi, k) \\ &> D(\bar{a}/L_j, \varphi, k) \geq D(\bar{a}/L_s, \varphi, k) = D(\bar{a}_s/A_s, \varphi, k), \end{aligned}$$

and $\mathrm{tp}(\bar{a}_r/A_r) \neq \mathrm{tp}(\bar{a}_s/A_s)$. As A_r is in the definable closure of two realizations of $\mathrm{tp}(\bar{a})$, there are uncountably many types over finite sets, contradicting smallness. ■

LEMMA 6.1.17 *Suppose P is a locally modular $\emptyset$-invariant family of types in a small theory, and $a \in \mathrm{cl}_P(\tilde{a}) - \mathrm{cl}_P(\emptyset)$ for some P-internal finitary $\tilde{a}$. Then there is some $a_0 \in \mathrm{cl}_P(a) - \mathrm{cl}_P(\emptyset)$ such that $\mathrm{tp}(a_0/\mathrm{cl}_P(\emptyset))$ is regular.*

PROOF: Put $p = \mathrm{tp}(a/\mathrm{cl}_P(\emptyset))$, and note that p is P-minimal. If $\mathcal{L}$ is the lattice of P-closed subclasses of $\mathrm{cl}_P(a)$ and $\bar{a}$ is a finite real tuple with $\tilde{a} \in \mathrm{dcl}(\bar{a})$, then $\mathcal{L}$ embeds into the lattice of P-closed subclasses of $\mathrm{cl}_P(\bar{a})$, so by Lemma 6.1.16 it does not embed a dense linear order. As $a \not\perp_{\mathrm{cl}_P(\emptyset)} \mathrm{cl}_P(a)$, there are L and L' in $\mathfrak{L}$ such that $L \subset L'$ and p is not foreign to $\mathrm{tp}(L'/L)$, but for any L'' with $L \subset L'' \subset L'$ either p is foreign to $\mathrm{tp}(L'/L'')$, or it is foreign to $\mathrm{tp}(L''/L)$. Choose some hyperimaginary element $b \in L'$ such that p is not foreign to $\mathrm{tp}(b/L)$ and some $B \supseteq L$ such that $p \not\perp_B \mathrm{tp}(b/B)$. Then p is non-orthogonal to $\mathrm{tp}(b/\mathrm{cl}_P(B))$; since the latter type is based on $\mathrm{cl}_P(b) \cap \mathrm{cl}_P(B) \subseteq L'$ by Corollary 3.5.16, we may replace L by $\mathrm{cl}_P(Lb) \cap \mathrm{cl}_P(B)$ and assume that $q := \mathrm{tp}(b/L)$ is non-orthogonal to p. Moreover, we can assume that L is closed under cl_P and cl_p, since p is P-minimal and p-minimal. Hence q is P-minimal and p-minimal as well by Lemma 3.5.3.

CLAIM. q is regular.

PROOF OF CLAIM: Suppose $b \not\perp_L c$. If $L'' = \mathrm{cl}_P(bL) \cap \mathrm{cl}_P(cL) \supseteq L$; then $b \perp_{L''} \mathrm{cl}_P(cL)$ by Corollary 3.5.16. As b and c fork over L, we have $b \not\perp_L L''$; since $L'' \subseteq \mathrm{cl}_P(bL) \subseteq L'$, either $\mathrm{tp}(L''/L)$ or $\mathrm{tp}(L'/L'')$ is co-foreign to p, whence to q by p-minimality. In the first case q would be foreign to $\mathrm{tp}(L''/L)$, which contradicts $b \not\perp_L L''$. So q is foreign to $\mathrm{tp}(L'/L'')$, and therefore to $\mathrm{tp}(b/\mathrm{cl}_P(cL))$; by P-minimality q is also foreign to $\mathrm{tp}(\mathrm{cl}_P(cL)/cL)$, and hence to $\mathrm{tp}(b/cL)$. As c was arbitrary, q is regular. ■

Since $p \not\perp q$, there is $a' \models p$, some $A \perp_{\mathrm{cl}_P(\emptyset)} a'$, and a non-forking extension $\mathrm{tp}(b/A)$ of q such that $a' \not\perp_A b$. Note that A contains L, and in particular $\mathrm{cl}_P(\emptyset)$. Put $a_0' = \mathrm{cl}_P(a') \cap \mathrm{cl}_P(bA)$. Then $a' \perp_{a_0'} bA$ by Corollary 3.5.16, whence $a' \not\perp_{\mathrm{cl}_P(\emptyset)} a_0'$; as $\mathrm{cl}_P(a') \perp_{\mathrm{cl}_P(\emptyset)} \mathrm{cl}_P(A)$, we have $a_0' \perp_{\mathrm{cl}_P(\emptyset)} \mathrm{cl}_P(A)$. Now $b \perp_A \mathrm{cl}_P(A)$ by P-minimality of q, so regularity of $\mathrm{tp}(b/A)$ implies regularity of $\mathrm{tp}(b/\mathrm{cl}_P(A))$, and of $\mathrm{tp}(\mathrm{cl}_P(bA)/\mathrm{cl}_P(A))$ by Lemma 5.2.10. Hence $\mathrm{tp}(a_0'/\mathrm{cl}_P(A))$ is regular, as is $\mathrm{tp}(a_0'/\mathrm{cl}_P(\emptyset))$ by Lemma 5.2.13.

Finally, we can choose a_0 to be the image of a_0' under an automorphism mapping a' to a. ■

THEOREM 6.1.18 *Suppose P is a locally modular $\emptyset$-invariant family of types in a small theory, and $a \in \mathrm{cl}_P(\tilde{a}) - \mathrm{cl}_P(\emptyset)$ for some P-internal finitary $\tilde{a}$. Then there are $n < \omega$ and an independent sequence $(a_i : i < n) \in \mathrm{cl}_P(a)$ over $\mathrm{cl}_P(\emptyset)$, such that $\mathrm{tp}(a_i/\mathrm{cl}_P(\emptyset))$ is regular for all $i < n$ and a is equidominant over $\mathrm{cl}_P(\emptyset)$ with $(a_i : i < n)$. In particular, $\mathrm{tp}(a/\mathrm{cl}_P(\emptyset))$ has finite weight.*

PROOF: First suppose $\mathrm{tp}(a/\mathrm{cl}_P(\emptyset))$ has infinite weight, as exemplified by a set $A \perp_{\mathrm{cl}_P(\emptyset)} a$ and some sequence $\{b_i : i \in \mathbb{Q}\}$ independent over $\mathrm{cl}_P(A)$ such that $a \not\perp_{\mathrm{cl}_P(A)} b_i$ for all $i \in \mathbb{Q}$. By Lemma 3.5.5 the family $\{\mathrm{cl}_P(Ab_i) : i \in \mathbb{Q}\}$ is independent over $\mathrm{cl}_P(A)$, and clearly $a \not\perp_{\mathrm{cl}_P(A)} \mathrm{cl}_P(Ab_i)$ for all $i \in \mathbb{Q}$. But this yields $a \not\perp_{\mathrm{cl}_P(Ab_i : i \leq q)} \mathrm{cl}_P(Ab_i : i \leq r)$ for any rational $q < r$. However, by Corollary 3.5.16

$$a \mathop{\perp}_{\mathrm{cl}_P(a) \cap \mathrm{cl}_P(Ab_i : i \leq q)} \mathrm{cl}_P(Ab_i : i \leq q),$$

so the intersections $\mathrm{cl}_P(a) \cap \mathrm{cl}_P(Ab_i : i \leq q)$ must all be different for varying $q \in \mathbb{Q}$, contradicting Lemma 6.1.16 applied to $\mathrm{cl}_P(\bar{a})$ for some real tuple $\bar{a}$ with $\tilde{a} \in \mathrm{dcl}(\bar{a})$.

It follows that there is a maximal sequence $(a_i : i < n) \in \mathrm{cl}_P(a)$ of hyperimaginary elements independent over $\mathrm{cl}_P(\emptyset)$, such that $\mathrm{tp}(a_i/\mathrm{cl}_P(\emptyset))$ is regular for all $i < n$.

CLAIM. $(a_i : i < n)$ dominates a over $\mathrm{cl}_P(\emptyset)$.

PROOF OF CLAIM: Suppose not, so there is b with $b \not\perp_{\mathrm{cl}_P(\emptyset)} a$, but $b \perp_{\mathrm{cl}_P(\emptyset)} (a_i : i < n)$. Put $b' = \mathrm{cl}_P(a) \cap \mathrm{cl}_P(b)$, then $a \perp_{b'} b$ by Corollary 3.5.16, and $a \not\perp_{\mathrm{cl}_P(\emptyset)} b'$; moreover $(a_i : i < n) \perp_{\mathrm{cl}_P(\emptyset)} \mathrm{cl}_P(b')$. By Lemma 6.1.17 there is some $a_n \in \mathrm{cl}_P(b') - \mathrm{cl}_P(\emptyset)$ such that $\mathrm{tp}(a_n/\mathrm{cl}_P(\emptyset))$ is regular. Hence $(a_i : i \leq n)$ would be a longer sequence, a contradiction. ■

As a trivially dominates $(a_i : i < n)$ over $\mathrm{cl}_P(\emptyset)$, we are done by Lemma 5.2.11 and 5.2.4, and Remark 5.2.7. ■

Theorem 6.1.18 is in particular applicable to a locally modular type over a finite set in a small theory.

COROLLARY 6.1.19 *Let T be small and one-based. Then every finitary type has finite weight, and is equidominant over its domain with a finite product of regular types.*

PROOF: Any type in a one-based theory is finitely based; we may therefore apply Theorem 6.1.18. ■

6.1.3 THEORIES WITH FINITE CODING

DEFINITION 6.1.20 A simple theory T admits *finite coding* if every real type is based on a finite set.

In other words, for every real type p there is a finite tuple $\bar{a}$ such that $\mathrm{Cb}(p) \in \mathrm{bdd}(\bar{a})$. One-based theories admit finite coding (any realization of p is a base for p), as do supersimple theories. We have already studied regular types and weight in those theories; it should not come too surprising that many results can also be proved in the context of finite coding.

PROPOSITION 6.1.21 *A small simple theory which admits finite coding has elimination of hyperimaginaries.*

PROOF: By Lemma 3.6.4 is is sufficient to show that the canonical base of every finitary Lascar strong type is interdefinable with a sequence of imaginary elements. If p is a finitary Lascar strong type, then p is based on some finite set A by finite coding. Consider two independent realizations a and a' of p over A. Then $\mathrm{tp}(a'/Aa) \vdash \mathrm{lstp}(a'/A)$, so $\mathrm{Cb}(p) \in \mathrm{dcl}(Aa)$. By Theorem 6.1.9 the canonical base of p is interdefinable with a sequence of imaginary elements, and we are done. ■

DEFINITION 6.1.22 A simple theory T has *no dense forking chain* if there is no chain of finitary types p_q for $q \in \mathbb{Q}$ such that p_r is a forking extension of p_q for any $q < r$.

Clearly, a theory has no dense forking chain if and only if it has no dense forking chain of real types.

PROPOSITION 6.1.23 *Small simple theories admitting finite coding have no dense forking chain.*

PROOF: Suppose there is a dense forking chain, i.e. there is a real tuple $\bar{a}$ and sets A_q for $q \in \mathbb{Q}$ such that $A_q \subset A_r$ and $\bar{a} \not\perp_{A_q} A_r$ for any $q < r$. Consider, for $r \in \mathbb{R}$, the types $p_r = \text{tp}(\bar{a}/\bigcup_{q<r} A_q)$. For any $q < r$ there are rational i, j with $q \leq i < j \leq r$, and there are some formula φ and some $k < \omega$ with

$$D(p_q, \varphi, k) = D(\bar{a}/\bigcup_{s<q} A_s, \varphi, k) \geq D(\bar{a}/A_i, \varphi, k)$$
$$> D(\bar{a}/A_j, \varphi, k) \geq D(\bar{a}/\bigcup_{s<r} A_s) = D(p_r, \varphi, k).$$

Hence p_q and p_r have no common non-forking extension; it follows from finite coding that there are continuum many types over finite sets, contradicting smallness. ■

LEMMA 6.1.24 *If T has no dense forking chain, then no finitary type has infinite weight.*

PROOF: Suppose $(b_i : i \in \mathbb{Q})$ is an independent family over A, and a a finitary hyperimaginary with $a \not\perp_A b_i$ for all $i \in \mathbb{Q}$. Put $B_q = A \cup \{b_i : i \leq q\}$ for each $q \in \mathbb{Q}$. Then $a \not\perp_{B_p} B_q$ for any two $p < q$, and we get a dense forking chain. ■

We shall now define dimension and SU_α–rank:

DEFINITION 6.1.25 The *dimension* $\dim(p; q)$ is the smallest function from pairs $(p; q)$, where q is a forking extension of p, to On^+ satisfying

> $\dim(p; q) \geq \alpha + 1$ if there are non-forking extensions p' of p and q' of q with $p' \subseteq q'$, and types p_i for $i < \omega$, such that $p' \subseteq p_0 \subseteq p_1 \subseteq \cdots \subseteq q'$ and $\dim(p_i; p_{i+1}) \geq \alpha$ for all $i < \omega$.

We put $\dim(p; q) = -1$ if q is a non-forking extension of p, and $\dim(p) = \dim(p; q)$, where q is any algebraic extension of p (i.e. if $p = \text{tp}(a/A)$, we can take $q = \text{tp}(a/Aa)$).

We shall write $\dim(a/A; a/B)$ for $\dim(\text{tp}(a/A); \text{tp}(a/B))$.

REMARK 6.1.26 1. Dimension is invariant under automorphisms.

2. If $p' \subseteq p$ and $q' \supseteq q$, then $\dim(p; q) \leq \dim(p'; q')$.

3. T is superstable if and only if $\dim(p) \leq 0$ for all finitary p.

4. If p' is a non-forking extension of p and q' is a non-forking extension of q, with $p \subseteq q$ and $p' \subseteq q'$, it is not clear that $\dim(p; q) = \dim(p'; q')$.

LEMMA 6.1.27 *For any elements a, b, sets $A \subseteq B$, and $\alpha \geq 0$:*

1. *If* $A' \perp_A Ba$ *and* $\dim(a/A; a/B) \geq \alpha$, *then* $\dim(a/AA'; a/BA') \geq \alpha$.
2. *If* $\dim(ab/A; ab/B) \geq \alpha$, *then either* $\dim(b/Aa; b/Ba) \geq \alpha$ *or* $\dim(a/A; a/B) \geq \alpha$.
3. *If* $\dim(a/A; a/B) \geq \alpha$, *then* $\dim(ab/A; ab/B) \geq \alpha$.

PROOF: We use induction on α for all three assertions.

1. If $\alpha = 0$ the assertion says that $a \not\perp_A B$ implies $a \not\perp_{AA'} B$, which follows from $a \perp_A A'$. So suppose $\dim(a/A; a/B) \geq \alpha + 1$, and consider a sequence $A \subseteq A_0 \subseteq A_1 \subseteq \cdots$ with $\dim(a/A_i; a/A_{i+1}) \geq \alpha$ for all $i < \omega$, and $a \perp_B \bigcup_{i<\omega} A_i$. By automorphism invariance we may choose $\bigcup_{i<\omega} A_i \perp_{Ba} A'$. Then $A' \perp_A Ba \bigcup_{i<\omega} A_i$; by inductive hypothesis $\dim(a/A_iA'; a/A_{i+1}A') \geq \alpha$ for all $i < \omega$. As $a \perp_B A' \bigcup_{i<\omega} A_i$, we get $a \perp_{BA'} \bigcup_{i<\omega} A_i$, and $\dim(a/AA'; a/BA') \geq \alpha + 1$.
2. If $\alpha = 0$, the assertion says that $ab \not\perp_A B$ implies $b \not\perp_{Aa} B$ or $a \not\perp_A B$, which is true. So suppose $\dim(ab/A; ab/B) \geq \alpha + 1$, and consider a sequence $A \subseteq A_0 \subseteq A_1 \subseteq \cdots$ with $ab \perp_B \bigcup_{i<\omega} A_i$ and $\dim(ab/A_i; ab/A_{i+1}) \geq \alpha$ for all $i < \omega$. Put $I = \{i < \omega : \dim(b/A_ia; b/A_{i+1}a) \geq \alpha\}$ and $J = \{i < \omega : \dim(a/A_i; a/A_{i+1}) \geq \alpha\}$. By inductive hypothesis $I \cup J = \omega$. Hence one of the two is infinite; as $b \perp_{Ba} \bigcup_{i<\omega} A_i$ and $a \perp_B \bigcup_{i<\omega} A_i$, the result follows.
3. If $\alpha = 0$, the assertion says that $a \not\perp_A B$ implies $ab \not\perp_A B$, which is true. So suppose $\dim(a/A; a/B) \geq \alpha + 1$, and consider a sequence $A \subseteq A_0 \subseteq A_1 \subseteq \cdots$ with $a \perp_B \bigcup_{i<\omega} A_i$ and $\dim(a/A_i; a/A_{i+1}) \geq \alpha$ for all $i < \omega$. We may assume $\bigcup_{i<\omega} A_i \perp_{Ba} b$, whence $ab \perp_B \bigcup_{i<\omega} A_i$. By inductive hypothesis $\dim(ab/A_i; ab/A_{i+1}) \geq \alpha$ for all $i < \omega$, and hence $\dim(ab/A; ab/B) \geq \alpha + 1$. ■

LEMMA 6.1.28 *T has a dense forking chain if and only if for some $p \subset q$, $\dim(p; q) = \infty$.*

PROOF: Suppose $(p_q : q \in \mathbb{Q})$ is a dense forking chain. We see immediately that if for all $q < r$ we have $\dim(p_q; p_r) \geq \alpha$, then for all $q < r$ we have $\dim(p_q; p_r) \geq \alpha + 1$. But this means $\dim(p_q) \geq \alpha$ for all α and all $q \in \mathbb{Q}$.

Conversely, as $\dim(p; q) \geq 0$ implies $D(p, \varphi, k) < D(q, \varphi, k)$ for some formula φ and some $k < \omega$, there is some ordinal δ such that if $\dim(p; q) \geq \delta$, then $\dim(p; q) = \infty$. Hence if $\dim(p; q) = \infty$, then $\dim(p; q) \geq \delta + 1$, so there are non-forking extensions p' of p and q' of q and a type r with $p' \subset r \subset q'$

and $\dim(p;r) \geq \delta$ and $\dim(r;q) \geq \delta$. Hence $\dim(p';r) = \dim(r;q') = \infty$, and we inductively construct a dense forking chain. ■

DEFINITION 6.1.29 For $\alpha \geq 0$ the SU_α*–rank* is the smallest function from the class of all types to On^+ satisfying:

> $SU_\alpha(p) \geq \beta + 1$ if p has an extension q such that $\dim(p;q) \geq \alpha$ and $SU_\alpha(q) \geq \beta$.

As usual, we write $SU_\alpha(a/A)$ for $SU_\alpha(\mathrm{tp}(a/A))$.

REMARK 6.1.30 1. $SU_0(p)$ is the usual SU–rank of p.

2. $SU_\alpha(p) = 0 \Leftrightarrow \dim(p) < \alpha$, and $SU_\alpha(p) = \infty \Leftrightarrow \dim(p) > \alpha$.

3. A theory has no dense forking chains if and only if for every type p there is some $\alpha < \infty$ with $SU_\alpha(p) < \infty$.

LEMMA 6.1.31 *If* $a \mathop{\smile\!\!\!\!\!\vert}_A A'$*, then* $SU_\alpha(a/A) = SU_\alpha(a/AA')$.

PROOF: Clearly $SU_\alpha(a/A) \geq SU(a/AA')$, so we show inductively on β that $SU_\alpha(a/A) \geq \beta$ implies $SU_\alpha(a/AA') \geq \beta$. This is clear for $\beta = 0$. Now if $SU_\alpha(a/A) \geq \beta + 1$, there is some $B \supset A$ with $\dim(a/A; a/B) \geq \alpha$ and $SU_\alpha(a/B) \geq \beta$; we may assume $B \mathop{\smile\!\!\!\!\!\vert}_{Aa} A'$, whence $A' \mathop{\smile\!\!\!\!\!\vert}_A Ba$. By Lemma 6.1.27.1 we get $\dim(a/AA'; a/BA') \geq \alpha$; the inductive assumption yields $SU_\alpha(a/BA') \geq \beta$, whence $SU(a/AA') \geq \beta + 1$. ■

PROPOSITION 6.1.32 SU_α–RANK INEQUALITIES

1. $SU_\alpha(b/Aa) + SU_\alpha(a/A) \leq SU_\alpha(ab/A) \leq SU_\alpha(b/Aa) \oplus SU_\alpha(a/A)$.

2. *Suppose* $SU_\alpha(a/Ab) < \infty$ *and* $SU_\alpha(a/A) \geq SU_\alpha(a/Ab) \oplus \beta$. *Then* $SU_\alpha(b/A) \geq SU_\alpha(b/Aa) + \beta$.

3. *Suppose* $SU_\alpha(a/Ab) < \infty$ *and* $SU_\alpha(a/A) \geq SU_\alpha(a/Ab) + \omega^\beta n$. *Then* $SU_\alpha(b/A) \geq SU_\alpha(b/Aa) + \omega^\beta n$.

4. *If* $a \mathop{\smile\!\!\!\!\!\vert}_A b$*, then* $SU_\alpha(ab/A) = SU_\alpha(a/A) \oplus SU_\alpha(b/A)$.

PROOF:

1. For the left-hand inequality, we show by induction on β that $SU_\alpha(a/A) \geq \beta$ implies $SU_\alpha(b/Aa) + \beta \leq SU_\alpha(ab/A)$. Clearly we may assume that $SU_\alpha(ab/A) < \infty$.

   For $\beta = 0$, the implication means $SU_\alpha(b/Aa) \leq SU_\alpha(ab/A)$, which again we prove inductively. So suppose $SU_\alpha(b/Aa) \geq \gamma + 1$, and consider $B \supseteq Aa$ such that $\dim(b/Aa; b/B) \geq \alpha$ and $SU_\alpha(b/B) \geq \gamma$. Then

$\dim(ab/Aa; ab/B) \geq \alpha$ by Lemma 6.1.27.3, whence $\dim(ab/A; ab/B) \geq \alpha$. As $B = Ba$, the inductive hypothesis implies $SU_\alpha(ab/B) \geq \gamma$, and $SU_\alpha(ab/A) \geq \gamma + 1$.

Suppose $SU_\alpha(a/A) \geq \beta+1$, and choose $B \supset A$ with $\dim(a/A; a/B) \geq \alpha$ and $SU_\alpha(a/B) \geq \beta$. We may take $B \mathop{\perp}_{Aa} b$, whence $SU_\alpha(b/Ba) = SU_\alpha(b/Aa)$ by Lemma 6.1.31. By Lemma 6.1.27.3 $\dim(ab/A; ab/B) \geq \alpha$, and hence $SU_\alpha(ab/A) \geq SU_\alpha(ab/B) + 1$. So the inductive hypothesis implies

$$\begin{aligned} SU_\alpha(b/Aa) + \beta + 1 &= SU_\alpha(b/Ba) + \beta + 1 \\ &\leq SU_\alpha(ab/B) + 1 \leq SU_\alpha(ab/A). \end{aligned}$$

For the right–hand inequality, we show by induction on β that $SU_\alpha(ab/A) \geq \beta$ implies $SU_\alpha(b/Aa) \oplus SU_\alpha(a/A) \geq \beta$. Again we may assume that both summands on the right–hand side have ordinal SU_α–rank. Suppose $SU_\alpha(ab/A) \geq \beta+1$, and consider $B \supset A$ with $\dim(ab/A; ab/B) \geq \alpha$ and $SU_\alpha(ab/B) \geq \beta$. By induction hypothesis $SU_\alpha(b/Ba) \oplus SU_\alpha(a/B) \geq \beta$. Lemma 6.1.27.2 implies that either $SU_\alpha(b/Ba) < SU_\alpha(b/Aa)$ or $SU_\alpha(a/B) < SU_\alpha(a/A)$, and hence $SU_\alpha(b/Aa) \oplus SU_\alpha(a/A) \geq \beta + 1$.

2. By induction on $SU_\alpha(a/Ab) \oplus \beta$, the cases $\beta = 0$ and β a limit ordinal being trivial. Suppose $SU_\alpha(a/A) \geq SU_\alpha(a/Ab) \oplus \beta + 1$, and consider some $B \supseteq A$ with $\dim(a/A; a/B) \geq \alpha$ and $SU_\alpha(a/B) \geq SU_\alpha(a/Ab) \oplus \beta$. We may choose $B \mathop{\perp}_{Aa} b$, whence $SU_\alpha(b/Ba) = SU_\alpha(b/Aa)$ by Lemma 6.1.31. If $\dim(b/A; b/B) \geq \alpha$, as $SU_\alpha(a/B) \geq SU_\alpha(a/Bb) \oplus \beta$, the inductive hypothesis implies

$$SU_\alpha(b/A) \geq SU_\alpha(b/B)+1 \geq SU_\alpha(b/Ba)+\beta+1 = SU_\alpha(b/Aa)+\beta+1.$$

Otherwise $\dim(a/Ab; a/Bb) \geq \alpha$ by Lemma 6.1.27.3 and 6.1.27.2. Therefore $SU_\alpha(a/Ab) \geq SU_\alpha(a/Bb) + 1$ and

$$SU_\alpha(a/B) \geq SU_\alpha(a/Ab) \oplus \beta \geq SU_\alpha(a/Bb) \oplus (\beta + 1),$$

whence again by inductive hypothesis

$$SU_\alpha(b/A) \geq SU_\alpha(b/B) \geq SU_\alpha(b/Ba) + \beta + 1 = SU_\alpha(b/Aa) + \beta + 1.$$

3. This follows by continuity from the previous part.

4. If, say, $SU_\alpha(a/A) \geq \beta + 1$, then there is a superset B of A such that $\dim(a/A; a/B) \geq \alpha$ and $SU_\alpha(a/B) \geq \beta$. We may choose $B \mathop{\perp}_{Aa} b$,

whence $Ba \mathop{\smile\!\!\!\!\!\!\,|}_A b$, and $SU_\alpha(b/B) = SU_\alpha(b/A)$ by Lemma 6.1.31. Moreover, $a \mathop{\smile\!\!\!\!\!\!\,|}_B b$, and $\dim(ab/A; ab/B) \geq \alpha$ by Lemma 6.1.27.3. By inductive hypothesis

$$\begin{aligned} SU_\alpha(ab/A) &\geq SU_\alpha(ab/B) + 1 \geq [SU_\alpha(a/B) \oplus SU_\alpha(b/B)] + 1 \\ &\geq (\beta + 1) \oplus SU_\alpha(b/A). \quad \blacksquare \end{aligned}$$

REMARK 6.1.33 It is not true that $SU_\alpha(a/B) = \inf\{SU_\alpha(a/\bar{b}) : \bar{b} \in B \text{ finite}\}$. Conversely, it is possible that $SU_\alpha(\bar{b}) = 0$ for all $\bar{b} \in B$, but $SU_\alpha(B) > 0$. This distinguishes SU_α-rank from the SU_P-rank introduced in Remark 5.1.19, which satisfies those continuity properties.

PROOF: Consider an elementary abelian 2-group G with predicates $(H_i : i < \omega)$ for a descending chain of subgroups, each of infinite index in its predecessor. This is an abelian structure and thus stable; in fact $SU_1(G) = 1$. Let a be a generic element (i.e. $a \in G - H_0$), and put $B = \{aH_i : i < \omega\}$ (a set of imaginary elements). Then $SU_1(a/\bar{b}) = 1$ for all finite $\bar{b} \in B$, but $SU_1(a/B) = 0$. Moreover, $SU_1(\bar{b}) = 0$ for all finite $\bar{b} \in B$, and $SU_1(B) = 1$, in accordance with Proposition 6.1.32.

In a similar vein, if B is an infinite independent set in a supersimple theory, then $SU_1(\bar{b}) = 0$ for all finite $\bar{b} \in B$, and $SU_1(B) = \infty$. ■

PROPOSITION 6.1.34 *In a small theory which admits finite coding, every type is non-orthogonal to some regular type.*

PROOF: By Proposition 6.1.23 and Lemma 6.1.28 all types have ordinal SU_α-rank for some $\alpha < \infty$. Consider a type $\mathrm{tp}(a/A)$, and choose $c \not\mathop{\smile\!\!\!\!\!\!\,|}_B a$ for some $B \mathop{\smile\!\!\!\!\!\!\,|}_A a$, such that first $\alpha = \dim(c/B)$ and then $\beta = SU_\alpha(c/B)$ are minimal. By local and finite character, we may increase B and assume that a dominates c over B.

We claim that $\mathrm{tp}(c/B)$ is regular. So consider $C \not\mathop{\smile\!\!\!\!\!\!\,|}_B c$, and suppose that $\mathrm{tp}(c/C)$ is non-orthogonal to $\mathrm{tp}(c/B)$. As $a \rhd_B c$, we see that $\mathrm{tp}(c/C)$ must be non-orthogonal to $\mathrm{tp}(a/A)$; by our minimal choice of α and β we have $SU_\alpha(c/C) = \beta$. Let $(c_i : i < \omega)$ be a Morley sequence in $\mathrm{lstp}(c/C)$. By elimination of hyperimaginaries, there is an imaginary element $d \in \mathrm{Cb}(c/C)$ with $c \not\mathop{\smile\!\!\!\!\!\!\,|}_B d$; since $\mathrm{Cb}(c/C) \in \mathrm{dcl}(c_i : i < \omega)$ by Corollary 3.3.13, there is some $n < \omega$ with $d \in \mathrm{dcl}(c_i : i < n)$. Then

$$\begin{aligned} \beta = SU_\alpha(c_i/B) &\geq SU_\alpha(c_i/B, c_j : j < i) \\ &\geq SU_\alpha(c_i/B, d, c_j : j < i) \geq SU_\alpha(c_i/C) = \beta \end{aligned}$$

for all $i < n$, and equality holds. By Proposition 6.1.32 we have

$$\begin{aligned} SU_\alpha(c_i : i < n/B) &\leq \beta \oplus \cdots \oplus \beta \quad (n \text{ summands}), \text{ and} \\ SU_\alpha(c_i : i < n/Bd) &\geq \beta + \cdots + \beta \quad (n \text{ summands}). \end{aligned}$$

As $SU_\alpha(d/B, c_i : i < n) = 0$, Proposition 6.1.32 yields

$$\begin{aligned} SU_\alpha(c_i : i < n/Bd) + SU_\alpha(d/B) &\leq SU_\alpha(d, c_i : i < n/B) \\ &= SU_\alpha(c_i : i < n/B), \end{aligned}$$

whence $SU_\alpha(d/B) < \beta$. Contradiction, as $c \not\perp_B d$ and hence $a \not\perp_B d$. ■

COROLLARY 6.1.35 *A finitary type in a small theory which admits finite coding is equidominant with a finite product of regular types (and in particular has finite weight).*

PROOF: By Proposition 6.1.23, Lemma 6.1.24 and Theorem 6.1.34, the assumptions of Theorem 5.2.18 are satisfied. ■

EXERCISE 6.1.36 In a theory without dense forking chain, every finitary type is equidominant with a finite product of regular types.

HINT. If we want to adapt the proof of Proposition 6.1.34, we have to deal with the fact that $\mathrm{Cb}(c/C)$ may be hyperimaginary, and thus not allow a restriction to a finite segment of the Morley sequence.

If P is an $\emptyset$-invariant family of types, relativize Definitions 6.1.25 and 6.1.29 to obtain P-dimension $\dim_P(p/q)$ and $SU_{\alpha,P}$-rank in the manner of Remark 5.1.19, and generalize 6.1.27–6.1.32.

Now, given a type p in a theory without dense forking chains, assume that $p \in S(\emptyset)$ and consider a p-internal type $\mathrm{tp}(c/B)$ dominated by p of least p-dimension α and $SU_{\alpha,p}$-rank β. If $\mathrm{tp}(c/B)$ is not regular, find $C' \supseteq C \supseteq B$ with $c \not\perp_C C'$ such that $C = \mathrm{dcl}(Cc) \cap \mathrm{cl}_p(C)$ and $C' = \mathrm{dcl}(C'c) \cap \mathrm{cl}_p(C')$, and both $\mathrm{tp}(c/C)$ and $\mathrm{tp}(c/C')$ are non-orthogonal to p, but whenever $C \subset D \subset D'$ with $c \not\perp_C D$ and $c \not\perp_D D'$, $\mathrm{dcl}(Dc) \cap \mathrm{cl}_p(D) = D$, and $c \perp_{C'} D'$, then $\mathrm{tp}(c/D)$ is orthogonal to p. By minimality, $SU_{\alpha,p}(c/C) = SU_{\alpha,p}(c/C') = \beta$; moreover one may assume that $C' = C \cup \{c_i : i < \omega\}$ for some Morley sequence $(c_i : i < \omega)$ in $\mathrm{lstp}(c/C')$.

There are $m, n, n' < \omega$, a tuple $(c'_i : i \leq n)$ of independent realizations of $\mathrm{tp}(c/C')$, and a tuple $(a_i : i < n')$ of independent realizations of a non-forking extension of p to $C' \cup \{c'_i : i < n\}$, such that $(a_i : i < n')$ forks with c_n over $C \cup \{c_j : j < m, c'_i : i < n\}$. Put

$$D = \mathrm{dcl}(C, c_j : j < m, c'_i : i \leq n) \subset \mathrm{cl}_p(C, c_j : j < m, c'_i : i < n).$$

Then p is non-orthogonal to $\mathrm{tp}(c_n/D)$, and $c_n \perp_{C'} D$; the choice of C and C' yields $c_n \perp_D C'$ and $\mathrm{Cb}(c_n/D) \in \mathrm{cl}_p(C, c_j : j < m, c'_i : i < n)$. Compute

$$\begin{aligned} &SU_{\alpha,p}(c_j : j < m, c'_i : i < n/C) \leq \beta \oplus \cdots \oplus \beta \quad (m+n \text{ summands}), \text{ and} \\ &SU_{\alpha,p}(c_j : j < m, c'_i : i < n/Cd) \geq \beta + \cdots + \beta \quad (m+n \text{ summands}). \end{aligned}$$

As $SU_{\alpha,p}(d/C, c_j : j < m, c'_i : i < n) = SU_{\alpha,p}(d/D) = 0$, deduce from the Lascar inequalities for $SU_{\alpha,p}$-rank that $SU_{\alpha,p}(d/C) < \beta$. But then $d \in \mathrm{cl}_p(C)$ and $c \perp_C d$, a contradiction.

So every type is non-orthogonal to a regular type, and we may finish by Theorem 5.2.18.

PROBLEM 6.1.37 Does a simple theory without dense forking chains have elimination of hyperimaginaries?

6.1.4 LACHLAN'S CONJECTURE

Lachlan's Conjecture states:

CONJECTURE 6.1.38 *A stable countable theory is either ω-categorical, or has infinitely many non-isomorphic countable models.*

The assertion would be false without the stability assumption.

EXAMPLE 6.1.39 Let $T = \mathrm{Th}\langle \mathbb{Q}, <, n : n \in \mathbb{N}\rangle$. Then T has exactly three non-isomorphic countable models: the prime model $\mathbb{Q}$, a model where $\{n : n \in \mathbb{N}\}$ has a supremum, and the saturated model where it has upper bounds but no supremum.

Similarly, for any $n > 2$ there is a theory which has exactly n non-isomorphic countable models; these theories are unstable, and in fact not simple. It is not possible, however, for a theory to have exactly two non-isomorphic countable models (Vaught [158]).

One may generalize Lachlan's Conjecture to simple theories:

CONJECTURE 6.1.40 *A simple countable theory is either ω-categorical, or has infinitely many non-isomorphic countable models.*

In this subsection we shall prove the above conjecture for simple theories with finite coding (and in particular for supersimple and for small one-based simple theories).

DEFINITION 6.1.41 Let $\bar{a}$ and $\bar{b}$ be real (or imaginary) tuples. Then $\bar{b}$ is *semi-isolated* over $\bar{a}$ if there is a formula $\varphi(\bar{x}, \bar{a}) \in \mathrm{tp}(\bar{b}/\bar{a})$ such that $\varphi(\bar{x}, \bar{a}) \vdash \mathrm{tp}(\bar{b})$.

LEMMA 6.1.42 *If $\mathrm{tp}(\bar{b}/\bar{a})$ is isolated, then $\bar{b}$ is semi-isolated over $\bar{a}$. If $\bar{c}$ is semi-isolated over $\bar{b}$ and $\bar{b}$ is semi-isolated over $\bar{a}$, then $\bar{c}$ is semi-isolated over $\bar{a}$.*

PROOF: The first statement is obvious. For the second assertion, if $\varphi(\bar{x}, \bar{b}) \vdash \mathrm{tp}(\bar{c})$ and $\psi(\bar{x}, \bar{a}) \vdash \mathrm{tp}(\bar{b})$, then $\exists \bar{y}\, [\varphi(\bar{x}, \bar{y}) \wedge \psi(\bar{y}, \bar{a})] \vdash \mathrm{tp}(\bar{c})$. ■

EXAMPLE 6.1.43 In a nonstandard model of $\mathrm{Th}(\mathbb{Z}, S)$, where S is the successor function, consider a and b in different S-chains. Then b is semi-isolated over a, but $\mathrm{tp}(b/a)$ is not isolated.

LEMMA 6.1.44 *If $\mathrm{tp}(\bar{b}/\bar{a})$ is isolated and $\bar{a}$ is semi-isolated over $\bar{b}$, then $\mathrm{tp}(\bar{a}/\bar{b})$ is isolated.*

PROOF: Choose $\varphi(\bar{x}, \bar{b}) \in \mathrm{tp}(\bar{a}/\bar{b})$ such that $\varphi(\bar{x}, \bar{b}) \vdash \mathrm{tp}(\bar{a})$, and let $\psi(\bar{y}, \bar{a})$ isolate $\mathrm{tp}(\bar{b}/\bar{a})$. Put $\vartheta(\bar{x}, \bar{b}) = \varphi(\bar{x}, \bar{b}) \wedge \psi(\bar{b}, \bar{x})$. If $\bar{a}' \models \vartheta$, then $\mathrm{tp}(\bar{a}') =$

$\text{tp}(\bar{a})$, so $\psi(\bar{y}, \bar{a}')$ isolates a complete type over $\bar{a}'$; it follows that $\text{tp}(\bar{a}', \bar{b}) = \text{tp}(\bar{a}, \bar{b})$, and $\text{tp}(\bar{a}/\bar{b})$ is isolated. ■

PROPOSITION 6.1.45 *If $\bar{a}$ and $\bar{b}$ are two independent realizations of a complete type $p \in S(\emptyset)$ in a simple theory, such that $\bar{b}$ is semi-isolated over $\bar{a}$, then $\bar{a}$ is semi-isolated over $\bar{b}$.*

PROOF: Pick $\bar{c} \downarrow \bar{a}\bar{b}$ such that $\text{tp}(\bar{c}\bar{b}) = \text{tp}(\bar{b}\bar{a})$. Choose an independent sequence $(\bar{a}_i\bar{b}_i\bar{c}_i : i < \omega)$ in $\text{tp}(\bar{a}\bar{b}\bar{c})$ with $\bar{a}_0\bar{b}_0\bar{c}_0 = \bar{a}\bar{b}\bar{c}$ and $\text{tp}(\bar{a}_{i+1}\bar{c}_i) = \text{tp}(\bar{b}\bar{a})$. By transitivity of semi-isolation, $\bar{a}_j$ is semi-isolated over $\bar{a}_i$ for all $j \geq i$.

If $\varphi(\bar{x}, \bar{a}) \in \text{tp}(\bar{b}/\bar{a})$ implies $\text{tp}(\bar{b})$, put $\vartheta(\bar{x}, \bar{a}, \bar{c}) = \varphi(\bar{c}, \bar{x}) \wedge \varphi(\bar{x}, \bar{a})$. Now if $\vartheta(\bar{x}, \bar{a}_i, \bar{c}_i) \wedge \vartheta(\bar{x}, \bar{a}_j, \bar{c}_j)$ were inconsistent for all $i \neq j$, then Ramsey's Theorem and compactness would yield an infinite indiscernible sequence $(\bar{a}'_i\bar{c}'_i : i < \omega)$ in $\text{tp}(\bar{a}\bar{c})$ such that $\vartheta(\bar{x}, \bar{a}'_i, \bar{c}'_i) \wedge \vartheta(\bar{x}, \bar{a}'_j, \bar{c}'_j)$ is inconsistent for all $i \neq j$. So $\vartheta(\bar{x}, \bar{a}, \bar{c})$ forks over $\emptyset$, contradicting $\bar{b} \models \vartheta(\bar{x}, \bar{a}, \bar{c})$ and $\bar{b} \downarrow \bar{c}\bar{a}$. Hence there is $i > j$ such that $\varphi(\bar{c}_j, \bar{x}) \wedge \varphi(\bar{x}, \bar{a}_i)$ is consistent, say it is realized by $\bar{d}$. Again by transitivity of semi-isolation, $\bar{c}_j$ is semi-isolated over $\bar{a}_i$. As $\bar{a}_i$ is semi-isolated over $\bar{a}_{j+1}$, so is $\bar{c}_j$ by transitivity; since $\text{tp}(\bar{a}_{j+1}\bar{c}_j) = \text{tp}(\bar{b}\bar{a})$, we see that $\bar{a}$ is semi-isolated over $\bar{b}$. ■

COROLLARY 6.1.46 *If $\bar{a}$ and $\bar{b}$ are two independent realizations of a complete type $p \in S(\emptyset)$ such that $\bar{b}$ is isolated over $\bar{a}$, then $\bar{a}$ is isolated over $\bar{b}$.* ■

REMARK 6.1.47 Simplicity of T is essential in Proposition 6.1.45. Consider the theory given in Example 6.1.39, and choose a, b with $n < a < b$ for all $n \in \mathbb{N}$. Then $\text{tp}(a) = \text{tp}(b)$ and $tp(b/a)$ is isolated, whereas $tp(a/b)$ is not isolated. But neither $\text{tp}(a/b)$ nor $\text{tp}(b/a)$ fork over $\emptyset$. However, for any c with $\text{tp}(cb) = \text{tp}(ba)$, the type $\text{tp}(b/ac)$ forks over $\emptyset$.

THEOREM 6.1.48 *A countable simple theory T with finite coding has either one or infinitely many non-isomorphic countable models.*

PROOF: If T is not small, it has continuum many non-isomorphic countable models. So we may assume that T is small; by Corollary 6.1.35 every type has finite weight. Clearly, we may also assume that T is not ω-categorical. So suppose for a contradiction that T has only finitely many non-isomorphic countable models.

CLAIM. There is a tuple $\bar{a}$ and a prime model $\mathfrak{M}$ over $\bar{a}$, such that $tp(\bar{a})$ is not isolated and every complete n-type (for all n) over $\emptyset$ is realized in $\mathfrak{M}$. Moreover there is a tuple $\bar{a}' \models \text{tp}(\bar{a})$ in $\mathfrak{M}$ such that $\bar{a}$ is not semi-isolated over $\bar{a}'$.

PROOF OF CLAIM: Let $(p_0, p_1, p_2, \ldots)$ be an enumeration of all real types in $S(\emptyset)$, choose realizations $\bar{c}_i \models p_i$, and put $\bar{d}_i = (\bar{c}_0, \ldots, \bar{c}_i)$ for all $i < \omega$. By Corollary 6.1.5 there is an atomic prime model $\mathfrak{M}_i$ over $\bar{d}_i$; since there are only finitely many non-isomorphic countable models, there is a countable model $\mathfrak{M}$ which is isomorphic to $\mathfrak{M}_i$ for infinitely many $i < \omega$. Hence $\mathfrak{M}$ realizes all types in $S(\emptyset)$; since $\mathfrak{M}$ must also realize a non-isolated type, it is not prime over $\emptyset$. So if $\mathfrak{M}$ is prime over $\bar{a}$, then $\mathrm{tp}(\bar{a})$ is not isolated.

As $\mathrm{Th}(\mathfrak{M}, \bar{a})$ is not ω-categorical either, there is some tuple $\bar{b}$ such that $\mathrm{tp}(\bar{b}/\bar{a})$ is not isolated. If $\bar{a}'\bar{b}'$ realizes $\mathrm{tp}(\bar{a}\bar{b})$ in $\mathfrak{M}$, then $\mathfrak{M}$ cannot be prime (and hence atomic) over $\bar{a}'$. Since M is prime over $\bar{a}$, the type $tp(\bar{a}/\bar{a}')$ cannot be isolated. Hence $\bar{a}$ is not semi-isolated over $\bar{a}'$ by Lemma 6.1.44. ■

Put $p = \mathrm{tp}(\bar{a})$.

CLAIM. There are two independent realizations $\bar{a}_0$ and $\bar{a}_0'$ in $\mathfrak{M}$ of p such that $\bar{a}_0'$ is not semi-isolated over $\bar{a}_0$.

PROOF OF CLAIM: Let $\bar{a}'' \perp \bar{a}\bar{a}'$ be a realization of p. As $\bar{a}$ is not semi-isolated over $\bar{a}'$, either $\bar{a}$ is not semi-isolated over $\bar{a}''$, or $\bar{a}''$ is not semi-isolated over $\bar{a}'$. Now choose a realization $\bar{a}_0'\bar{a}_0$ of $\mathrm{tp}(\bar{a}\bar{a}'')$ or $\mathrm{tp}(\bar{a}''\bar{a}')$ inside $\mathfrak{M}$. ■

Since $\bar{a}_0, \bar{a}_0'$ are both in $\mathfrak{M}$, they must be semi-isolated over $\bar{a}$. Choose inductively $\bar{a}_i, \bar{a}_i' \models p$ such that $\mathrm{tp}(\bar{a}_{i+1}\bar{a}_{i+1}'\bar{a}_i') = \mathrm{tp}(\bar{a}_0\bar{a}_0'\bar{a})$, and $\bar{a}_{i+1}\bar{a}_{i+1}' \perp_{\bar{a}_i'} (\bar{a}_j : j \leq i)$ for all $i < \omega$. By transitivity of semi-isolation, $\bar{a}_i$ and $\bar{a}_i'$ are semi-isolated over $\bar{a}$ for all $i < \omega$.

CLAIM. $\bar{a}$ is not semi-isolated over $\bar{a}_i$.

PROOF OF CLAIM: Suppose $\bar{a}$ is semi-isolated over $\bar{a}_i$. Then $\bar{a}_i'$ is semi-isolated over $\bar{a}_i$ by transitivity, a contradiction. ■

Hence $\bar{a} \not\perp \bar{a}_i$ by Proposition 6.1.45.

CLAIM. $(\bar{a}_i : i < \omega)$ is independent.

PROOF OF CLAIM: We show that $(\bar{a}_0, \ldots, \bar{a}_n, \bar{a}_n')$ is independent for all $n < \omega$. This is true for $n = 0$, so assume it holds for n. Then $(\bar{a}_i : i \leq n) \perp \bar{a}_n'$; since $\bar{a}_{n+1}\bar{a}_{n+1}' \perp_{\bar{a}_n'} (\bar{a}_i : i \leq n)$, we get $\bar{a}_{n+1}\bar{a}_{n+1}' \perp (\bar{a}_i : i \leq n)$. The assertion now follows from $\bar{a}_{n+1} \perp \bar{a}_{n+1}'$. ■

But this contradicts finiteness of $w(p)$. ■

6.2 ω-CATEGORICAL THEORIES

An ω-categorical superstable theory is well understood: it is one-based and has finite SU-rank (even finite Morley rank). Much less is known about ω-categorical supersimple structures (and very little about simple ω-categorical theories), except that things are not as nice. In this section, we shall first present Hrushovski's amalgamation method and obtain an ω-categorical structure of SU-rank 1 which is not one-based. We shall then deal with the known cases:

ω-categorical supersimple groups are finite-by-abelian-by-finite of finite SU-rank, and ω-categorical one-based theories have finite SU-rank. In fact, the latter result allows a generalization to a slightly more complicated class of theories, the *CM-trivial* theories (see Subsection 6.2.3). However, in general we cannot solve:

CONJECTURE 6.2.1 *A supersimple ω-categorical theory has finite SU-rank.*

6.2.1 AN AMALGAMATION CONSTRUCTION

Let $\mathfrak{L}$ be a countable relational language $\{R_i : i < \omega\}$, such that there are only finitely many relations of any given arity, and fix non-negative real numbers $(\alpha_i : i < \omega)$; we call α_i the *weight* of the relation R_i. If A is an $\mathfrak{L}$-structure, we call a tuple $\bar{a} \in A$ with $A \models R_i(\bar{a})$ an *i-edge* of A, and denote the set of i-deges of A by $R_i(A)$. Put

$$\delta(A) = |A| - \sum_{i<\omega} \alpha_i |R_i(A)|,$$

i.e. $\delta(A)$ is the number of points of A minus the number of weighted edges of A. Let $\mathcal{K}$ be the universal class of finite $\mathfrak{L}$-structures such that a relation R_i holds only on distinct elements, for all $i < \omega$, and put

$$\mathcal{K}_0 = \{A \in \mathcal{K} : \delta(B) > 0 \text{ for all non-empty } B \subseteq A\},$$

again a universal class. As there are only finitely many relations of arity at most $|A|$, there are only finitely many non-isomorphic structures in $\mathcal{K}$ of any given finite cardinality, and $\sum_{i<\omega} \alpha_i |R_i(A)|$ is well-defined for all $A \in \mathcal{K}$.

DEFINITION 6.2.2 Let B be an $\mathfrak{L}$-structure, $A_1, A_2 \subseteq B$, and $A_0 = A_1 \cap A_2$. Then A_1 and A_2 are *freely amalgamated* over A_0 if the only edges in $A_1 \cup A_2$ which are not edges of A_1 or of A_2 have weight 0. If A_1 and A_2 are freely amalgamated over A_0, the amalgam is *canonical* if there are no edges of $A_1 \cup A_2$ which are not edges of A_1 or of A_2. We denote the canonical free amalgam of A_1 and A_2 over A_0 by $A_1 \otimes_{A_0} A_2$.

REMARK 6.2.3 If A_1 and A_2 are (canonically) freely amalgamated over their intersection A_0 and $B \subseteq A_1 \cup A_2$, we may put $B_i = B \cap A_i$ for $i = 0, 1, 2$. Then B_1 and B_2 are (canonically) freely amalgamated over their intersection B_0.

Clearly, if A_1 and A_2 are $\mathfrak{L}$-structures with substructures $A_0 \subseteq A_1$ and $A_0' \subseteq A_2$ such that $A_0 \cong A_0'$, we can identify A_0 and A_0' and form the canonical free amalgam $A_1 \otimes_{A_0} A_2$.

LEMMA 6.2.4 *Let $B \in \mathcal{K}$, and $A_1, A_2 \subseteq B$. If $A_0 = A_1 \cap A_2$, then*

$$\delta(A_1 \cup A_2) + \delta(A_0) \leq \delta(A_1) + \delta(A_2),$$

and equality holds if and only if A_1 and A_2 are freely amalgamated over A_0.

PROOF: Clearly $|A_1 \cup A_2| + |A_0| = |A_1| + |A_2|$. Now an edge of A_1 or of A_2 is an edge of $A_1 \cup A_2$; if it is both in A_1 and in A_2, it also is in $A_1 \cap A_2$. The assertion follows. ■

DEFINITION 6.2.5 If $B \in \mathcal{K}$ and $A \subseteq B$, we say that A is *closed* in B, denoted $A \leq B$, if $\delta(A') > \delta(A)$ for all A' with $A \subset A' \subseteq B$.
If $\sigma : A \to B$ is an embedding, σ is *closed* if its image is closed in B.

Note that $\emptyset \leq A$ for all $A \in \mathcal{K}_0$.

LEMMA 6.2.6 *Let $B \in \mathcal{K}$ and $A \leq B$. If $B' \subseteq B$, then $A \cap B' \leq B'$.*

PROOF: For any A'' with $A \cap B' \subset A'' \subseteq B'$ we have $A \subset A \cup A'' \subseteq B$, whence $\delta(A) < \delta(A \cup A'')$. Moreover $A \cap B' = A \cap A''$, and by Lemma 6.2.4

$$\delta(A \cap B') = \delta(A \cap A'') \leq \delta(A) + \delta(A'') - \delta(A \cup A'') < \delta(A''). \quad ■$$

LEMMA 6.2.7 *If $C \in \mathcal{K}$ and $A \leq B \leq C$, then $A \leq C$.*

PROOF: Consider A' with $A \subset A' \subseteq C$, and put $B' = A' \cap B$. If $A' \subseteq B$, easily $\delta(A) < \delta(A')$ since $A \leq B$. Otherwise $B' \subset A'$; as $B' \leq A'$ by Lemma 6.2.6, we get $\delta(B') < \delta(A')$. Finally, $A \leq B$ implies $\delta(A) \leq \delta(B')$, whence $\delta(A) < \delta(A')$. ■

COROLLARY 6.2.8 *If $B \in \mathcal{K}$ and $A \subseteq B$, then there is a unique minimal closed superstructure of A in B.*

PROOF: By Lemma 6.2.4 the union of all supersets of A in B of minimal δ-rank is closed in B, and clearly minimal and unique such. ■
We denote the minimal closed superset of A in B by $\mathrm{cl}_B(A)$.

LEMMA 6.2.9 *Let $B \in \mathcal{K}$, and A_1, A_2 be substructures of B which are in $\mathcal{K}_0$. If A_1 and A_2 are freely amalgamated over their intersection A_0 and $A_0 \leq A_1$, then $A_1 \cup A_2 \in \mathcal{K}_0$, and $A_2 \leq A_1 \cup A_2$.*

PROOF: Consider any B with $A_2 \subset B \subseteq A_1 \cup A_2$, and put $B_1 = B \cap A_1$. Then $A_0 \subset B_1 \subseteq A_1$, so $\delta(A_0) < \delta(B_1)$. Moreover, B is a free amalgam of A_2 and B_1 over A_0 by Remark 6.2.3. Hence

$$\delta(B) = \delta(A_2) + \delta(B_1) - \delta(A_0) > \delta(A_2),$$

and $A_2 \leq A_1 \cup A_2$. Finally, if $C \subseteq A_1 \cup A_2$ is non-empty, then either $C \subseteq A_1$ and $\delta(C) > 0$, or $C \cap A_2$ is non-empty and $C \cap A_2 \leq C$ by Lemma 6.2.6, whence $0 < \delta(C \cap A_2) \leq \delta(C)$. Thus $A_1 \cup A_2 \in \mathcal{K}_0$. ■

DEFINITION 6.2.10 Let $f : \mathbb{R}^+ \to \mathbb{R}^+$ be a monotone increasing unbounded function. Put

$$\mathcal{K}_f = \{A \in \mathcal{K} : \delta(B) \geq f(|B|) \text{ for all } B \subseteq A\}.$$

In order to avoid case distinctions, for such a function f we shall put $f(0) = 0$, and assume $\emptyset \in \mathcal{K}_f$. Note that $\mathcal{K}_f \subseteq \mathcal{K}_0$. We call $\mathcal{K}_f$ *trivial* if $\mathcal{K}_f = \{\emptyset\}$. If M is an infinite $\mathfrak{L}$-structure whose finite substructures are all in $\mathcal{K}$, write $A \leq M$ for some finite $A \subset M$ if $A \leq A'$ for all finite $A' \subset M$ containing A.

DEFINITION 6.2.11 $\mathcal{K}_f$ is *closed under free amalgamation* if whenever $A_0 \leq A_1, A_2 \in \mathcal{K}_f$, then $A_1 \otimes_{A_0} A_2 \in \mathcal{K}_f$.

Note that any free amalgam B of A_1 and A_2 over A_0 has the same δ-rank as the canonical one. So $B \in \mathcal{K}_f$ if and only if $A_1 \otimes_{A_0} A_2 \in \mathcal{K}_f$.

DEFINITION 6.2.12 A countably infinite $\mathfrak{L}$-structure $\mathfrak{M}$ is a *generic* model for $\mathcal{K}_f$ if every finite substructure of $\mathfrak{M}$ is in $\mathcal{K}_f$, and whenever $A \leq \mathfrak{M}$ and $A \leq B \in \mathcal{K}_f$, then there is a closed embedding σ of B into $\mathfrak{M}$ over A.

THEOREM 6.2.13 *Suppose $\mathcal{K}_f$ is non-trivial, and closed under free amalgamation. Then there is a generic model for $\mathcal{K}_f$.*

PROOF: As $\mathcal{K}_f$ is non-trivial and closed under free amalgamation, it is countably infinite. Let $(A_i : i < \omega)$ be an enumeration of $\mathcal{K}_f$ (up to isomorphism). We shall inductively construct an increasing chain $(M_n : n < \omega)$ of finite $\mathfrak{L}$-structures in $\mathcal{K}_f$, and an enumeration $(B_i : i < \omega)$ of all finite closed substructures of $\bigcup_{n<\omega} M_n$, such that $B_n \leq M_n \leq M_{n+1}$ for all $n < \omega$, and whenever $B_i \leq C$ for some $C \cong A_j$, then there is a closed embedding of C into M_{i+j} over B_i. We put $M_0 = A_0$.

Suppose M_n has been constructed, together with an enumeration $(B_i : i < k_n)$ of the closed substructures of M_n. Enumerate all pairs $B \leq C$ with $B = B_i$ for some $i < k_n$ and $C \cong A_j$ for some $j \leq n + 1 - i$ as $(B_jC_j : j < k)$ for some $k < \omega$. Define inductively $X_0 = M_n$ and $X_{j+1} = X_j \otimes_{B_j} C_j$, for $j \leq k$; since $\mathcal{K}_f$ is closed under free amalgamation, $X_j \in \mathcal{K}_f$ for all $j \leq k$. Put $M_{n+1} = X_k$; Lemmas 6.2.7 and 6.2.9 imply that M_n has the required properties. Now extend the enumeration $(B_i : i < k_n)$ of the closed subsets of M_n (which are still closed in M_{n+1} by Lemma 6.2.7) to an enumeration $(B_i : i < k_{n+1})$ of the closed subsets of M_{n+1}.

Put $\mathfrak{M} = \bigcup_{n<\omega} M_n$; it is easy to see that $\mathfrak{M}$ is generic. ∎

From now on, we fix a monotone increasing unbounded function f such that $\mathcal{K}_f$ is non-trivial and closed under free amalgamation.

LEMMA 6.2.14 *Let $\mathfrak{M}$ be a generic model, and $A \subset \mathfrak{M}$ a finite substructure. Then there is a unique minimal finite set $B \leq \mathfrak{M}$ containing A.*

PROOF: Since f is unbounded increasing, there is some $n < \omega$ such that $f(n) > \delta(A)$. Hence $\delta(B) > \delta(A)$ for all $B \in \mathcal{K}_f$ with $|B| \geq n$; it follows that there is a set $B_0 \subseteq \mathfrak{M}$ containing A such that every superset of B_0 in $\mathfrak{M}$ has bigger δ-rank. Thus $B_0 \leq \mathfrak{M}$; by Lemma 6.2.7 we may take $\text{cl}_{\mathfrak{M}}(A) = \text{cl}_{B_0}(A)$. ∎

DEFINITION 6.2.15 If $\mathfrak{M}$ is a generic model and $A \subset \mathfrak{M}$, the *closure* $\text{cl}_{\mathfrak{M}}(A)$ is the unique minimal closed superset of A in $\mathfrak{M}$.

We omit the subscript $\mathfrak{M}$ it the ambient model is clear from the context.

THEOREM 6.2.16 *Let $\mathfrak{M}$ and $\mathfrak{M}'$ be two generic models. If σ_A is an isomorphism from $A \leq \mathfrak{M}$ to $A' \leq \mathfrak{M}'$, then σ_A extends to an isomorphism between $\mathfrak{M}$ and $\mathfrak{M}'$. In particular, a generic model $\mathfrak{M}$ is unique up to isomorphism, the type of a closed set is determined by its isomorphism type, and* $\text{Th}(\mathfrak{M})$ *is ω-categorical.*

PROOF: As $\mathfrak{M}$ and $\mathfrak{M}'$ are both countable, they can be written as $\mathfrak{M} = \{m_i : i < \omega\}$ and $\mathfrak{M}' = \{m'_i : i < \omega\}$. Clearly, it is enough to show that we can extend σ_A one step, as we then construct the isomorphism between $\mathfrak{M}$ and $\mathfrak{M}'$ by a back-and-forth argument. So put $B = \text{cl}_{\mathfrak{M}}(A, m_0)$. Then $A \leq B \in \mathcal{K}_f$; as $A \cong A'$, genericity of $\mathfrak{M}'$ yields $B' \leq \mathfrak{M}'$ such that (A, B) and (A', B') are isomorphic. Next, put $C' = \text{cl}_{\mathfrak{M}'}(B', m'_0)$. Then $B' \leq C'$, and genericity of $\mathfrak{M}$ provides $C \leq \mathfrak{M}$ with $(B', C') \cong (B, C)$. Thus we may extend σ_A by mapping C to C'. It follows that σ_A can be extended to an isomorphism $\sigma : \mathfrak{M} \to \mathfrak{M}'$.

As $\emptyset \leq \mathfrak{M}$ and $\emptyset \leq \mathfrak{M}'$ are isomorphic, any two generic models are isomorphic. Finally, the condition $A \leq B$ is first-order definable in (A, B), as K_f has only finitely many isomorphism types of a given size; moreover even the condition $A \leq \mathfrak{M}$ is first-order definable in A, as we only have to check supersets of A of size at most $f^{-1}(\delta(A))$. Since "$\delta(A) \geq f(|A|)$ for all finite $A \subset \mathfrak{M}$" is a universal type-definable condition (we have to exclude certain finite $\mathfrak{L}$-structures), genericity of an $\mathfrak{L}$-structure is first-order definable, and $\text{Th}(\mathfrak{M})$ is ω-categorical. ∎

LEMMA 6.2.17 *Let $\mathfrak{M}$ be a generic model, and $A \subset \mathfrak{M}$ a finite subset. Then* $\text{acl}(A) = \text{cl}_{\mathfrak{M}}(A)$.

PROOF: Since $\text{cl}_{\mathfrak{M}}(A)$ is unique over A and finite, clearly $\text{cl}_{\mathfrak{M}}(A) \subseteq \text{acl}(A)$. On the other hand, if $\text{cl}_{\mathfrak{M}}(A) \subset B \leq \mathfrak{M}$, one sees inductively that the free amalgam B_n of n copies of B over $\text{cl}_{\mathfrak{M}}(A)$ is in $\mathcal{K}_f$, for any $n < \omega$. As B_n embeds closedly into $\mathfrak{M}$ over $\text{cl}_{\mathfrak{M}}(A)$ by genericity, B has at least n automorphic conjugates over $\text{cl}_{\mathfrak{M}}(A)$ for all $n < \omega$, and $B \not\subseteq \text{acl}(A)$. ∎

Let T_f be the theory of the generic model. Note that Lemmas 6.2.14 and 6.2.17 hold for any model of T_f.

DEFINITION 6.2.18 Let $\mathcal{K}' \subseteq \mathcal{K}_0$ be closed under substructures, and $n < \omega$. Then $\mathcal{K}'$ has the $\mathcal{P}(n)^-$-amalgamation property if for every family $\{A, A_i, A_i' : i < n\}$ of structures in $\mathcal{K}'$ with $A \leq A_i$ for all $i < n$, and closed A-embeddings $\sigma_{ij} : A_i \to A_j'$ for $i \neq j$ satisfying for all $i < n$

1. for all $j \neq i$ there is an A-isomorphism

$$\tau_{ij} : \mathrm{cl}_{A_i'}(\bigcup_{k \neq i,j} \sigma_{ki} A_k) \to \mathrm{cl}_{A_j'}(\bigcup_{k \neq i,j} \sigma_{kj} A_k),$$

   such that $\tau_{jk} \circ \tau_{ij}$ agrees with τ_{ik} on $\mathrm{cl}_{A_i'}(\bigcup_{l \neq i,j,k} \sigma_{li}(A_l))$ and $\tau_{jk} \circ \sigma_{ij} = \sigma_{ik}$ for all $k \neq i, j$, and

2. $A_i' = \mathrm{cl}_{A_i'}(\bigcup_{j \neq i} \sigma_{ji} A_j)$, and $\delta(A_i') = \delta(A) + \sum_{j \neq i}[\delta(A_j) - \delta(A)]$,

there are some $B \in \mathcal{K}'$ and closed A-embeddings $\sigma_i : A_i' \to B$, with $\sigma_j \circ \tau_{ij} = \sigma_i \restriction_{\mathrm{dom}(\tau_{ij})}$ for all distinct $i, j < n$, such that $\delta(\mathrm{cl}_B(\bigcup_{i<n} \sigma_i A_i')) = \delta(A) + \sum_{i<n}[\delta(A_i) - \delta(A)]$.

REMARK 6.2.19 Condition 1. is a compatibility condition. Condition 2. implies (by Lemma 6.2.4) that $\{\sigma_{ji} A_j : j \neq i\}$ is in free amalgamation over A inside A_i'. Since $\mathcal{K}'$ is closed under substructures, we may assume that $B = \mathrm{cl}_B(\bigcup_{i<n} \sigma_i A_i')$. Putting $\sigma_i' = \sigma_j \circ \sigma_{ij} : A_i \to B$ for some (any) $j \neq i$, the requirement on B in particular implies that $\{\sigma_i' A_i : i < n\}$ is in free amalgamation over A inside B.

REMARK 6.2.20 $\mathcal{P}(0)^-$ and $\mathcal{P}(1)^-$ are trivial. $\mathcal{P}(2)^-$ is the existence of free amalgamation: if $A \leq A_0$ and $A \leq A_1$ for structures $A, A_0, A_1 \in \mathcal{K}'$, then there exists some $B \in \mathcal{K}'$ and closed A-embeddings $\sigma_i : A_i \to B$ such that $\sigma_0(A_0)$ and $\sigma_1(A_1)$ are freely amalgamated over A, and $\delta(\mathrm{cl}_B(\sigma_0 A_0 \cup \sigma_1 A_1)) = \delta(A_0) + \delta(A_1) - \delta(A_0)$. As $\mathcal{K}'$ is closed under substructures, some free amalgam of A_0 and A_1 (not necessarily the canonical one) over A is in $\mathcal{K}'$.

DEFINITION 6.2.21 Let $\mathfrak{M}$ be a model of T_f. For finite $\bar{a}, A \subset \mathfrak{M}$ put $d(\bar{a}) = \delta(\mathrm{acl}(\bar{a}))$, and $d(\bar{a}/A) = d(\bar{a}A) - d(A)$; if $A \subset \mathfrak{M}$ is infinite, put $d(\bar{a}/A) = \inf\{d(\bar{a}/A_0) : A_0 \subset A \text{ finite}\}$. Put $A \mathop{\smile\!\!\!\!\!\mid}^f_C B$ if $\mathrm{acl}(AC) \cap \mathrm{acl}(BC) = \mathrm{acl}(C)$ and for all finite $\bar{a} \in \mathrm{acl}(AC)$ and $\bar{b} \in \mathrm{acl}(BC)$ we have $d(\bar{a}/\bar{b}\bar{c}) = d(\bar{a}/\bar{c})$, where $\bar{c} = \mathrm{acl}(\bar{a}\bar{b}) \cap \mathrm{acl}(C)$.

REMARK 6.2.22 Clearly, $\mathop{\smile\!\!\!\!\!\mid}^f$ is automorphism invariant and has the finite character property. If A and B are two finite closed sets, Lemma 6.2.4 immediately yields $d(AB) + d(A \cap B) \leq d(A) + d(B)$.

LEMMA 6.2.23 *1.* $\mathop{\smile\!\!\!\!\!\mid}^f$ *is symmetric.*

2. *If* $\mathfrak{M} \models T_f$ *and* $A \subseteq B \subseteq \mathfrak{M}$, *then* $d(\bar{a}/A) \geq d(\bar{a}/B)$ *for any* $\bar{a} \in \mathfrak{M}$.

3. *If* $A, B, C \subset \mathfrak{M}$ *are finite with* $d(A/B) = d(A/BC)$, *then* $A \perp^f_B C$.

4. *If* $A \subseteq B \subseteq \mathfrak{M}$ *and* $\bar{a} \in \mathfrak{M}$, *then* $\bar{a} \perp^f_A B$ *if and only if* $d(\bar{a}/A) = d(\bar{a}/B)$ *and* $\mathrm{acl}(\bar{a}A) \cap \mathrm{acl}(B) = \mathrm{acl}(A)$.

PROOF:

1. $d(\bar{a}/\bar{b}\bar{c}) = d(\bar{a}/\bar{c})$ is equivalent to $d(\bar{a}\bar{b}\bar{c}) + d(\bar{c}) = d(\bar{a}\bar{c}) + d(\bar{b}\bar{c})$.

2. It is enough to show the assertion for finite algebraically closed A, B. But since $\delta(\mathrm{acl}(\bar{a}B)) \leq \delta(\mathrm{acl}(\bar{a}A) \cup B)$ and $A \leq \mathrm{acl}(\bar{a}A) \cap B$,

$$\begin{aligned} d(\bar{a}/B) - d(\bar{a}/A) &= \delta(\mathrm{acl}(\bar{a}B)) - \delta(B) - \delta(\mathrm{acl}(\bar{a}A)) + \delta(A) \\ &\leq \delta(\mathrm{acl}(\bar{a}A) \cup B) + \delta(\mathrm{acl}(\bar{a}A) \cap B) - \delta(\mathrm{acl}(\bar{a}A)) - \delta(B) \leq 0 \end{aligned}$$

by Lemma 6.2.4.

3. We may assume A, B, C are closed and $B \leq A \cap C$. Then

$$\begin{aligned} d(A/C) = \delta(\mathrm{acl}(AC)) - \delta(C) &\leq \delta(A \cup C) - \delta(C) \\ &\leq \delta(A) - \delta(A \cap C) \leq \delta(A) - \delta(B) = d(A/B), \end{aligned}$$

by Lemma 6.2.4. As $d(A/C) = d(A/B)$, we must have equality; since B is closed, this implies $B = A \cap C$. Moreover, A and C are freely amalgamated over B.

Suppose $\bar{a} \in A$ and $\bar{c} \in C$, and put $\bar{b} = \mathrm{acl}(\bar{a}\bar{c}) \cap B$. Let $\bar{a}_0 = A \cap \mathrm{acl}(\bar{a}\bar{c})$ and $\bar{c}_0 = C \cap \mathrm{acl}(\bar{a}\bar{c})$. Then $A \cap C = B$ implies $\bar{a}_0 \cap \bar{c}_0 = \bar{b}$. We claim that $d(\bar{a}_0/\bar{b}) = d(\bar{a}_0/\bar{b}\bar{c}_0)$. Note first that $\bar{a}_0$ and $\bar{c}_0$ must be freely amalgamated over $\bar{b}$, since A and C are freely amalgamated over B. Next, if $d(\bar{a}_0\bar{c}_0) < \delta(\bar{a}_0\bar{c}_0)$, then there is some $\bar{e} \in \mathrm{acl}(\bar{a}_0\bar{c}_0)$ disjoint from $\bar{a}_0\bar{c}_0$, such that

$$|\bar{e}| < \sum_{i<\omega} \alpha_i |R_i(\bar{a}_0\bar{c}_0\bar{e}) - [R_i(\bar{a}_0) \cup R_i(\bar{c}_0)]|.$$

But this $\bar{e}$ is also disjoint from $A \cup C$, and would force $d(A/C) < d(A/B)$, a contradiction.

Therefore $d(\bar{a}_0/\bar{b}) = d(\bar{a}_0/\bar{b}\bar{c}_0)$. Since $\bar{c} \in \bar{c}_0$, part 2. implies

$$d(\bar{a}_0/\bar{b}) \geq d(\bar{a}_0/\bar{b}\bar{c}) \geq d(\bar{a}_0/\bar{b}\bar{c}_0) = d(\bar{a}_0/\bar{b})$$

and equality holds. By symmetry, $d(\bar{c}/\bar{b}) = d(\bar{c}/\bar{b}\bar{a}_0)$, whence by part 2.

$$d(\bar{c}/\bar{b}) \geq d(\bar{c}/\bar{b}\bar{a}) \geq d(\bar{c}/\bar{b}\bar{a}_0) = d(\bar{c}/\bar{b})$$

and equality holds again. So $d(\bar{a}/\bar{b}) = d(\bar{a}/\bar{b}\bar{c})$ by symmetry.

4. If $d(\bar{a}/A) > d(\bar{a}/B)$, then this must hold for sufficiently big finite subsets of A and B, whence $\bar{a} \not\perp^f_A B$. So we only have to show "if". By part 3. it is sufficient to show $d(\bar{a}'/\bar{b}\bar{c}') = d(\bar{a}'/\bar{b})$ for any $\bar{c} \in \mathrm{acl}(B)$ and $\bar{a}' = \mathrm{acl}(\bar{a}\bar{c}) \cap \mathrm{acl}(\bar{a}A)$, $\bar{b} = \mathrm{acl}(\bar{a}\bar{c}) \cap \mathrm{acl}(A)$ and $\bar{c}' = \mathrm{acl}(\bar{a}\bar{c}) \cap \mathrm{acl}(B)$. Note that $\bar{b} = \bar{a}' \cap \bar{c}'$. But if $d(\bar{a}'/\bar{b}) > d(\bar{a}'/\bar{b}\bar{c}')$, there is $\bar{e} \in \mathrm{acl}(\bar{a}'\bar{c}')$ disjoint from $\bar{a}'\bar{c}'$, such that
$$|\bar{e}| - \sum_{i<\omega} \alpha_i |R_i(\bar{a}'\bar{c}'\bar{e}) - [R_i(\bar{a}') \cup R_i(\bar{c}')]| = -\epsilon < 0.$$
(If $\bar{a}'$ and $\bar{c}'$ are not freely amalgamated over $\bar{b}$, we may take $\bar{e} = \emptyset$.) So if $A_0 \subseteq A$ is big enough such that $\bar{a}'\bar{b} \in \mathrm{acl}(\bar{a}A_0)$ and $d(\bar{a}/A_0) \leq d(\bar{a}/A) + \epsilon/2$, then $\bar{e}$ forces
$$d(\bar{a}/B) \leq d(\bar{a}/A_0\bar{c}') \leq d(\bar{a}/A_0) - \epsilon \leq d(\bar{a}/A) - \epsilon/2,$$
a contradiction. ∎

PROPOSITION 6.2.24 *$\perp^f$ satisfies transitivity, extension and local character.*

PROOF: TRANSITIVITY: Suppose $A \perp^f_B C$ and $A \perp^f_{BC} D$; we want to show that $A \perp^f_B CD$. First,
$$\begin{aligned}\mathrm{acl}(AB) \cap \mathrm{acl}(BCD) &\subseteq \mathrm{acl}(AB) \cap [\mathrm{acl}(ABC) \cap \mathrm{acl}(BCD)] \\ &\subseteq \mathrm{acl}(AB) \cap \mathrm{acl}(BC) = \mathrm{acl}(B).\end{aligned}$$
But now $d(\bar{a}/B) = d(\bar{a}/BC) = d(\bar{a}/BCD)$ for all $\bar{a} \in A$.

Conversely, if $A \perp^f_B CD$, clearly $A \perp^f_B C$, and $d(\bar{a}/BC) = d(\bar{a}/BCD)$ for all $\bar{a} \in A$. We have to show that $\mathrm{acl}(ABC) \cap \mathrm{acl}(BCD) = \mathrm{acl}(BC)$. Consider $\bar{e} \in [\mathrm{acl}(ABC) \cap \mathrm{acl}(BCD)] - [\mathrm{acl}(AB) \cup \mathrm{acl}(BC)]$. So the number of weighted edges between $\bar{e}$ and $\mathrm{acl}(AB)$ or between $\bar{e}$ and $\mathrm{acl}(BC)$ is less than $|\bar{e}|$, but we may suppose that the number of weighted edges between $\bar{e}$ and $\mathrm{acl}(AB) \cup \mathrm{acl}(BC)$ is at least $|\bar{e}|$. Hence there are edges of non-zero weight between $\bar{e}$ and $\mathrm{acl}(AB) \cup \mathrm{acl}(BC)$ which are not just edges between $\bar{e}$ and $\mathrm{acl}(BC)$. Since $\bar{e} \in \mathrm{acl}(BCD)$, this forces $A \not\perp^f_B CD$, a contradiction.

EXTENSION: Suppose $A \leq B \leq \mathfrak{M}$ and $\bar{a} \in \mathfrak{M}$. As $\mathcal{K}_f$ is closed under free amalgamation, $\mathrm{acl}(\bar{a}A) \otimes_A B$ is a structure whose finite substructures are in $\mathcal{K}_f$, so by compactness and genericity we may embed it closedly over B into $\mathfrak{M}$ (assuming $\mathfrak{M}$ sufficiently saturated). If $\bar{a}'$ is the image of $\bar{a}$ under this embedding, it is easy to check that $d(\bar{a}'/A) = d(\bar{a}'/B)$ and $\mathrm{acl}(\bar{a}A') \cap B = A$, whence $\bar{a}' \perp^f_A B$; moreover $\mathrm{tp}(\bar{a}'/A) = \mathrm{tp}(\bar{a}/A)$ by Lemma 6.2.16.

LOCAL CHARACTER: If a finite tuple $\bar{a}$ and A are given, there is a countable $A_0 \subseteq A$ with $d(\bar{a}/A_0) = d(\bar{a}/A)$; if $A' \subseteq A$ is countable with $\mathrm{acl}(A') \supseteq \mathrm{acl}(\bar{a}A_0) \cap \mathrm{acl}(A)$, then $\bar{a} \mathop{\smile\!\!\!\!\!|}^{f}_{A'} A$. ∎

THEOREM 6.2.25 *If $\mathcal{K}_f$ satisfies $\mathcal{P}(3)^-$, then $\mathop{\smile\!\!\!\!\!|}^{f}$ satisfies the Independence Theorem over algebraically closed sets.*

PROOF: Suppose $A_0 \leq A_i \leq \mathfrak{M}$ for $i = 1,2$ with $A_1 \mathop{\smile\!\!\!\!\!|}^{f}_{A_0} A_2$, and $\bar{b}_i \mathop{\smile\!\!\!\!\!|}^{f}_{A_0} A_i$ for $i = 1,2$ with $\mathrm{tp}(\bar{b}_1/A_0) = \mathrm{tp}(\bar{b}_2/A_0)$. Choose finite $\bar{a}_i \in A_i$ for $i = 1,2$. Put $A'_i = A_i \cap \mathrm{acl}(\bar{a}_1\bar{a}_2\bar{b}_1\bar{b}_2)$, and $B_i = \mathrm{acl}(A'_0\bar{b}_i)$, for $i = 1,2$. Then $A'_0 \leq A'_i \leq \mathfrak{M}$, $A'_1 \mathop{\smile\!\!\!\!\!|}^{f}_{A'_0} A'_2$, $B_i \mathop{\smile\!\!\!\!\!|}^{f}_{A'_0} A'_i$ and $\mathrm{tp}(B_1/A'_0) = \mathrm{tp}(B_2/A'_0)$, for $i = 1,2$. By the $\mathcal{P}(3)^-$-condition applied to the family $\{A'_0, A'_1, A'_2, B'_1, \mathrm{acl}(A'_2B'_2), \mathrm{acl}(A'_1B'_1), \mathrm{acl}(A'_1A'_2)\}$, we find $D \in \mathcal{K}_f$ with $\mathrm{acl}(A'_1A'_2) \leq D$ and some $B \leq D$ with $\mathrm{tp}(B/A'_i) = \mathrm{tp}(B_i/A'_i)$ for $i = 1,2$, $\mathrm{acl}(A'_1A'_2B) = D$, and $d(B/A'_1A'_2) = d(B/A'_0)$. By genericity, we can embed B closedly over $\mathrm{acl}(A'_1A'_2)$ into $\mathfrak{M}$, whence $\mathrm{tp}(B/A'_i) = \mathrm{tp}(B_i/A'_i)$ for $i = 1,2$. By compactness, we find the required $\bar{b} \mathop{\smile\!\!\!\!\!|}_{A_0} A_1A_2$ realizing $\mathrm{tp}(\bar{b}_1/A_1) \cup \mathrm{tp}(\bar{b}_2/A_2)$. ∎

COROLLARY 6.2.26 *If $\mathcal{K}_f$ satisfies $\mathcal{P}(3)^-$, then the theory T_f of the generic model for $\mathcal{K}_f$ is simple, and $\mathop{\smile\!\!\!\!\!|}^{f}$ is forking independence. Moreover, T eliminates bounded hyperimaginaries.*

PROOF: By Theorem 2.6.1, forking independence is the same as $\mathop{\smile\!\!\!\!\!|}^{f}$ and T is simple. If E is a bounded type-definable equivalence relation over a set A with class $\bar{a}_E$ for some real tuple $\bar{a}$, let $\bar{A}$ be the algebraic closure of A in the home sort. As the Independence Theorem holds over $\bar{A}$, we get $\mathrm{tp}(\bar{a}/\bar{A}) \vdash \mathrm{lstp}(\bar{a}/A)$. So $\bar{a}_E \in \mathrm{dcl}(\bar{A})$, and we finish by Lemma 3.6.3 (with A named). ∎

EXAMPLE 6.2.27 Suppose $\mathfrak{L}$ consists of a single ternary relation R, and $f(x) = \log_3(x) + 1$. Then $\mathcal{K}_f$ is non-empty; it contains any finite set of unrelated points, and e.g. three R-related points. Suppose $A_0 \leq A_i \in \mathcal{K}_f$ for $i = 1,2$, and put $B = A_1 \otimes_{A_0} A_2$; clearly we may assume $A_1 \neq A_0 \neq A_2$, and $|A_1| \geq |A_2|$. Now $B \in \mathcal{K}_0$ by Lemma 6.2.9, and

$$\frac{\delta(B) - \delta(A_1)}{|B| - |A_1|} = \frac{\delta(A_2) - \delta(A_0)}{|A_2| - |A_0|} \geq \frac{1}{|A_2|} \geq f'(|A_2|) \geq f'(|A_1|);$$

since f is convex and $\delta(A_1) \geq f(|A_1|)$, this proves $B \in \mathcal{K}_f$. (In fact, any convex unbounded function f with $0 < f'(x) \leq \frac{1}{x}$ would do.)

Next, we verify that $\mathcal{K}_f$ has the $\mathcal{P}(3)^-$-amalgamation property. So suppose $\{A, A_i, A'_i : i = 0,1,2\}$ and closed embeddings $\{\sigma_{ij} : i \neq j\}$ are given as in Definition 6.2.18; for ease of notation we assume that the σ_{ij} are identity maps. Let $B_0 = A'_1 \otimes_{A_0} A'_2$, then $B_0 \in \mathcal{K}_f$, and $A_1 \cup A_2 \leq B_0$. Next, let $B = B_0 \otimes_{A_1 \cup A_2} A'_0$; so $B \in \mathcal{K}_0$ by Lemma 6.2.9, and $A'_0 \leq B$. (By symmetry, we also have $A'_1 \leq B$ and $A'_2 \leq B$.)

Let $C \subseteq B$; as $|\mathrm{cl}_B(C)| \geq |C|$ and $\delta(\mathrm{cl}_B(C)) \leq \delta(C)$, we may assume $C \leq B$. Put $C_i = C \cap A_i$ and $C'_i = C \cap A'_i$; then $C_i \leq A_i$ and $C'_i \leq A'_i$, whence $C_i, C'_i \in \mathcal{K}_f$ for $i = 0, 1, 2$. By symmetry, we may assume $\delta(C'_0) \geq \delta(C'_i)$ for $i = 1, 2$. If $\delta(C'_0) = \delta(C)$, then $C'_0 \leq A'_0 \leq B$ implies $C'_0 \leq B$ and $C = C'_0 \in \mathcal{K}_f$. Otherwise $\delta(C) \geq \delta(C'_0) + 1$, and

$$|C| \leq \sum_{i=0}^{2} |C'_i| \leq \sum_{i=0}^{2} 3^{\delta(C'_i)-1} \leq 3 \cdot 3^{\delta(C'_0)-1} \leq 3^{\delta(C)-1},$$

whence $\delta(C) \geq f(|C|)$ and $C \in \mathcal{K}_f$.

So T_f is simple. In fact, since $d(a/A)$ can either be 0 or 1 for a single element a, and $d(a/A) = 0$ implies $a \in \mathrm{acl}(A)$, the theory must have SU-rank 1. Note that there are two distinct two-types: a pair of points which is closed, and a pair of points whose closure contains a third point and one edge. So algebraic closure induces a non-homogeneous matroid on the generic model $\mathfrak{M}_f$. Moreover, it is easy to see that $\mathfrak{M}_f$ is not one-based.

EXERCISE 6.2.28 Put $g(x) = (x+1)!$ and $f = g^{-1}$. If again $\mathfrak{L}$ consists of a single ternary relation R and $\alpha = 1$, show that T_f has the $\mathcal{P}(n)^-$-amalgamation property for all $n < \omega$.

PROBLEM 6.2.29 Suppose $\mathfrak{M}$ is a countable, ω-categorical structure of SU-rank 1, such that there are constants c, d with $|\mathrm{acl}(A)| \leq cd^{SU(A)}$ for all finite $A \subset \mathfrak{M}$. If $\mathfrak{M}$ has the $\mathcal{P}(n)^-$-amalgamation property for all $n < \omega$, is $\mathfrak{M}$ necessarily one-based?

REMARK 6.2.30 Instead of amalgamating sets, one might also amalgamate finite elementary Abelian 2-groups, with additional relations $\{R_i : i < \omega\}$. In this case, one should take $\delta(A) = \log_2(|A|) - \sum_{i<\omega} \alpha_i |R_i(A)|$.

EXERCISE 6.2.31 Construct an ω-categorical group of SU-rank 1 which is not one-based.

6.2.2 ω-CATEGORICAL SUPERSIMPLE GROUPS

In this subsection, we shall show:

THEOREM 6.2.32 *An ω-categorical supersimple group is finite-by-Abelian-by-finite, and has finite SU-rank.*

EXAMPLE 6.2.33 Let V be a vector space over $\mathbb{F}_3$, and $\langle .,. \rangle$ a non-degenerate antisymmetric bilinear form on V. It is easy to see that V is ω-categorical supersimple, of SU-rank 1. Define a group G on $\mathbb{F}_3 \rtimes V$ as follows:

$$(n, v)(n', v') = (n + n' + \langle v, v' \rangle, v + v').$$

Then G is non-Abelian, with centre $\mathbb{F}_3 \times \{0\}$; since G is interpretable in V and a finite cover of V, it is again ω-categorical of SU-rank 1.

So Theorem 6.2.32 is the best we can hope for. Let us start by considering ω-categorical groups with simple theory. Note that by ω-categoricity, all characteristic subgroups of G (such as the derived subgroups, the subgroups in the upper central series, $\tilde{Z}(G)$, or the iterated centres of G) are $\emptyset$-definable.

REMARK 6.2.34 Suppose N is a normal subgroup of finite index in G with N' finite. If K is an automorphic image of N, then $N \cap K$ is central in NK modulo $N'K'$. It follows that $(NK)'$ is again finite. Hence if we choose a normal subgroup with finite derived subgroup of minimal possible index in G, it must be characteristic, and thus $\emptyset$-definable.

THEOREM 6.2.35 *An ω-categorical simple group G is nilpotent-by-finite.*

PROOF: We may assume that G is countably infinite. By ω-categoricity, we may suppose that G has no $\emptyset$-definable subgroup of finite index; moreover G is locally finite of finite exponent, so there is an ascending sequence $\{G_i : i < \omega\}$ of finite subgroups with $\bigcup_{i<\omega} G_i = G$. The classification of finite simple groups tells us that there are only finitely many simple groups of given exponent, so we may assume that the G_i have non-trivial proper normal subgroups A_i.

CASE 1: The A_i may be chosen of bounded size.

We may assume that all A_i have the same size. Put $H_j = \bigcap_{i \geq j} N_G(A_i)$, and let $\varphi(x, y, Y)$ be the formula $x \in yN_G(Y)$. If $k < \omega$ is such that $D^*(H_k, \varphi, 2)$ is maximal, then $N_G(A_i)$ can intersect H_j only in a subgroup of finite index for any $j \geq i \geq k$. If follows that H_j and H_k are commensurate. On the other hand, H_j is commensurate to a finite sub-intersection M_j of bounded size by Theorem 4.2.12, so all M_j must be commensurable for $j \geq k$; this commensurability is uniform by ω-categoricity (or Lemma 4.2.6). Now $G_i \leq N_G(A_j) \leq M_j$ for $i \leq j$. Since any finite collection $\bar{g}$ of coset representatives of M_k in G is contained in G_i for big $i < \omega$, the index $|G : M_k|$ is bounded by

$$\sup_{j \geq k} |M_j M_k : M_k| = \sup_{j \geq k} |M_j : (M_j \cap M_k)| < \omega.$$

Thus M_k, and hence $N_G(A_k)$, has finite index in G. As A_k is finite, $C_G(a)$ has finite index in G for all $a \in A_k$, so $A_k \leq \tilde{Z}(G)$. Since $G^0_\emptyset = G$, Proposition 4.4.10 yields $\tilde{Z}(G) \leq Z_2(G)$ and $Z_2(G)$ is non-trivial.

CASE 2: The A_i have increasing size.

If we choose A_i minimal normal in G_i, then A_i is either Abelian, or a direct product of finite simple groups $(S_i^j : j < n_i)$. Suppose the latter. The S_i^j are again bounded in size, so n_i must increase with i. By Theorem 4.2.12 there are $k, k' < \omega$ such that there are no elements $(a_i : i < k)$ with $|C_G(a_i : i < j) : C_G(a_i : i \leq j)| \geq k'$ for all $j < k$. But if $n_i \geq kk'$ we can choose non-central elements $b_j \in S_i^j$ and put $a_s = \prod_{j<k'(s+1)} b_j$ for $s < k$; we obtain

$$|C_G(a_s : s < j) : C_G(a_s : s \leq j) \geq |S_i^{k'j} \times \cdots \times S_i^{k'(j+1)-1} : C_G(a_j)| \geq 2^{k'}$$

for all $j < k$, a contradiction. Hence we may assume that all A_i are Abelian. By Theorem 4.2.12 for every $i < \omega$ there is a tuple $\bar{a}_i \in A_i$ of bounded size such that $C_G(\bar{a}_i)$ is commensurate to $C_G(A_i)$. Put $N_i = \tilde{N}_G(C_G(\bar{a}_i))$, then $G_i \leq N_i$ for all $i < \omega$. As in case 1. we see that N_i has finite index in G for sufficiently large $i < \omega$.

By ω-categoricity, we may assume that all $\bar{a}_i$ have the same type. Since A_i is Abelian, $A_i \leq \tilde{Z}(C_G(\bar{a}_i))$, so $\tilde{Z}(C_G(\bar{a}_i))$ has unbounded size. The N_i-conjugates of $\tilde{Z}(C_G(\bar{a}_i))$ are commensurable; by Proposition 4.2.7 there is a definable normal subgroup K_i of N_i commensurable with $\tilde{Z}(C_G(\bar{a}_i))$; it must be finite-by-Abelian-by-finite by Remark 4.4.11. By Remark 6.2.34 there is a characteristic normal subgroup L_i of finite index in K_i such that L_i' is finite; replacing K_i by $C_{L_i}(L_i')$, we may assume that K_i is nilpotent.

Replacing N_i by a subgroup of finite index, we may (for sufficiently large i) assume that N_i is normal in G. But then the finitely many G-conjugates of K_i generate a nilpotent normal subgroup of G, and G has an Abelian normal subgroup. By ω-categoricity the subgroup $\langle g^G \rangle$ is uniformly definable in g; as G does not have the strict order property, there must be a minimal normal Abelian subgroup A. But then the family of automorphic images of A generates a characteristic Abelian normal subgroup of G, which must be $\emptyset$-definable by ω-categoricity.

In either case we have found an $\emptyset$-definable non-trivial Abelian normal subgroup of G; by induction on the number of $\emptyset$-definable normal subgroups G must be soluble.

In order to show nilpotency, again by induction on the number of $\emptyset$-definable normal subgroups, we may assume that G' is nilpotent (note that G' is connected over $\emptyset$ by connectivity of G). For $a \in Z(G')$ put $A_a = \langle a^G \rangle$; by ω-categoricity A_a is definable and we may choose non-trivial $a \in Z(G')$ such that A_a is minimal. Put $M = G/C_G(A_a)$; this is an Abelian group acting on A_a. By definability of A_a, there is some $k < \omega$ such that for every $b \in A_a$ there are $m_0, \ldots, m_k \in M$ with $b = \sum_{i \leq k} a^{m_i}$.

CLAIM. If $\sum_{i<k'} b_0^{m_i} = 0$ for some non-trivial $b_0 \in A_a$, some $k' < \omega$ and $m_0, \ldots, m_{k'} \in M$, then $\sum_{i \leq k'} b^{m_i} = 0$ for all $b \in A_a$.

PROOF OF CLAIM: Consider $B = \{b \in A_a : \sum_{i\leq k'} b^{m_i} = 0\}$. Clearly, B is a subgroup of A_a; if $b \in B$ and $m \in M$, then

$$\sum_{i\leq k'} (b^m)^{m_i} = \sum_{i\leq k'} (b^{m_i})^m = (\sum_{i\leq k'} b^{m_i})^m = 0^m = 0,$$

so B is M- and hence G-invariant. Now $b_0 \in B$ implies $A_{b_0} \leq B$; as $B \leq A_a$ and A_a was chosen minimal, we get $B = A_a$. ■

It follows that the ring of endomorphisms of A_a generated by $\{a \mapsto a^g : g \in G\}$ is definable, commutative, and without zero divisors. So it embeds into a definable field K; by ω-categoricity K is finite. So the homomorphism $G \to K^\times$ given by the action of $g \in G$ on A_a has a kernel of finite index. Therefore $A_a \leq \tilde{Z}(G) \leq Z_2(G)$ and $Z_2(G)$ is non-trivial; as $G/Z_2(G)$ is nilpotent by inductive hypothesis, we are done. ■

As in the stable case, it is an open question whether we can improve Theorem 6.2.35:

PROBLEM 6.2.36 Is an ω-categorical group with simple theory finite-by-Abelian-by-finite?

In preparation for the proof of Theorem 6.2.32, we shall first study the structure of ω-categorical groups of finite SU-rank. Of course, *a posteriori* these results will be applicable in general to ω-categorical supersimple groups. The following definition differs from the stable case.

DEFINITION 6.2.37 A definable group is *minimal* if it is infinite and every definable subgroup of infinite index is finite.

Note that in a supersimple theory every definable group contains a definable minimal subgroup (any infinite definable subgroup of minimal SU-rank possible will do).

DEFINITION 6.2.38 Let A and B be definable Abelian minimal groups. A *virtual isogeny* f between A and B is a definable isomorphism $f : D/K \to I/C$, where D has finite index in A, I has finite index in B, and K and C are both finite. Two virtual isogenies f_1 and f_2 are *equivalent* if the derived maps from $D_1 \cap D_2$ to $(I_1 + I_2)/(C_1 + C_2)$ agree on a subgroup of finite index in A.

It is easy to check that equivalence of virtual isogenies is a congruence with respect to composition (whenever composition makes sense). Note that if f is defined over some parameters $\bar{a}$, say $f = f_{\bar{a}}$, then the equivalence relation on $\text{tp}(\bar{a})$ given by equivalence of $f_{\bar{x}}$ and $f_{\bar{y}}$ is definable by ω-categoricity. Thus the equivalence class $[f]$ of f is an imaginary element.

LEMMA 6.2.39 *Let A and B be $\emptyset$-definable Abelian minimal groups in an ω-categorical theory. Then the equivalence class of every virtual isogeny from A to B is algebraic over $\emptyset$.*

PROOF: Consider first the set R of virtual isogenies from a minimal group A to A itself, modulo equivalence. If $f : D/K \to I/C$ is in R (so D and I have finite index in A and K and C are finite), then $f^{-1} : I/C \to D/K$ is also a virtual isogeny, and every element in R has an inverse. Moreover, if $f_1 : D_1/K_1 \to I_1/K_1$ and $f_2 : D_2/K_2 \to I_2/C_2$ are two virtual isogenies in R, then their sum

$$f_1 + f_2 : x \mapsto f_1(x + K_1) + f_2(x + K_2) + (C_1 + C_2)$$

is a definable endomorphism from $D_1 \cap D_2$ to $A/(C_1+C_2)$; since A is minimal, it must either have finite image (in which case f_1 and $-f_2$ are equivalent), or have an image of finite index, and $f_1 + f_2$ induces another virtual isogeny in R; it is easy to see that the equivalence class of $f_1 + f_2$ only depends on the equivalence classes of f_1 and f_2. Clearly addition satisfies the group laws (with $x \mapsto -f(x)$ as additive inverse of f), and composition distributes over addition. Thus addition and composition endow $R \cup \{0\}$ with the structure of a division ring. If A is ω-categorical, R must be locally finite, whence a locally finite field, and every element of R is algebraic over $\emptyset$. Furthermore, if f and f' are two virtual isogenies between A and B, then f^{-1} is a virtual isogeny between B and A, and $f^{-1}f'$ is a virtual isogeny of A. Clearly f' is algebraic over $f^{-1}f'$ and f; as $[f^{-1}f']$ is algebraic over $\emptyset$, we see that $[f']$ is algebraic over $[f]$.

As f and f' were arbitrary virtual isogenies between A and B, we see that $[f''] \in \mathrm{acl}([f])$ for any $[f''] \models \mathrm{tp}([f'])$. Hence $\mathrm{tp}([f'])$ has only boundedly, whence finitely, many realizations, and $[f'] \in \mathrm{acl}(\emptyset)$. This proves the lemma. ■

PROPOSITION 6.2.40 *Let G be an ω-categorical finite-by-Abelian-by-finite group of finite SU-rank. Then any definable subgroup of G is commensurable with one definable over* $\mathrm{acl}(\emptyset)$.

PROOF: G has a characteristic normal subgroup N with finite commutator subgroup N' by Remark 6.2.34. Replacing a definable subgroup H of G by the definable subgroup $(H \cap N)N'/N'$ of N/N', we can reduce the problem to the case when G is Abelian.

For Abelian G, consider first a minimal subgroup A of G. By the finiteness of rank, there are only finitely many conjugates of A, say $(A_i : i < n)$, such that every conjugate of A intersects $A^0 := \sum_{i<n} A_i$ in a subgroup of finite index. We may choose the A_i almost linearly independent, i.e. $A_i \cap \sum_{j \neq i} A_j$ is finite for all $i < n$. (Then $SU(A^0) = SU(A) \cdot n$.)

Now consider another conjugate A' of A. Since $A' \cap A^0$ is infinite by maximality of n, there must be some $i = i(A') < n$ such that $A' + \sum_{k \neq i} A_k$ intersects A_i in an infinite subgroup, which must have finite index by minimality of A_i. For every $j \neq i$ such that $A_j \cap (A' + \sum_{k \neq j} A_k)$ is infinite, put

$$r(A', j)(x) = \{y \in A_j : x - y \in A' + \sum_{k \neq i,j} A_k\} = A_j \cap (x + A' + \sum_{k \neq i,j} A_k).$$

CLAIM. $r(A', j)$ is a virtual isogeny from A_i to A_j.

PROOF OF CLAIM: By almost linear independence, $(A_i + A_j) \cap \sum_{k \neq i,j} A_k$ is finite. We may therefore divide out by $\sum_{k \neq i,j} A_k$ and assume $n = 2$, $i = 0$ and $j = 1$. But then for every $x \in A_0 \cap (A_1 + A')$ the value $r(A', j)(x)$ is in $A_1/(A_1 \cap A')$, and in fact $r(A', j)$ defines an isomorphism between

$$[A_0 \cap (A_1 + A')]/(A_0 \cap A') \quad \text{and} \quad [A_1 \cap (A_0 + A')]/(A_1 \cap A').$$

This is a virtual isogeny from A_1 to A_2. ■

If $A_j \cap (A' + \sum_{k \neq j} A_k)$ is finite, we put $r(A', j) = 0$.

CLAIM. Suppose A'' is another conjugate of A such that $i(A'') = i(A')$, and $r(A', j)$ and $r(A'', j)$ are equivalent virtual isogenies for all $j \neq i(A')$ with $r(A', j) \neq 0$ or $r(A'', j) \neq 0$. Then A' and A'' are commensurable.

PROOF OF CLAIM: Put $i = i(A') = i(A'')$. If $r(A', j) = 0$, then $\sum_{k \neq j} A_k$ intersects A' in an infinite subgroup, and hence in a subgroup of finite index by minimality of A'. It is therefore sufficient to consider the intersection of A' with $\sum_{r(A',j) \neq 0} A_j$. So we may assume $r(A', j) \neq 0$ and $r(A'', j) \neq 0$ for all $j \neq i$.

Let B be a subgroup of finite index in A_i and B_j a finite subgroup of A_j for all $j < n$ such that $r(A', j)(x) + B_j = r(A'', j)(x) + B_j$ for all $x \in B$ and all $j \neq i$. Now

$$\{x - \sum_{j \neq i} x_i : x \in B, x_i \in r(A', j)(x) \text{ for all } j \neq i\}$$

is a subgroup of finite index in $A' + \sum_{j \neq i} [A_j \cap (A' + \sum_{k \neq i,j} A_k)]$; since the second summand is finite, it is commensurable with A'. Similarly

$$\{x - \sum_{j \neq i} x_i : x \in B, x_i \in r(A'', j)(x) \text{ for all } j \neq i\}$$

is commensurable with A''. But these two groups are equal modulo $\sum_{j \neq i} B_j$, i.e. modulo a finite group, and must be commensurable. ■

It follows that there are only finitely many commensurability classes among the conjugates of A. By Corollary 4.2.10 there is a locally connected group A^c commensurable with A (alternatively, the locally connected component A^c of A given by Corollary 4.5.16 is definable by ω-categoricity). It has only finitely many conjugates, corresponding to the commensurability classes of conjugates of A, and must therefore be definable over $\mathrm{acl}(\emptyset)$. This proves the assertion for minimal groups.

If $H \leq G$ is not minimal, then by supersimplicity it contains a minimal subgroup A which is commensurable with some $\mathrm{acl}(\emptyset)$-definable A^c. But HA^c/A^c is a subgroup of G/A^c of smaller SU-rank; by induction it is commensurable with an $\mathrm{acl}(\emptyset)$-definable group H_c/A^c, whose pre-image H^c in G is as required. ∎

PROPOSITION 6.2.41 *An ω-categorical group of finite SU-rank is finite-by-Abelian-by-finite.*

PROOF: Suppose not, and let G be a counterexample. By ω-categoricity, we may assume that G has no proper $\emptyset$-definable subgroups of finite index.

CLAIM. We may assume that $G'/(G' \cap Z(G))$ is finite.

PROOF OF CLAIM: G is nilpotent by Lemma 6.2.35. Since the subgroups $\gamma_m(G)$ in the lower central series are $\emptyset$-definable for all $m \geq 1$ by ω-categoricity, there is some minimal $n > 2$ such that $G'/\gamma_n(G)$ is infinite. (Note that G/G' must be infinite since G has no characteristic subgroup of finite index, and G' is infinite since G is not finite-by-Abelian.)

We claim that $G/\gamma_n(G)$ is not finite-by-Abelian-by-finite. So suppose otherwise, and let N be a normal subgroup of G of finite index containing $\gamma_n(G)$ such that $N'\gamma_n(G)/\gamma_n(G)$ is finite; by Remark 6.2.34 we may assume that N is characteristic in G. But since G has no proper characteristic subgroup of finite index, $N = G$, and $G'/\gamma_n(G)$ is finite, a contradiction.

$G'/\gamma_{n-1}(G)$ is finite by minimality of n, so $\gamma_{n-1}(G)/\gamma_n(G)$ is infinite; since $\gamma_{n-1}(G)/\gamma_n(G)$ is contained in $Z(G/\gamma_n(G))$, we may replace G by $G/\gamma_n(G)$ and are done. ∎

For $g \in G$ put $G_g = C_{G/Z(G)}(gZ(G))$. Since $G'/(G' \cap Z(G))$ is finite, G_g has finite index in $G/Z(G)$. Define $H_g = \{(hZ(G), [h,g]) : hZ(G) \in G_g\}$; this is a subgroup of $G/Z(G) \times Z(G)$, and is well-defined, as the commutator $[h,g]$ does not depend on the choice of h in $hZ(G)$. Since $G/Z(G) \times Z(G)$ is finite-by-Abelian, H_g is commensurable with an $\mathrm{acl}(\emptyset)$-definable group by Proposition 6.2.40; it follows by compactness that there are only finitely many commensurability classes in the family $\{H_g : g \in G\}$, and we may choose two independent generics $g, g' \in G$ such that H_g and $H_{g'}$ are commensurable. So the projection of $H_g \cap H_{g'}$ to the first co-ordinate is a subgroup of finite index in $G_g \cap G_{g'}$, and there is a generic element $h \in G$ over g, g' such that $hZ(G)$ is in that projection; then $[h,g] = [h,g']$ and $[h, g'g^{-1}] = 1$. But $g'g^{-1}$ is a generic element of G independent of h; it follows that $|G : C_G(h)|$ is finite, $h \in \tilde{Z}(G)$, and $\tilde{Z}(G)$ has finite index in G. So G is finite-by-Abelian-by-finite by Remark 4.4.11. ∎

PROPOSITION 6.2.42 *An ω-categorical supersimple group has finite SU-rank.*

PROOF: Let G be an ω-categorical supersimple group of infinite rank, and let p be a 1-type of SU-rank ω over a finite set, which we may assume to be $\emptyset$. By ω-categoricity, p is isolated, say by some formula $\varphi(x)$. As $|S_1(\emptyset)|$ is finite, there is a bound $n < \omega$ for the SU-rank of a 1-type over $\emptyset$ of finite SU-rank.

Take a forking extension q of p of SU-rank greater than n, and a Morley sequence $(a_i : i < \omega)$ in q. Let $\bar{a} = (a_i : i < k)$ be a maximal independent initial segment of $(a_i : i < \omega)$. Then $a_k \not\perp \bar{a}$, so

$$\omega = SU(p) = SU(a_k) > SU(a_k/\bar{a}) \geq SU(q) > n.$$

Let $\bar{p}$ be the type of $\bar{a}$, and let m be the maximal SU-rank of an extension of p of finite SU-rank over $2k$ elements (which exists by ω-categoricity); clearly $m > n$.

CLAIM. There are a finite set X of parameters and an extension $\bar{q} \in S_k(X)$ of $\bar{p}$ of finite SU-rank, such that the SU-rank of every co-ordinate of a realization of $\bar{q}$ over $X \cup$ (the other co-ordinates) is greater than m.

PROOF OF CLAIM: $SU(a_i) = SU(p) = \omega$ for all $i < k$, and $\{a_i : i < k\}$ is an independent set. For every $i < k$ choose a finite set A_i of parameters with $\omega > SU(a_i/A_i) > m$; we may assume $A_i \perp_{a_i} \bar{a} \cup \{A_j : j < i\}$ for all $i < k$. Suppose inductively that $\{a_j A_j : j < i\} \cup \{a_j : j \geq i\}$ is an independent set. Then $a_i \perp \{a_j A_j : j < i\} \cup \{a_j : j > i\}$, whence $a_i A_i \perp \{a_j A_j : j < i\} \cup \{a_j : j > i\}$, and $\{a_j A_j : j \leq i\} \cup \{a_j : j > i\}$ is independent. Therefore $\{a_i A_i : i < k\}$ is independent, and $a_i \perp_{A_i} \{a_j A_j : j \neq i\}$. Put $X = \{A_j : j < k\}$ and $\bar{q} = \mathrm{tp}(\bar{a}/X)$, then $SU(a_i/X, a_j : j \neq i) = SU(a_i/A_i) > m$. ■

Consider a Morley sequence $I = (\bar{a}^s : s < \omega)$ in $\bar{q}$, and consider the type of $\bar{a}^0\bar{a}^1$ over $\emptyset$. For every $i < k$

$$\begin{aligned} SU(a_i^1/\bar{a}^0, \bar{a}^1 - \{a_i^1\}) &\geq SU(a_i^1/X, \bar{a}^0, \bar{a}^1 - \{a_i^1\}) \\ &= SU(a_i^1/X, \bar{a}^1 - \{a_i^1\}) > m; \end{aligned}$$

so $SU(a_i^1/\bar{a}^0, \bar{a}^1 - \{a_i^1\}) = \omega$ by assumption on m. Similarly, for all $i < k$ we have $SU(a_i^0/\bar{a}^1, \bar{a}^0 - \{a_i^0\}) = \omega$. Hence $\bar{a}^0\bar{a}^1$ is a tuple of $2k$ independent realizations of p; in particular $\bar{a}^0 \perp \bar{a}^1$. By supersimplicity there is a finite initial segment $A = (\bar{a}^i : i < i_0)$ of I with $SU(\bar{a}^{i_0}/A) = SU(\bar{q}) < \omega$. Let $H = \{g \in G : SU(g/\bar{a}^{i_0}) < \omega\}$ and $K = \{g \in G : SU(g/A) < \omega\}$; note that $H \subseteq K$ by the Lascar inequality 5.1.6. By ω-categoricity, K is an A-definable subgroup of G and H is an $\bar{a}^{i_0}$-definable subgroup of K, with $SU(K) < \omega$. Note that $SU(H) \geq SU(a_k/\bar{a}) > n$, since $\mathrm{tp}(\bar{a}) = \mathrm{tp}(\bar{a}^{i_0})$.

As I is a Morley sequence, for every $j \geq i_0$ there is an automorphism σ_j which fixes A and maps $\bar{a}^{i_0}$ to $\bar{a}^j$. So σ_j stabilizes K setwise. By Propositions

6.2.40 and 6.2.41 there are only finitely many commensurability classes among K-conjugates of H, so there is $i_0 \leq i < j$ such that $\sigma_i(H)$ and $\sigma_j(H)$ are commensurable, whence $\sigma_i(H^c) = \sigma_j(H^c)$, where H^c is a locally connected definable group commensurable with H, as given by Corollary 4.2.10, or by Corollary 4.5.16 and ω-categoricity.

Now H^c is definable over $\bar{a}^{i_0}$, so $\sigma_i(H^c)$ is definable over $\mathrm{dcl}(\bar{a}^i) \cap \mathrm{dcl}(\bar{a}^j)$. But $\bar{a}^0 \downarrow \bar{a}^1$ implies $\bar{a}^i \downarrow \bar{a}^j$ by indiscernibility, and $\mathrm{dcl}(\bar{a}^i) \cap \mathrm{dcl}(\bar{a}^j) \subseteq \mathrm{acl}(\emptyset)$. Thus $\sigma_i(H^c)$ is a group of finite SU-rank definable over $\mathrm{acl}(\emptyset)$. If $r \in S_1(\mathrm{acl}(\emptyset))$ is a generic type for $\sigma_i(H^c)$, then $SU(r) = SU(\sigma_i(H^c)) = SU(H^c) = SU(H) > n$. Since $SU(r) = SU(H) < \omega$, this contradicts the maximal choice of n, as r does not fork over $\emptyset$. ■

Theorem 6.2.32 now follows from Propositions 6.2.41 and 6.2.42. ■

6.2.3 ω-CATEGORICAL CM-TRIVIAL THEORIES

There is another case in which we can prove finiteness of SU-rank for an ω-categorical supersimple theory, namely, if the underlying forking geometry is not too complicated.

DEFINITION 6.2.43 An $\emptyset$-invariant family P of types in a simple theory is *CM-trivial* if whenever $\mathrm{tp}(c/A)$ is P-internal and $A \subseteq B \subseteq C$ with $\mathrm{cl}_P(cB) \cap \mathrm{cl}_P(C) = \mathrm{cl}_P(B)$, then $\mathrm{Cb}(c/\mathrm{cl}_P(B)) \subseteq \mathrm{cl}_P(\mathrm{Cb}(c/\mathrm{cl}_P(C)), A)$.
A simple theory T is *CM-trivial* if the family of all types is CM-trivial.

CM-triviality prohibits the interpretation of a *pseudo-planar space:* a set of points, lines and planes which behaves essentially like a sufficiently rich collection of points, curves and surfaces. If P is the family of all types, then P-closure is bounded closure. So in a CM-trivial theory, $\mathrm{Cb}(c/A) \subseteq \mathrm{bdd}(\mathrm{Cb}(c/B))$ whenever $\mathrm{bdd}(cA) \cap \mathrm{bdd}(B) = \mathrm{bdd}(A)$. If T eliminates hyperimaginaries (e.g. if T is small or superstable) and is CM-trivial, we may replace the bounded by the algebraic closure.

REMARK 6.2.44 A locally modular family P is CM-trivial.

PROOF: Suppose P is locally modular, $A \subseteq B \subseteq C$, $\mathrm{tp}(c/A)$ is P-internal and $\mathrm{cl}_P(cB) \cap \mathrm{cl}_P(C) = \mathrm{cl}_P(B)$. Then $\mathrm{Cb}(c/\mathrm{cl}_P(C)) \in \mathrm{cl}_P(cA) \subseteq \mathrm{cl}_P(cB)$ by Lemma 3.5.13 (with A named); since $\mathrm{Cb}(c/\mathrm{cl}_P(C)) \subseteq \mathrm{cl}_P(C)$, we get $\mathrm{Cb}(c/\mathrm{cl}_P(C)) \subseteq \mathrm{cl}_P(B)$. Therefore $c \downarrow_{\mathrm{cl}_P(B)} \mathrm{cl}_P(C)$ and the result follows. ■

REMARK 6.2.45 The ω-categorical simple structures produced by Hrushovski's amalgamation method from Subsection 6.2.1 are CM-trivial.

PROOF: Suppose $f : \mathbb{R}^+ \to \mathbb{R}^+$ is monotone increasing and $\mathcal{K}_f$ is closed under free amalgamation and has the $\mathcal{P}(3)^-$-amalgamation property. Let $\mathfrak{M}$

be a model of the generic theory T_f, and consider first a finite closed $A \leq \mathfrak{M}$ and a finite tuple $\bar{c}$.

CLAIM. Let $A_1, A_2 \leq B \leq \mathfrak{M}$. If $\bar{c} \mathop{\smile\!\!\!\!\!\mid}^f_{A_1} B$ and $\mathrm{acl}(cA_2) \cap \mathrm{acl}(B) = A_2$, then $\bar{c} \mathop{\smile\!\!\!\!\!\mid}^f_{A_1 \cap A_2} A_2$.

PROOF OF CLAIM: Put $A_0 = A_1 \cap A_2$. By definition of $\mathop{\smile\!\!\!\!\!\mid}^f$, for any finite $\bar{b} \in B$ we have $d(\bar{c}/\bar{a}\bar{b}) = d(\bar{c}/\bar{a})$, where $\bar{a} = \mathrm{acl}(\bar{c}\bar{b}) \cap A_1$. If $\bar{b} \in A_2$, then

$$\bar{a} \in \mathrm{acl}(\bar{c}A_2) \cap A_1 \subseteq \mathrm{acl}(\bar{c}A_2) \cap \mathrm{acl}(B) \cap A_1 \subseteq A_2 \cap A_1 = A_0.$$

Hence $\bar{a} = \mathrm{acl}(\bar{b}\bar{c}) \cap A_0$; moreover $\mathrm{acl}(\bar{c}\bar{a}) \cap \mathrm{acl}(\bar{b}\bar{a}) = \mathrm{acl}(\bar{a})$. Therefore $\mathrm{acl}(\bar{c}A_0) \cap A_2 = A_0$, and $\bar{c} \mathop{\smile\!\!\!\!\!\mid}^f_{A_0} A_2$. ■

Now suppose $A \leq B \leq \mathfrak{M}$, and $\bar{c} \in \mathfrak{M}$ satisfies $\mathrm{acl}(\bar{c}A) \cap \mathrm{acl}(B) = \mathrm{acl}(A)$. Choose $B' \leq \mathfrak{M}$ with $\mathrm{Cb}(\bar{c}/B) \in \mathrm{bdd}(B')$ and $B' \mathop{\smile\!\!\!\!\!\mid}_{\mathrm{Cb}(\bar{c}/B)} B\bar{c}$. Then $\bar{c} \mathop{\smile\!\!\!\!\!\mid}_B B'$ and $\bar{c} \mathop{\smile\!\!\!\!\!\mid}_{\mathrm{Cb}(\bar{c}/B)} BB'$, whence $\bar{c} \mathop{\smile\!\!\!\!\!\mid}_{B'} B$. By Corollary 6.2.26 we get $\bar{c} \mathop{\smile\!\!\!\!\!\mid}^f_{B'} \mathrm{acl}(BB')$.

On the other hand, $\bar{c} \mathop{\smile\!\!\!\!\!\mid}_B B'$ implies $\mathrm{acl}(\bar{c}B) \cap \mathrm{acl}(BB') = \mathrm{acl}(B) = B$, whence $\mathrm{acl}(\bar{c}A) \cap \mathrm{acl}(BB') = \mathrm{acl}(\bar{c}A) \cap \mathrm{acl}(B) = \mathrm{acl}(A) = A$. By the claim above $\bar{c} \mathop{\smile\!\!\!\!\!\mid}^f_{A\cap B'} BB'$. So $\bar{c} \mathop{\smile\!\!\!\!\!\mid}_{A\cap B'} A$, whence $\mathrm{Cb}(\bar{c}/A) \in \mathrm{bdd}(B')$; since $B' \mathop{\smile\!\!\!\!\!\mid}_{\mathrm{Cb}(\bar{c}/B)} A$, we get $\mathrm{Cb}(\bar{c}/A) \in \mathrm{bdd}(\mathrm{Cb}(\bar{c}/B))$. ■

A superstable ω-categorical structure is one-based [29]. This is no longer true in the supersimple case, as Hrushovski's example 6.2.27 shows; however, a significant modification (or a new construction) seems to be required to answer the following question negatively:

PROBLEM 6.2.46 Is a supersimple ω-categorical structure CM-trivial?

As far as merely simple ω-categorical structures are concerned, Hrushovski has used an irrational weight α and a slightly more complicated function f than that used in Subsection 6.2.1 to construct a stable ω-categorical pseudo-plane, which cannot be one-based; of course it is still CM-trivial.

PROBLEM 6.2.47 Is a stable, or simple, ω-categorical structure CM-trivial?

THEOREM 6.2.48 *If $p \in S(\emptyset)$ is a regular type in a supersimple ω-categorical CM-trivial theory, then p is non-orthogonal to a type of SU-rank 1.*

PROOF: Let C be the set of realizations of p in $\mathfrak{C}^{eq}$. For any $x \in C$ and $X \subseteq C$ put $\mathrm{cl}(X) = \{x \in C : x \not\mathop{\smile\!\!\!\!\!\mid} X\}$; since p is regular, this is a closure operator; by ω-categoricity, $\mathrm{cl}(X)$ is definable for any finite set X. For any $x, y \in C$ we have $\mathrm{cl}(x) = \mathrm{cl}(y)$ if and only if $x \not\mathop{\smile\!\!\!\!\!\mid} y$, so $\not\mathop{\smile\!\!\!\!\!\mid}$ yields a definable equivalence relation E on C. Put $p_E = \mathrm{tp}(x_E)$ (for any $x \models p$) and $C_E = \{x_E : x \in C\}$.

Then C_E is the set of realizations of p_E. If $x_E \models p_E$ and $x_E \not\perp A$, then $x \not\perp A$, so $\text{tp}(x/A)$ and hence $\text{tp}(x_E/A)$ is orthogonal to p, and therefore also to p_E. Thus p_E is again regular, and $p \doteq p_E$.

CLAIM. If the geometry of p_E is locally finite, then $SU(p_E) \leq 1$.

PROOF OF CLAIM: Suppose the geometry is locally finite, $x \models p_E$ and A is a set of parameters with $x \not\perp A$. We claim that $x \in \text{acl}(A)$; this will clearly prove the assertion.

So suppose $x \notin \text{acl}(A)$, and let $(x_i : i < \omega)$ be a Morley sequence in $\text{tp}(x/A)$. By local finiteness of the geometry, we may replace the sequence by an infinite subsequence such that $(x_i : i < \omega)$ is independent over $\emptyset$. But $\text{Cb}(x/A) = \text{Cb}(x_0/A) \subseteq \text{dcl}(x_i : i > 0)$, whence $x_0 \perp A$, a contradiction. ■

Suppose now that p is orthogonal to all types of SU-rank 1. Then there is a minimal $n < \omega$ such that there are $n+1$ points $\{a_i : i \leq n\} \subset C_E$ with $\text{cl}_E(a_i : i \leq n)$ infinite; since $\text{cl}(a_0) = \{a_0\}$ by definition of p_E, we have $n \geq 1$. Put $A = \{a_0, \ldots, a_{n-2}\}$ (so $A = \emptyset$ for $n = 1$), and add A to the language.

Put $C'_E = C_E - \text{cl}_E(A)$. For any $a \in C'_E$ we call $\text{cl}_E(A, a)$ a *point*. By the choice of n, points are finite subsets of C_E; note that by our definitions, any two dependent points must be equal. For any two $a, a' \in C'_E$ with $a' \notin \text{cl}_E(A, a)$ (and hence $a \notin \text{cl}_E(A, a')$) we call $\text{cl}_E(A, a, a')$ a *line*; by the choice of n there is at least one line which is an infinite subset of C'_E; it must therefore contain infinitely many points. Finally, for any $a, a', a'' \in C'_E$ with $a'' \notin \text{cl}(A, a, a')$ and $a' \notin \text{cl}_E(A, a)$ we call $\text{cl}_E(A, a, a', a'')$ a *plane*; incidence between points, lines and planes means containment of the respective sets. By ω-categoricity, points, lines and planes are elements of $(C'_E)^{eq}$. We denote the set of points by P, the set of lines by L, and the set of planes by S.

CLAIM. There are incident $a \in P, b \in L, c \in S$ with $b \notin \text{acl}(ac)$, $a \notin \text{acl}(bc)$ and $c \perp_b a$.

PROOF OF CLAIM: Let b be a line containing infinitely many points. By compactness there is a point $a^0 \notin \text{acl}(b)$ incident with b. Let k be the maximum number of points in the algebraic closure of any two lines, and let $a^0, \ldots, a^k$ be part of a Morley sequence of $\text{tp}(a^0/\text{acl}(b))$. Let b' realize a non-forking extension of $\text{tp}(b/\text{acl}(a^0))$ to $\text{acl}(a^0, \ldots, a^k, b)$. Note that $b \notin \text{acl}(a^0)$ so b' is a line through a^0 distinct from b, and we may assume that $a^1 \notin \text{acl}(bb')$ by the choice of k. Let c be the unique plane containing b', a^1. As $c \in \text{acl}(bb')$ we therefore have $a^1 \notin \text{acl}(bc)$.

Next, $b' \perp_{a^0} a^1$ and $a^1 \perp a^0$, so $b'a^0 \perp a^1$, whence $a^0 \perp_{b'} a^1$. As $a^0 \notin \text{acl}(b')$, we get $a^0 \notin \text{acl}(a^1 b')$; so $a^0 \in \text{acl}(bb')$ implies $b \notin \text{acl}(a^1 b')$, whence $b \notin \text{acl}(a^1 c)$.

Finally let $a_2 = \text{cl}_E(x, A)$ be a point of $b' - b$. Then $x \notin b = \text{cl}_E(A, a^0, a^1)$, whence $x \perp a^0 a^1$ by definition of the closure, and $a_2 \perp a^1 a^0$. Hence $a_2 \perp a^1 b$

and $a_2 \mathop{\smile\!\!\!\!\!\!\mid}_b a^1$, so $c \in \mathrm{acl}(a_2 b)$ implies $c \mathop{\smile\!\!\!\!\!\!\mid}_b a^1$. Taking $a = a^1$ gives what we want. ■

CLAIM. If $X = \mathrm{acl}(\mathrm{Cb}(a/\mathrm{acl}(b)))$, then $b \in X$.

PROOF OF CLAIM: Suppose otherwise. Then there exists $b' \neq b$ with $\mathrm{tp}(b/X) = \mathrm{tp}(b'/X)$ and $b \mathop{\smile\!\!\!\!\!\!\mid}_X b'$. Let $r = \mathrm{tp}(a/\mathrm{acl}(b))$ and r' a translate of this (over X) to $\mathrm{acl}(b')$. These have the same restrictions to X (which is a Lascar strong type), and neither forks over X. By the Independence Theorem 2.5.20 there exists $a' \mathop{\smile\!\!\!\!\!\!\mid}_X bb'$ realizing $r \cup r'$. But a' is a point on both lines b and b', whence $a' \in \mathrm{acl}(bb')$. Therefore $a' \in \mathrm{acl}(X)$, and $a \in \mathrm{acl}(X) \subseteq \mathrm{acl}(b)$, a contradiction. ■

CLAIM. Let d be a canonical parameter for $\mathrm{acl}(ac) \cap \mathrm{acl}(bc)$. Then c is the unique plane in $\mathrm{acl}(d)$; moreover $\mathrm{acl}(ad) \cap \mathrm{acl}(bd) = \mathrm{acl}(d)$, and $\mathrm{acl}(\mathrm{Cb}(a/\mathrm{acl}(bd))) = \mathrm{acl}(b)$.

PROOF OF CLAIM: It is easy to see that c is the unique plane in $\mathrm{acl}(bc)$, so the first statement follows. The second is trivial, and the third follows from the last claim, together with the fact that $a \mathop{\smile\!\!\!\!\!\!\mid}_b c$ and $\mathrm{acl}(bd) = \mathrm{acl}(bc)$. ■

CLAIM. $\mathrm{Cb}(a/d) \not\subseteq \mathrm{acl}(b)$.

PROOF OF CLAIM: Let $Y = \mathrm{acl}(\mathrm{Cb}(a/d))$. If $c \in Y$, we are done as $c \notin \mathrm{acl}(b)$. Hence we may assume $c \notin Y$, so $d \notin Y$. Let $(d_i : i < \omega)$ be a Morley sequence in $\mathrm{lstp}(d/Y)$, with $d = d_0$. By the Independence Theorem 2.5.20 over Y twice, there exists a point $a' \mathop{\smile\!\!\!\!\!\!\mid}_Y d_0 d_1 d_2$ with $\mathrm{tp}(a' d_i) = \mathrm{tp}(ad)$ for $i < 3$. Let c_i be the unique plane in $\mathrm{acl}(d_i)$, for $i < 3$. As a is on the plane $c = c_0$, the point a' is incident with all three planes c_0, c_1 and c_2. Moreover, a' is not algebraic over $c_0 c_1 c_2$: if it were, then $a' \in \mathrm{acl}(Y)$, whence $a \in \mathrm{acl}(Y) \subseteq \mathrm{acl}(d) \subseteq \mathrm{acl}(bc)$. Thus the three planes have infinitely many points in common, and must intersect in a line b_1. As d_0, d_1, d_2 are independent over Y and b_1 lies in the algebraic closure of any pair of them, we get that $b_1 \in Y$. As $b \notin \mathrm{acl}(ac)$, we have $b \notin \mathrm{acl}(d)$, whence $b_1 \neq b$. Thus $Y \not\subseteq \mathrm{acl}(b)$, as otherwise $\mathrm{acl}(b)$ would contain two distinct lines of the plane c, and hence c itself, a contradiction. ■

So now we have $\mathrm{acl}(ad) \cap \mathrm{acl}(bd) = \mathrm{acl}(d)$ and $\mathrm{Cb}(a/d) \not\subseteq \mathrm{Cb}(a/bd)$, contradicting CM-triviality. ■

COROLLARY 6.2.49 *Let T be an ω-categorical supersimple CM-trivial theory. Then T has finite SU-rank.*

PROOF: If there were a type of infinite SU-rank, we could find a finitely based type p of SU-rank ω. Then p is regular, and orthogonal to all types of SU-rank 1; adding the parameters of its domain to the language, we contradict Theorem 6.2.48. ■

REMARK 6.2.50 In [29] finiteness of the rank of a superstable ω-categorical theory is proved by showing first that a type of finite rank is locally modular, and then considering forking extensions of a type of rank ω, obtaining a contradiction by using local modularity for types of finite rank. Our proof of Theorem 6.2.32 proceeds similarly by describing the structure of subgroups of finite rank, and then finding suitable subgroups of finite rank in a group of infinite rank (and in fact Proposition 6.2.40 is very close to local modularity, see Proposition 4.8.3). In contrast, our proof of Theorem 6.2.48 uses CM-triviality for a type of infinite rank, and thus does not appear to be too helpful for proving Conjecture 6.2.1. Moreover, it cannot be localized to a CM-trivial type of SU-rank ω, as it uses local finiteness of algebraic closure in an essential way.

6.3 SIMPLE EXPANSIONS OF SIMPLE THEORIES

6.3.1 AMALGAMATING SIMPLE THEORIES

In this section we shall try to amalgamate two simple theories T_1 and T_2 in disjoint languages (apart from equality) into a simple theory $T \supseteq T_1 \cup T_2$.

DEFINITION 6.3.1 A theory T *eliminates infinite quantifiers* if for every formula $\varphi(x, \bar{y})$ there is a formula $\psi(\bar{y})$ such that in any model $\mathfrak{M} \models T$ and for any $\bar{a} \in \mathfrak{M}$ the formula $\varphi(x, \bar{a})$ defines an infinite set if and only if $\mathfrak{M} \models \psi(\bar{a})$.

Note that the elements and tuples in this definition are real, not imaginary.

LEMMA 6.3.2 *Let T be a theory which eliminates infinite quantifiers. For a formula $\varphi(\bar{x}, \bar{y})$ let $\exists^{na}\bar{x}\,\varphi(\bar{x}, \bar{y})$ denote the condition*

$$\exists \bar{x}\,[\bigwedge_{i<j} x_i \neq x_j \wedge \bar{x} \cap \mathrm{acl}(\bar{y}) = \emptyset].$$

Then there is a formula $\vartheta(\bar{y})$ equivalent to $\exists^{na}\bar{x}\,\varphi(\bar{x}, \bar{y})$ modulo T.

PROOF: For an arbitrary formula $\psi(\bar{x}, \bar{y})$ and a tuple $\bar{a}$ of parameters, let a-dim$(\psi(\bar{x}, \bar{a}))$ be the maximal $n < |\bar{x}|$ such that there are $\bar{b} \models \psi(\bar{x}, \bar{a})$ and some permutation σ with $b_{\sigma(i)} \notin \mathrm{acl}(\bar{a}, b_{\sigma(0)}, \ldots, b_{\sigma(i-1)})$ for all $i < n$.
CLAIM. The set $\{\bar{a} : \text{a-dim}(\psi(\bar{x}, \bar{a})) = n\}$ is definable for all $n < \omega$.
PROOF OF CLAIM: Let Σ be the group of permutations of $|\bar{x}|$. Then

$$\bigvee_{\sigma \in \Sigma} \exists^\infty x_{\sigma(0)} \cdots \exists^\infty x_{\sigma(n-1)} \exists x_{\sigma(n)} \cdots \exists x_{\sigma(|\bar{x}|-1)} \psi(\bar{x}, \bar{y})$$
$$\wedge \neg \bigvee_{\sigma \in \Sigma} \exists^\infty x_{\sigma(0)} \cdots \exists^\infty x_{\sigma(n)} \exists x_{\sigma(n+1)} \cdots \exists x_{\sigma(|\bar{x}|-1)} \psi(\bar{x}, \bar{y})$$

is equivalent to a first-order formula by elimination of infinite quantifiers, and defines the required set. ■

Given a formula $\psi(\bar{x},\bar{y})$, let $\psi'(\bar{x},\bar{y})$ be the condition

$$\psi(\bar{x},\bar{y}) \wedge \bigwedge_{i<|\bar{x}|} \text{a-dim}(\psi(\bar{x}-x_i,\bar{y}x_i)) < \text{a-dim}(\psi(\bar{x},\bar{y})).$$

As $\text{a-dim}(\psi(\bar{x},\bar{y}))$ is bounded by $|\bar{x}|$, this is first-order definable by the claim above. Now suppose $\models \psi(\bar{b},\bar{a})$ and $\bar{b}\cap \text{acl}(\bar{a}) = \emptyset$. For any $b_i \in \bar{b}$ we have $b_i \notin \text{acl}(\bar{a})$, so $\text{a-dim}(\psi(\bar{x}-x_i,\bar{a}b_i)) < \text{a-dim}(\psi(\bar{x},\bar{a}))$ and $\models \psi'(\bar{b},\bar{a})$. Hence $\exists^{na}\bar{x}\,\psi(\bar{x},\bar{a})$ if and only if $\exists^{na}\bar{x}\,\psi'(\bar{x},\bar{a})$.

CLAIM. If $\text{a-dim}(\psi(\bar{x},\bar{a})) = \text{a-dim}(\psi'(\bar{x},\bar{a}))$, then $\exists^{na}\bar{x}\,\psi(\bar{x},\bar{a})$ if and only if $\exists\bar{x}\,[\psi'(\bar{x},\bar{a}) \wedge \bigwedge_{i<j} x_i \neq x_j]$.

PROOF OF CLAIM: We only have to show "if". So suppose $\models \psi'(\bar{b},\bar{a}) \wedge \bigwedge_{i<j} b_i \neq b_j$; we may choose it such that for some permutation σ we have $b_{\sigma(i)} \notin \text{acl}(\bar{a}, b_{\sigma(0)},\ldots,b_{\sigma(i-1)})$ for all $i < \text{a-dim}(\psi(\bar{x},\bar{a}))$. Suppose $i \geq \text{a-dim}(\psi(\bar{x},\bar{a}))$ and $b_{\sigma(i)} \in \text{acl}(\bar{a})$. Then $\bar{b}-b_{\sigma(i)}$ witnesses $\text{a-dim}(\psi(\bar{x}-x_{\sigma(i)},\bar{a}b_{\sigma(i)})) \geq \text{a-dim}(\psi(\bar{x},\bar{a}))$, contradicting $\bar{b}\models\psi'(\bar{x},\bar{a})$. Hence $\bar{b}\cap \text{acl}(\bar{a}) = \emptyset$. ■

As $|\bar{x}| \geq \text{a-dim}(\varphi(\bar{x},\bar{y})) \geq \text{a-dim}(\varphi'(\bar{x},\bar{y})) \geq \cdots$, the assertion follows. ■

REMARK 6.3.3 It is clear that for any formula $\varphi(\bar{x},\bar{a})$ over some model $\mathfrak{M}$ we have $\mathfrak{M} \models \exists^{na}\bar{x}\,\varphi(\bar{x},\bar{a})$ if and only if there are a model $\mathfrak{N} \succ \mathfrak{M}$ and a tuple $\bar{b}\in\mathfrak{N}-\mathfrak{M}$ of distinct elements, such that $\mathfrak{N}\models\varphi(\bar{b},\bar{a})$.

THEOREM 6.3.4 *Let T_1 and T_2 be two model-complete theories in relational languages $\mathfrak{L}_1$ and $\mathfrak{L}_2$ respectively, with $\mathfrak{L}_1\cap\mathfrak{L}_2 = \{=\}$. Suppose both theories eliminate quantifiers and infinite quantifiers. Then $T_1\cup T_2$ has a model-companion T. Moreover, T eliminates infinite quantifiers, and* acl(.) *is the transitive closure of algebraic closure in the sense of T_1 and in the sense of T_2.*

PROOF: For every $\mathfrak{L}_1$-formula $\varphi_1(\bar{x},\bar{y})$ and $\varphi_2(\bar{x},\bar{y})$, consider the following sentence

$$\forall\bar{y}\,\{[\exists^{na}\bar{x}\,\varphi_1(\bar{x},\bar{y}) \wedge \exists^{na}\bar{x}\,\varphi_2(\bar{x},\bar{y})] \rightarrow \exists\bar{x}\,[\varphi_1(\bar{x},\bar{y})\wedge\varphi_2(\bar{x},\bar{y})]\}.$$

By Lemma 6.3.2, modulo $T_1\cup T_2$ this is a first-order sentence $\sigma(\varphi_1,\varphi_2)$. Put

$$T = T_1\cup T_2\cup\{\sigma(\varphi_1,\varphi_2) : \varphi_1\in\mathfrak{L}_1, \varphi_2\in\mathfrak{L}_2\}.$$

CLAIM. Any model of $T_1\cup T_2$ can be embedded into a model of T.

PROOF OF CLAIM: Suppose $\mathfrak{M}\models T_1\cup T_2$. If $\varphi_i(\bar{x},\bar{a})$ is an $\mathfrak{L}_i(\mathfrak{M})$-formula with $\mathfrak{M}\models\exists^{na}\bar{x}\,\varphi_i(\bar{x},\bar{a})$ for $i=1,2$, then there are $\mathfrak{L}_i$-elementary $\mathfrak{L}_i$-superstructures $\mathfrak{M}_i$ of $\mathfrak{M}$, for $i=1,2$, such that there is a tuple $\bar{m}_i\in\mathfrak{M}_i-\mathfrak{M}$ of distinct elements satisfying $\varphi_i(\bar{x},\bar{a})$. We may choose $\mathfrak{M}_1$ and $\mathfrak{M}_2$ of the same cardinality, greater than $|\mathfrak{M}|$. But then we may identify the universes of $\mathfrak{M}_1$ and $\mathfrak{M}_2$ over $\mathfrak{M}$ in such a way that $\bar{m}_1$ is identified with $\bar{m}_2$. We thus obtain a model $\mathfrak{M}'$ of $T_1 \cup T_2$, an $\mathfrak{L}_1$- and $\mathfrak{L}_2$-elementary superstructure of $\mathfrak{M}$, such that $\mathfrak{M}' \models \exists\bar{x}\,[\varphi_1(\bar{x},\bar{a}) \wedge \varphi_2(\bar{x},\bar{a})]$. Repeating this construction with unions at the limit stages, we obtain an $\mathfrak{L}_1$- and $\mathfrak{L}_2$-elementary superstructure $\mathfrak{M}^1 \supseteq \mathfrak{M}$, and by iteration an $\mathfrak{L}_1$- and $\mathfrak{L}_2$-elementary chain of models $\mathfrak{M} = \mathfrak{M}^0 \subseteq \mathfrak{M}^1 \subseteq \cdots$ of $T_1 \cup T_2$, such that for every $i < \omega$, every $\mathfrak{L}_1$-formula $\varphi_1(\bar{x},\bar{y})$, every $\mathfrak{L}_2$-formula $\varphi_2(\bar{x},\bar{y})$, and every $\bar{a} \in \mathfrak{M}_i$, if $\exists^{na}\bar{x}\,\varphi_1(\bar{x},\bar{a})$ and $\exists^{na}\bar{x}\,\varphi_2(\bar{x},\bar{a})$ are both true in $\mathfrak{M}_i$ (and hence in every $\mathfrak{M}_j$ for $j \geq i$), then there is $\bar{m} \in \mathfrak{M}_{i+1}$ satisfying $\varphi_1(\bar{x},\bar{a}) \wedge \varphi_2(\bar{x},\bar{a})$. Then $\bigcup_{i<\omega}\mathfrak{M}^i \models T$. ■

CLAIM. Every model of T is existentially closed.

PROOF OF CLAIM: Let $\mathfrak{M} \subset \mathfrak{N}$ be two models of T, and $\varphi(\bar{x},\bar{a})$ a quantifier-free $\mathfrak{L}_1 \cup \mathfrak{L}_2$-formula with parameters $\bar{a}$ from $\mathfrak{M}$, which has a solution in $\mathfrak{N} - \mathfrak{M}$. As $\mathfrak{L}_1$ and $\mathfrak{L}_2$ are both relational, we may assume that $\varphi(\bar{x},\bar{a})$ has the form $\varphi_1(\bar{x},\bar{a}) \wedge \varphi_2(\bar{x},\bar{a})$, where $\varphi_i(\bar{x},\bar{y})$ is a quantifier-free $\mathfrak{L}_i$-formula, for $i = 1, 2$. As $\varphi_i(\bar{x},\bar{a})$ has a solution in $\mathfrak{N} - \mathfrak{M}$, we get $\exists^{na}\bar{x}\,\varphi_i(\bar{x},\bar{a})$; since $\mathfrak{M} \models \sigma(\varphi_1,\varphi_2)$, the conjunction $\varphi_1(\bar{x},\bar{a}) \wedge \varphi_2(\bar{x},\bar{a})$ must have a solution in $\mathfrak{M}$. Thus $\mathfrak{M}$ is existentially closed. ■

By the Robinson test 1.2.15, T is model-complete.

For the last part, consider again an $(\mathfrak{L}_1 \cup \mathfrak{L}_2)$-formula $\psi(x,\bar{y})$; by model-completeness we may assume that it is of the form $\exists\bar{x}\,[\varphi_1(x,\bar{x},\bar{y}) \wedge \varphi_2(x,\bar{x},\bar{y})]$, where φ_i is an $\mathfrak{L}_i$-formula which implies $\bigwedge_{i<|\bar{x}|} x_i \neq x \wedge \bigwedge_{i<j} x_i \neq x_j$, for $i = 1, 2$. Suppose $\psi(x,\bar{a})$ defines an infinite set for some $\bar{a}$ in a model $\mathfrak{M} \models T$. Let $b, \bar{b} \in \mathfrak{M}$ be such that $b \notin \mathrm{acl}(\bar{a})$ and $\mathfrak{M} \models \varphi_1(b,\bar{b},\bar{a}) \wedge \varphi_2(b,\bar{b},\bar{a})$. Then $\bar{b} = \bar{b}_0\bar{b}_1$, where $\bar{b}_0 \cap \mathrm{acl}(\bar{a}) = \emptyset$ and $\bar{b}_1 \in \mathrm{acl}(\bar{a})$. Hence $b\bar{b}_0 \cap \mathrm{acl}(\bar{b}_1\bar{a}) = \emptyset$, and $\mathfrak{M} \models \exists^{na} x\bar{x}_0\,\varphi_i(x,\bar{x}_0,\bar{b}_1,\bar{a})$ for $i = 1, 2$.

Conversely, if there is a division $\bar{x} = \bar{x}_0\bar{x}_1$ and $\bar{b}_1 \in \mathrm{acl}(\bar{a})$ such that $\mathfrak{M} \models \exists^{na} x\bar{x}_0\,\varphi_i(x,\bar{x}_0,\bar{b}_1,\bar{a})$ for $i = 1, 2$, then the axioms of T imply that $\varphi_1(x,\bar{x}_0,\bar{b}_1,\bar{a}) \wedge \varphi_2(x,\bar{x}_0,\bar{b}_1,\bar{a})$ has an infinite sequence of solutions $(b^i\bar{b}_0^i : i < \omega)$ with $b^i\bar{b}_0^i \cap \mathrm{acl}(\bar{b}_1,\bar{a},b^j\bar{b}_0^j : j < i) = \emptyset$ for all $i < \omega$. Hence $\psi(x,\bar{a})$ defines an infinite set.

It follows that $\exists^\infty x\,\psi(x,\bar{a})$ is equivalent modulo T to a (possibly infinite) disjunction; since it is obviously equivalent to an infinite conjunction, it is definable by compactness.

Finally, consider $a \in \mathrm{acl}(A)$ (inside some model $\mathfrak{M} \models T$); we may assume that A is algebraically closed in the sense of T_1 and of T_2. Suppose $\varphi(x,A)$ isolates $\mathrm{tp}(a/A)$; by model-completeness and relationality of the languages we may assume that $\varphi(x,A) = \exists\bar{x}\,[\varphi_1(x,\bar{x},A) \wedge \varphi_2(x,\bar{x},A)]$, where φ_i is an $\mathfrak{L}_i$-formula for $i = 1, 2$. We choose it such that the length $|\bar{x}|$ is minimal; if $\bar{a}$ is such that $\mathfrak{M} \models \varphi_1(a,\bar{a},A) \wedge \varphi_2(a,\bar{a},A)$, then $\bar{a}$ is a tuple of distinct elements disjoint from $A \cup \{a\}$ by minimality. Hence $a \notin A$ would imply

$\exists^{na} x\bar{x}\, \varphi_i(x,\bar{x},A)$ for $i = 1,2$, and $\varphi_1(x,\bar{x},A) \wedge \varphi_2(x,\bar{x},A)$ would have infinitely many solutions in $\mathfrak{M}$, contradicting our choice of φ to isolate an algebraic type. Thus $a \in A$ and we are done. ∎

We shall now see that, in a way, elimination of infinite quantifiers is a necessary condition.

DEFINITION 6.3.5 A set is *strongly minimal* if every definable subset is finite or co-finite.

Strongly minimal sets are ω-stable of finite rank, and uncountably categorical.

LEMMA 6.3.6 *Let T_1 and T_2 be model-complete theories in languages $\mathfrak{L}_1$ and $\mathfrak{L}_2$, respectively, with $\mathfrak{L}_1 \cap \mathfrak{L}_2 = \{=\}$. Suppose T_1 does not eliminate infinite quantifiers, and T_2 is not strongly minimal. Then $T_1 \cup T_2$ has no model-companion.*

PROOF: Suppose $T_1 \cup T_2$ has a model-companion T, and let $\varphi(x,\bar{y})$ be a formula such that $\exists^\infty x\, \varphi(x,\bar{y})$ is not equivalent to a first-order formula. Then there is a model $\mathfrak{M} \models T_1$ and a sequence $(\bar{a}_i : i < \omega)$, such that $\varphi(x,\bar{a}_i)$ defines a finite set of size at least i in $\mathfrak{M}$. If $\psi(x)$ is an infinite co-infinite $\mathfrak{L}_2$-formula (possibly with parameters), choose $\mathfrak{N} \succ \mathfrak{M}$ of greater cardinality and turn $\mathfrak{N}$ into a model of $T_1 \cup T_2$ such that $\mathfrak{N} \models \psi(m)$ for all $m \in \mathfrak{M}$. We may assume that in fact $\mathfrak{N} \models T$.

Let $\mathcal{U}$ be a non-principal ultrafilter on ω, and $\mathfrak{N}'$ the ultrapower $\prod_\omega \mathfrak{N}/\mathcal{U}$. Put $\bar{a} = (\bar{a}_i : i < \omega)/\mathcal{U}$. Then $\mathfrak{N}' \models \varphi(x,\bar{a}) \rightarrow \psi(x)$. But since $\varphi(x,\bar{a})$ and $\neg\psi(x)$ both define infinite sets, $\mathfrak{N}'$ embeds into a model of T satisfying $\exists x\, [\varphi(x,\bar{a}) \wedge \neg\psi(x)]$. So the embedding is not elementary, contradicting model-completeness of T. ∎

DEFINITION 6.3.7 A formula $\varphi(x,y)$ has the *independence property* (modulo a theory T) if there are a model $\mathfrak{M} \models T$, a sequence $(a_i : i < \omega)$ and elements $(b_I : I \subseteq \omega)$ in $\mathfrak{M}$, such that $\mathfrak{M} \models \varphi(a_i, b_I)$ if and only if $i \in I$. A theory T has the *independence property* if there is a formula which has it (modulo T).

It is easy to see that a theory with the independence property must be unstable. It can, however, still be simple. Under some weak condition, the model companion of $T_1 \cup T_2$ will have the independence property.

PROPOSITION 6.3.8 *Let T_1 and T_2 be model-complete theories in languages $\mathfrak{L}_1$ and $\mathfrak{L}_2$, respectively, with $\mathfrak{L}_1 \cap \mathfrak{L}_2 = \{=\}$, such that $T_1 \cup T_2$ has a model-companion T. Suppose T_1 is simple, T_2 is not strongly minimal, and (possibly after adding parameters) there are independent a, b in a model $\mathfrak{M}$ of T_1 with* $\mathrm{acl}(a,b) \neq \mathrm{acl}(a) \cup \mathrm{acl}(b)$. *Then T has the independence property.*

PROOF: Put $p(x,y) = \mathrm{tp}(a,b)$. Let $(b_i : i < \omega)$ be a Morley sequence in $\mathrm{lstp}(b/a)$, and $(a_I : I \subseteq \omega)$ a sequence of independent realizations of $\bigcup_{i<\omega} p(x,b_i)$, all in some model $\mathfrak{M} \models T_1$. Then the set $\{a_I, b_i : i < \omega, I \subseteq \omega\}$ is independent, and $\mathrm{acl}(a_I, b_i) \cap \mathrm{acl}(a_J, b_j : J \neq I, j < \omega) = \mathrm{acl}(b_i)$. So if $\varphi(x,a,b)$ isolates $\mathrm{tp}(e/ab)$ for some $e \in \mathrm{acl}(ab) - [\mathrm{acl}(a) \cup \mathrm{acl}(b)]$, then $\varphi(x,a_I,b_i) \wedge \varphi(x,a_J,b_j)$ is inconsistent unless $I = J$ and $i = j$.

Choose an infinite co-infinite $\mathfrak{L}_2$-formula $\psi(x)$. We can then expand $\mathfrak{M}$ to a model of $T_1 \cup T_2$ in such a way that $\forall x\,[\varphi(x,a_I,b_i) \to \psi(x)]$ holds if and only if $i \in I$. Now $\mathfrak{M}$ embeds into a model $\mathfrak{N}$ of T; by model-completeness of T_1 and T_2 we also have $\mathfrak{N} \models \forall x\,[\varphi(x,a_I,b_i) \to \psi(x)]$ if and only if $i \in I$ (the universal quantifier is over a finite set). This witnesses the independence property. ■

PROPOSITION 6.3.9 *Let T_1 and T_2 be simple theories in relational languages $\mathfrak{L}_1$ and $\mathfrak{L}_2$, respectively, with $\mathfrak{L}_1 \cap \mathfrak{L}_2 = \{=\}$. Suppose T_1 and T_2 are model-complete, and $T_1 \cup T_2$ has a model-companion T. For a model $\mathfrak{M}$ of T and subsets A, B, C of $\mathfrak{M}$ define $A \mathop{\smile\!\!\!\!\!\mid}^0_B C$ if $\mathrm{acl}(AB)$ is independent of $\mathrm{acl}(BC)$ over $\mathrm{acl}(B)$ in the sense of T_1 and in the sense of T_2 (as subsets of the $\mathfrak{L}_1$- and $\mathfrak{L}_2$-reducts of $\mathfrak{M}$), where $\mathrm{acl}(.)$ is taken with respect to T. Then $\mathop{\smile\!\!\!\!\!\mid}^0$ is invariant under automorphisms, and satisfies extension, symmetry, finite character, and local character. Moreover, it satisfies* partial *transitivity: if $A \subseteq B \subseteq C$ and $a \mathop{\smile\!\!\!\!\!\mid}^0_A B$ and $a \mathop{\smile\!\!\!\!\!\mid}^0_B C$, then $a \mathop{\smile\!\!\!\!\!\mid}^0_A C$.*

PROOF: Clearly $\mathop{\smile\!\!\!\!\!\mid}^0$ is invariant under automorphisms, and inherits symmetry, partial transitivity and finite character from independence in the sense of T_1 and T_2.

CLAIM. $\mathop{\smile\!\!\!\!\!\mid}^0$ satisfies local character.

PROOF OF CLAIM: Consider a tuple a and a set A in some model of T. By local character in T_1 and in T_2, we find an increasing chain $(A_i : i < \omega)$ of subsets of A of size at most $|T|$ (with $A_0 = \emptyset$) such that $\mathrm{acl}(aA_i)$ is independent of $\mathrm{acl}(A)$ over $\mathrm{acl}(A_{i+1})$ in the sense of T_1 and of T_2 for all $i < \omega$. Put $B = \bigcup_{i<\omega} A_i$. Then $|B| \leq |T|$ and $\mathrm{acl}(aB)$ is independent of $\mathrm{acl}(A)$ over $\mathrm{acl}(B)$ in the sense of T_1 and of T_2. Hence $a \mathop{\smile\!\!\!\!\!\mid}^0_B A$. ■

CLAIM. $\mathop{\smile\!\!\!\!\!\mid}^0$ satisfies the extension axiom.

PROOF OF CLAIM: Let $\mathfrak{M} \models T$, consider subsets $A \subseteq B \subseteq \mathfrak{M}$ with $\mathrm{acl}(A) = A$, and take some $\bar{a} \in \mathfrak{M}$. By finite character we may assume that $B = \mathfrak{M}$. Denote independence in the sense of T_i by $\mathop{\smile\!\!\!\!\!\mid}^i$, for $i = 1, 2$. By the extension axiom for $\mathop{\smile\!\!\!\!\!\mid}^i$ there are $\mathfrak{L}_i$-elementary $\mathfrak{L}_i$-superstructures $\mathfrak{N}_i$ of $\mathfrak{M}$ containing automorphic copies $\mathfrak{M}_i$ of $\mathfrak{M}\restriction_{\mathfrak{L}_i}$, with $\mathfrak{M}_i \mathop{\smile\!\!\!\!\!\mid}^i_A \mathfrak{M}$ (for $i = 1, 2$); we may choose them such that $\mathfrak{N}_1$ and $\mathfrak{N}_2$ have the same cardinality, greater than $|\mathfrak{M}|$. As $\mathfrak{M}_1 \cap \mathfrak{M} = \mathfrak{M}_2 \cap \mathfrak{M} = \mathrm{acl}(A) = A$, we may identify the universes of

$\mathfrak{N}_1$ and $\mathfrak{N}_2$ over A in such a way that $\mathfrak{M}_1$ is identified with $\mathfrak{M}_2$ as an isomorphic copy $\mathfrak{M}'$ of $\mathfrak{M}$, which must be a model of T. We thus obtain a model of $T_1 \cup T_2$, which can be embedded into a model $\mathfrak{N}$ of T; by model-completeness of T we have $\mathfrak{M} \prec \mathfrak{N}$ and $\mathfrak{M}' \prec \mathfrak{N}$. Note that by model-completeness of T_1 and T_2 we also have $\mathfrak{N}_1 \prec \mathfrak{N}$ in the sense of T_1 and $\mathfrak{N}_2 \prec \mathfrak{N}$ in the sense of T_2. Thus $\mathfrak{M}'$ (which equals $\mathfrak{M}_1 = \mathfrak{M}_2$ as a set) is independent of $\mathfrak{M}$ over A in the sense of T_1 (as it is so in $\mathfrak{N}_1$) and of T_2 (as it is so in $\mathfrak{N}_2$); since $\mathfrak{M}$ and $\mathfrak{M}'$ are models of T containing A, we get $\mathrm{acl}(\mathfrak{M}A) = \mathfrak{M}$ and $\mathrm{acl}(\mathfrak{M}'A) = \mathfrak{M}'$, whence $\mathfrak{M}' \perp^0_A \mathfrak{M}$.

Finally, if $\bar{a}' \in \mathfrak{M}'$ corresponds to $\bar{a} \in \mathfrak{M}$, then $\mathfrak{M}' \prec \mathfrak{N}$ and $\mathfrak{M} \prec \mathfrak{N}$ imply

$$\mathrm{tp}_{\mathfrak{N}}(\bar{a}'/A) = \mathrm{tp}_{\mathfrak{M}'}(\bar{a}'/A) = \mathrm{tp}_{\mathfrak{M}}(\bar{a}/A) = \mathrm{tp}_{\mathfrak{N}}(\bar{a}/A). \quad \blacksquare$$

PROPOSITION 6.3.10 *Let T_1 and T_2 be simple theories in relational languages $\mathfrak{L}_1$ and $\mathfrak{L}_2$, respectively, with $\mathfrak{L}_1 \cap \mathfrak{L}_2 = \{=\}$. Suppose T_1 and T_2 are model-complete, and $T_1 \cup T_2$ has a model-companion T. Let $\perp^0$ be the conjunction of T_1- and T_2-independence. Then $\perp^0$ satisfies the Independence Theorem over a model.*

PROOF: Let $\mathfrak{M} \models T$ contain a submodel $\mathfrak{M}_0$, sets A_1 and A_2 with $A_1 \perp^0_{\mathfrak{M}_0} A_2$, and tuples $a_1 \perp^0 \mathfrak{M}_0 A_1$ and $a_2 \perp^0_{\mathfrak{M}_0} A_2$, with $\mathrm{tp}(a_1/\mathfrak{M}_0) = \mathrm{tp}(a_2/\mathfrak{M}_0)$. Let $\mathfrak{N}_1$ be any model containing $\mathfrak{M}_0 a_1$; by extension we may assume $\mathfrak{N}_1 \perp^0_{\mathfrak{M}_0 a_1} A_1$, whence $A_1 \perp^0_{\mathfrak{M}_0} \mathfrak{N}_1$ by partial transitivity. As $\mathrm{tp}(a_1/\mathfrak{M}_0) = \mathrm{tp}(a_2/\mathfrak{M}_0)$, there is $\mathfrak{N}_2$ with $\mathrm{tp}(\mathfrak{N}_2 a_2/\mathfrak{M}_0) = \mathrm{tp}(\mathfrak{N}_1 a_1/\mathfrak{M}_0)$; by extension we may assume $\mathfrak{N}_2 \perp^0_{\mathfrak{M}_0 a_2} A_2$, whence $A_2 \perp^0_{\mathfrak{M}_0} \mathfrak{N}_2$ by partial transitivity. Finally, by extension there are models $\mathfrak{M}_1 \supseteq \mathfrak{M}_0 A_1$ and $\mathfrak{M}_2 \supseteq \mathfrak{M}_0 A_2$ with $\mathfrak{M}_1 \perp^0_{\mathfrak{M}_0 A_1} \mathfrak{N}_1 A_2$ and $\mathfrak{M}_2 \perp^0_{\mathfrak{M}_0 A_2} \mathfrak{M}_1 \mathfrak{N}_2$. By partial transitivity and finite character, $\mathfrak{N}_1 \perp^0_{\mathfrak{M}_0} \mathfrak{M}_1$, $\mathfrak{N}_2 \perp^0_{\mathfrak{M}_0} \mathfrak{M}_2$, and $\mathfrak{M}_1 \perp^0_{\mathfrak{M}_0} \mathfrak{M}_2$. We shall show that there is $\mathfrak{N} \perp^0_{\mathfrak{M}_0} \mathfrak{M}_1 \mathfrak{M}_2$ realizing $\mathrm{tp}(\mathfrak{N}_1/\mathfrak{M}_1) \cup \mathrm{tp}(\mathfrak{N}_2/\mathfrak{M}_2)$; choosing some $a \in \mathfrak{N}$ which corresponds to $a_1 \in \mathfrak{N}_1$ and $a_2 \in \mathfrak{N}_2$, we are done by finite character.

Denote independence in the sense of T_i by $\perp^i$. By the Independence Theorem applied to the simple theories T_1 and T_2, there are $\mathfrak{L}_i$-structures $\bar{\mathfrak{N}}^i \models T_i$ containing $\mathfrak{M}$ for $i = 1, 2$, which contain $\mathfrak{L}_i$-elementarily copies of the $\mathfrak{L}_i$-reduct of $\mathfrak{N}_i$, say $\mathfrak{N}^i$, such that $\mathfrak{N}^i \mathfrak{M}_j$ is $\mathfrak{L}_i$-equivalent to $\mathfrak{N}_j \mathfrak{M}_j$ for $j = 1, 2$, and $\mathfrak{N}^i \perp^i_{\mathfrak{M}_0} \mathfrak{M}$. Choosing $\bar{\mathfrak{N}}^i$ big enough, we can also find two further copies of the $\mathfrak{L}_i$-reduct of $\mathfrak{M}$ inside $\bar{\mathfrak{N}}^i$, say $\mathfrak{M}^i_1$ and $\mathfrak{M}^i_2$, such that $\mathfrak{M}^i_j \mathfrak{M}_j \mathfrak{N}^i$ is $\mathfrak{L}_i$-equivalent to $\mathfrak{M}\mathfrak{M}_j \mathfrak{N}_j$, for $j = 1, 2$. We choose them such that $\mathfrak{M}^i_1 \perp^i_{\mathfrak{M}_1 \mathfrak{N}^i} \mathfrak{M}$ and $\mathfrak{M}^i_2 \perp^i_{\mathfrak{M}_2 \mathfrak{N}^i} \mathfrak{M}^i_1 \mathfrak{M}$. As $\mathfrak{N}^i \perp^i_{\mathfrak{M}_1} \mathfrak{M}$ by transitivity, $\mathfrak{M}^i_1 \cap \mathfrak{M} = \mathrm{acl}_i(\mathfrak{M}_1 \mathfrak{N}^i) \cap \mathfrak{M} = \mathfrak{M}_1$; similarly $\mathfrak{M}^i_2 \cap \mathfrak{M} = \mathrm{acl}_i(\mathfrak{M}_2 \mathfrak{N}^i) \cap \mathfrak{M} = \mathfrak{M}_2$ (where $\mathrm{acl}_i(.)$ is algebraic closure in the sense of T_i). Finally $\{\mathfrak{M}_1, \mathfrak{M}_2, \mathfrak{N}^i\}$

forms an T_i-independent family over $\mathfrak{M}_0$, so $\mathfrak{M}_1 \mathop{\smile\!\!\!\!\!\mid}^i_{\mathfrak{N}^i} \mathfrak{M}_2$, and

$$\mathfrak{M}_1^i \cap \mathfrak{M}_2^i = \mathrm{acl}_i(\mathfrak{M}_2\mathfrak{N}^i) \cap \mathfrak{M}_1^i = \mathrm{acl}_i(\mathfrak{M}_2\mathfrak{N}^i) \cap \mathrm{acl}_i(\mathfrak{M}_1\mathfrak{N}^i) = \mathfrak{N}^i.$$

If we choose $\bar{\mathfrak{N}}^1$ and $\bar{\mathfrak{N}}^2$ of the same cardinality greater than $|\mathfrak{M}|$, we can identify $\bar{\mathfrak{N}}^1$ and $\bar{\mathfrak{N}}^2$ in such a way that $\mathfrak{N}^1$ and $\mathfrak{N}^2$, as well as $\mathfrak{M}_j^1$ and $\mathfrak{M}_j^2$ for $j = 1, 2$, become identified as isomorphic copies of $\mathfrak{N}_1$ (or $\mathfrak{N}_2$) and $\mathfrak{M}$, respectively, and hence form models of T, which we just call $\mathfrak{N}$ and $\bar{\mathfrak{M}}_j$ (for $j = 1, 2$), respectively. Clearly $\bar{\mathfrak{N}}^1 = \bar{\mathfrak{N}}^2$ is a model of $T_1 \cup T_2$, which we can embed into a model $\mathfrak{N}'$ of T.

By model-completeness of T_1, T_2 and T, all embeddings are elementary. Now $\mathfrak{N}^i$ is T_i-independent of $\mathfrak{M}$ over $\mathfrak{M}_0$ in $\bar{\mathfrak{N}}^i$ (and hence in $\mathfrak{N}'$, which is an $\mathfrak{L}_i$-elementary superstructure of $\bar{\mathfrak{N}}^i$) for $i = 1, 2$; moreover $\mathfrak{M}$ and $\mathfrak{N}$ are models of T containing $\mathfrak{M}_0$, whence $\mathrm{acl}(\mathfrak{N}\mathfrak{M}_0) = \mathfrak{N}$ and $\mathrm{acl}(\mathfrak{M}\mathfrak{M}_0) = \mathfrak{M}$. Therefore $\mathfrak{N} \mathop{\smile\!\!\!\!\!\mid}^0_{\mathfrak{M}_0} \mathfrak{M}$, and $\mathfrak{N} \mathop{\smile\!\!\!\!\!\mid}^0_{\mathfrak{M}_0} \mathfrak{M}_1\mathfrak{M}_2$. Moreover, by construction

$$\mathrm{tp}_{\mathfrak{N}'}(\mathfrak{N}/\mathfrak{M}_j) = \mathrm{tp}_{\bar{\mathfrak{M}}_j}(\mathfrak{N}/\mathfrak{M}_j) = \mathrm{tp}_{\mathfrak{M}}(\mathfrak{N}_j/\mathfrak{M}_j).$$

So we have found the required $\mathfrak{N}$; this finishes the proof. ■

The question thus becomes under what condition $\mathop{\smile\!\!\!\!\!\mid}^0$ satisfies *backwards transitivity*, i.e. $a \mathop{\smile\!\!\!\!\!\mid}^0_A BC$ implies $a \mathop{\smile\!\!\!\!\!\mid}^0_{AB} C$ (as $a \mathop{\smile\!\!\!\!\!\mid}^0_A B$ follows from finite character). The following example shows that this is easily violated:

EXAMPLE 6.3.11 Let T_1 and T_2 be the theories of elementary Abelian 2-groups, with additions $+$ and $+'$, respectively. Then T_1 and T_2 are strongly minimal. Suppose $T_1 \cup T_2$ has a simple model-companion T. Let $\mathfrak{M}$ be a model of T with $0 = 0'$, and containing distinct elements a, b, c, c' with $c = a + b$ and $c' = a +' b$ (we can construct such an object and then embed it into a model of T). Then $\mathrm{acl}(.)$ is the iterative closure of algebraic closure in the sense of T_1 and of T_2; in particular $\mathrm{acl}(\emptyset) = \{0\}$ and $\mathrm{acl}(x) = \{0, x\}$ for any x. Now $\exists^{na} xy\, x + y = c$ and $\exists^{na} xy\, x +' y = c'$ both hold, so we may assume $a \notin \mathrm{acl}(c, c')$, by existential closedness of $\mathfrak{M}$. But T_1 and T_2 both have SU-rank 1, so $\mathrm{acl}(a)$ is independent of $\mathrm{acl}(c, c')$ in the sense of T_1 and of T_2; similarly $\mathrm{acl}(c)$ is independent of $\mathrm{acl}(c')$ in the sense of T_1 and T_2. However, if $(c'_i : i < \omega)$ is an indiscernible sequence in $\mathrm{tp}(c'/c)$ and a' satisfies $\mathrm{tp}(a'cc'_i) = \mathrm{tp}(acc')$ for all $i < \omega$, then it is easy to see that $c'_i = c'_j$ for all $i, j < \omega$. But there are indiscernible sequences with $c'_0 \neq c'_1$, so $\mathrm{tp}(a/cc')$ divides over cc', and conjunction of T_1-independence and T_2-independence is not the same as non-dividing. By Theorem 2.6.1 backwards transitivity for $\mathop{\smile\!\!\!\!\!\mid}^0$ cannot hold (as this is the only missing assumption).

DEFINITION 6.3.12 A theory has *identical algebraic closure* if $\mathrm{acl}(A) = A$ for all sets A. It has *trivial algebraic closure* if $\mathrm{acl}(A) = \bigcup\{\mathrm{acl}(a) : a \in A \text{ a singleton}\}$ for all sets A.

So a theory with identical algebraic closure has trivial algebraic closure, and functions in a theory with trivial algebraic closure are essentially unary.

PROPOSITION 6.3.13 *Let T_1 and T_2 be model-complete simple theories which eliminate infinite quantifiers, in languages $\mathfrak{L}_1$ and $\mathfrak{L}_2$, respectively, with $\mathfrak{L}_1 \cap \mathfrak{L}_2 = \{=\}$. Suppose T_1 and T_2 have trivial algebraic closure. Then the model-companion T of $T_1 \cap T_2$ is simple.*

PROOF: We may assume that both T_1 and T_2 are relational. By Propositions 6.3.9 and 6.3.10 and Theorem 2.6.1 it is sufficient to show that $\mathop{\smile\hspace{-0.9em}\mid}^{0}$ satisfies backwards transitivity. Recall that by Theorem 6.3.4 algebraic closure in T is the iteration of algebraic closures in T_1 and in T_2.

Since T_1 and T_2 both have trivial algebraic closure, so does T. Hence if $A \subseteq B \subseteq C$, and $a \mathop{\smile\hspace{-0.9em}\mid}^{0}_{A} C$, then $\mathrm{acl}(aB) = \mathrm{acl}(aA) \cup \mathrm{acl}(B)$; backwards transitivity of independence in T_1 and in T_2 now yields $\mathrm{acl}(aB) \mathop{\smile\hspace{-0.9em}\mid}^{0}_{\mathrm{acl}(B)} \mathrm{acl}(C)$, whence $a \mathop{\smile\hspace{-0.9em}\mid}^{0}_{B} C$. ∎

However, supersimplicity is in general not preserved.

EXAMPLE 6.3.14 Let T_1 be the theory of an equivalence relation E with infinitely many infinite classes, and T_2 the theory of a bijection S without cycles. Then T_1 and T_2 are model-complete, eliminate infinite quantifiers, and have trivial algebraic closure; T_1 has SU-rank 2 and T_2 has SU-rank 1. Let $\mathfrak{M}$ be a model of the model-companion for $T_1 \cup T_2$. If $a \in \mathfrak{M}$ is such that all $S^i(a)$ lie in different classes Ea_i for $i \in \mathbb{Z}$, then $a \not\mathop{\smile\hspace{-0.9em}\mid} a_i$ for all $i \in \mathbb{Z}$. Since we can choose the $(a_i : i \in \mathbb{Z})$ independent, $w(a) = \omega$ and T is not supersimple by Theorem 5.2.5.

PROPOSITION 6.3.15 *Let T_1 and T_2 be model-complete simple theories which eliminate infinite quantifiers, in languages $\mathfrak{L}_1$ and $\mathfrak{L}_2$, respectively, with $\mathfrak{L}_1 \cap \mathfrak{L}_2 = \{=\}$. Suppose T_2 has identical algebraic closure and SU-rank 1. Then the model-companion T of $T_1 \cap T_2$ is simple. Moreover, algebraic closure and forking (for algebraically closed sets) in T are the same as in T_1. In particular, if T_1 is supersimple of rank α, so is T.*

PROOF: By Theorem 6.3.4 algebraic closure in T is the iteration of algebraic closure in T_1 and T_2, and hence equals algebraic closure in T_1.

CLAIM. In T_2 we have $A \mathop{\smile\hspace{-0.9em}\mid}^{2}_{B} C$ if and only if $A \cap C \subseteq B$.

PROOF OF CLAIM: Clearly $A \mathop{\smile\hspace{-0.9em}\mid}^{2}_{B} C$ implies the containment, as $\mathrm{acl}_2(B) = B$. On the other hand, if $A \not\mathop{\smile\hspace{-0.9em}\mid}^{2}_{B} C$, choose a minimal $\bar{c} \in C$ with $A \not\mathop{\smile\hspace{-0.9em}\mid}^{2}_{B} \bar{c}$. If $\bar{c} = (c_0, \ldots, c_n)$ and $\bar{c}' = (c_0, \ldots, c_{n-1})$, this means $A \mathop{\smile\hspace{-0.9em}\mid}^{2}_{B} \bar{c}'$ and

$$c_n \in \mathrm{acl}_2(AB\bar{c}') - \mathrm{acl}_2(B\bar{c}') \subseteq A - B,$$

as algebraic closure is identical, whence $c_n \in [A \cap C] - B$. ■

Suppose now $A \mathop{\smile\!\!\!\!\!\!\mid}^1_B C$, where $B \subseteq A \cap C$ are algebraically closed sets (in the sense of T_1 or T). Hence $A \cap C = B$ and $A \mathop{\smile\!\!\!\!\!\!\mid}^2_B C$, whence $A \mathop{\smile\!\!\!\!\!\!\mid}^0_B C$. The assertion follows. ■

EXERCISE 6.3.16 Show that the model-companion of the union of a simple theory and a simple theory of SU-rank 1 with trivial algebraic closure (where both theories eliminate infinite quantifiers) is not necessarily simple.

HINT. Consider the theory of an elementary Abelian 2-group and a bijection without cycles, and adapt Example 6.3.11. So even the model-companion of two uncountably categorical theories need not be simple.

COROLLARY 6.3.17 *Let T be a simple theory with elimination of infinite quantifiers in a language $\mathfrak{L}$. Suppose $\mathcal{P}$ is a family of new predicates (of arbitrary arity). Then T has a simple model-companion T^* in $\mathfrak{L} \cup \mathcal{P}$; moreover T^* is supersimple if and only if T is, and has the same SU-rank.*

PROOF: Let T' be the theory of a family $\mathcal{P}$ of generic predicates, i.e. the theory which states that for any $k < \omega$, and any finite sequences $(P_i : i < k) \subseteq \mathcal{P}$ and $(\bar{a}_i, \bar{b}_i : i < k)$ (with repetitions) the formula

$$\bigwedge_{i<k} [P_i(x, \bar{a}_i) \wedge \neg P_i(x, \bar{b}_i)]$$

defines an infinite set unless it contains a subformula $P_i(x, \bar{a}) \wedge \neg P_i(x, \bar{a})$. Moreover, T' either contains the axiom saying that P_i is symmetric, or versions of the above axiom scheme for all relevant permutations of the variables. Then it is easy to see that T' eliminates quantifiers. So T' is simple of SU-rank 1, eliminates infinite quantifiers, and has identical algebraic closure. The assertion now follows from Proposition 6.3.15. ■

6.3.2 SIMPLE THEORIES WITH AN AUTOMORPHISM

In this subsection, we shall show that if a stable theory, expanded by a new function for an automorphism, allows a model-companion, then that model-companion is simple. We shall also give some criteria as to when such a model-companion exists.

Throughout this subsection, let T be a complete theory in a language $\mathfrak{L}$, σ a new unary function symbol not in $\mathfrak{L}$, and T_σ the expansion of T which asserts that σ is an automorphism.

LEMMA 6.3.18 *If T is stable, T_σ has the amalgamation property. More precisely, if $(\mathfrak{M}_1, \sigma_1)$ and $(\mathfrak{M}_2, \sigma_2)$ are two models of T_σ which intersect in a* *common submodel $(\mathfrak{M}_0, \sigma_0)$, such that $\mathfrak{M}_1 \mathop{\smile\!\!\!\!\!\mid}_{\mathfrak{M}_0} \mathfrak{M}_2$ (in the sense of T) inside some $|\mathfrak{M}_1 \cup \mathfrak{M}_2|$-homogeneous model $\mathfrak{N} \models T$, then there is an automorphism σ of $\mathfrak{N}$ extending $\sigma_1 \cup \sigma_2$.*

PROOF: Since types over a model in a stable theory have a unique non-forking extension over any superset, for any two $\mathfrak{L}$-types $p_1(\bar{x})$ and $p_2(\bar{y})$ over $\mathfrak{M}_0$ there is a unique completion $p(\bar{x}, \bar{y})$ which implies $\bar{x} \mathop{\smile\!\!\!\!\!\mid}_{\mathfrak{M}_0} \bar{y}$; we denote it by $p_1 \otimes_{\mathfrak{M}_0} p_2$. Then

$$\begin{aligned} &\mathrm{tp}_{\mathfrak{L}}(\sigma_1(\mathfrak{M}_1), \sigma_2(\mathfrak{M}_2)/\sigma_0(\mathfrak{M}_0)) \\ &\quad = \mathrm{tp}_{\mathfrak{L}}(\sigma_1(\mathfrak{M}_1)/\sigma_0(\mathfrak{M}_0)) \otimes_{\sigma_0(\mathfrak{M}_0)} \mathrm{tp}_{\mathfrak{L}}(\sigma_2(\mathfrak{M}_2)/\sigma_0(\mathfrak{M}_0)) \\ &\quad = \sigma_0(\mathrm{tp}_{\mathfrak{L}}(\mathfrak{M}_1/\mathfrak{M}_0)) \otimes_{\sigma_0(\mathfrak{M}_0)} \sigma_0(\mathrm{tp}_{\mathfrak{L}}(\mathfrak{M}_1/\mathfrak{M}_0)) \\ &\quad = \sigma_0(\mathrm{tp}_{\mathfrak{L}}(\mathfrak{M}_1/\mathfrak{M}_0) \otimes_{\mathfrak{M}_0} \mathrm{tp}_{\mathfrak{L}}(\mathfrak{M}_2/\mathfrak{M}_0)) \\ &\quad = \sigma_0(\mathrm{tp}_{\mathfrak{L}}(\mathfrak{M}_1, \mathfrak{M}_2/\mathfrak{M}_0)). \end{aligned}$$

Hence $\sigma_1 \cup \sigma_2$ is an $\mathfrak{L}$-elementary partial map on $\mathfrak{N}$, and can be extended to an automorphism of $\mathfrak{N}$ by homogeneity. ∎

REMARK 6.3.19 In fact, the proof does not require $\mathfrak{M}_0$ to be a model, but merely to be algebraically closed in T^{eq} and to be closed under σ. Hence in a stable theory we can amalgamate automorphisms over algebraically and σ-closed sets.

If $\mathfrak{M} \models T_\sigma$ and $A \subset \mathfrak{M}$, let $\mathrm{acl}_{\mathfrak{L}}(A)$ denote the algebraic closure of A in the $\mathfrak{L}$-reduct of $\mathfrak{M}$, and $\mathrm{acl}_\sigma(A)$ the closure of $\mathrm{acl}_{\mathfrak{L}}(A)$ under σ and σ^{-1}. We shall need the following general Lemma:

LEMMA 6.3.20 *Let T be a stable theory. If $a \in \mathrm{dcl}(bc)$ and $c' = \mathrm{Cb}(c/ab)$, then $a \in \mathrm{dcl}(bc')$.*

PROOF: Let a' realize $\mathrm{tp}(a/bc')$, and suppose c'' realizes the unique non-forking extension of $\mathrm{tp}(c/c')$ to $c'baa'$. Then $\mathrm{tp}(a'c''bc') = \mathrm{tp}(ac''bc') = \mathrm{tp}(acbc')$, so $a \in \mathrm{dcl}(bc)$ yields $a = a' \in \mathrm{dcl}(bc')$. ∎

REMARK 6.3.21 Note that example 3.4.4 shows that Lemma 6.3.20 need not hold in a simple theory.

THEOREM 6.3.22 *Suppose T is stable and T_σ has a model-companion T^*. For subsets A, B, C of a model $\mathfrak{M} \models T^*$ put $A \mathop{\smile\!\!\!\!\!\mid}^*_B C$ if $\mathrm{acl}_\sigma(AB)$ is independent of $\mathrm{acl}_\sigma(BC)$ over $\mathrm{acl}_\sigma(B)$ in the sense of T. Then $\mathop{\smile\!\!\!\!\!\mid}^*$ is invarinat under automorphisms, and satisfies symmetry, transitivity, extension, local and finite character, and the Independence Theorem. In particular, T^* is simple. Moreover, if T is superstable, then so is (every completion of) T^*.*

PROOF: Invariance under automorphisms is obvious, as are symmetry, local and finite character. For transitivity, note that $\mathrm{acl}_\sigma(ABC) = \mathrm{acl}_\mathfrak{L}(\mathrm{acl}_\sigma(AB), \mathrm{acl}_\sigma(BC))$, since σ is an $(\mathfrak{L} \cup \{\sigma\})$-automorphism of $\mathfrak{M}$.

Extension: Suppose $\mathfrak{M} \models T^*$, $A \subseteq \mathfrak{M}$, and $\bar{a} \in \mathfrak{M}$. Clearly we may assume that $A = \mathrm{acl}_\sigma(A)$, so A is algebraically closed in the sense of T and closed under σ. Since T is stable, we find an $|\mathfrak{M}|^+$-homogeneous and $|\mathfrak{M}|^+$-saturated T-elementary extension $\mathfrak{N}$ of $\mathfrak{M}$ containing a copy $\mathfrak{M}'$ of $\mathfrak{M}$ with $\mathfrak{M}' \mathop{\smile\!\!\!\!\!|}_A \mathfrak{M}$. If σ' is the corresponding copy of σ, then $\sigma \cup \sigma'$ is an elementary map by Lemma 6.3.18, and can be extended to an automorphism τ of $\mathfrak{N}$. Thus $(\mathfrak{N}, \tau) \models T_\sigma$, and has an extension $(\mathfrak{N}', \tau') \models T^*$. If $\bar{a}' \in \mathfrak{M}'$ corresponds to $\bar{a} \in \mathfrak{M}$, model-completeness implies

$$\mathrm{tp}_{\mathfrak{N}'}(\bar{a}'/A) = \mathrm{tp}_{\mathfrak{M}'}(\bar{a}'/A) = \mathrm{tp}_{\mathfrak{M}}(\bar{a}/A);$$

and clearly $\mathrm{acl}_\sigma(\bar{a}A) \subseteq \mathfrak{M}'$ is independent of $\mathfrak{M}$ over A in the sense of T.

In order to prove the Independence Theorem, assume $\mathfrak{M} \prec \mathfrak{N} \models T^*$, $A_1 \mathop{\smile\!\!\!\!\!|}^*_{\mathfrak{M}} A_2$, $\bar{a}_1 \mathop{\smile\!\!\!\!\!|}^*_{\mathfrak{M}} A_1$, $\bar{a}_2 \mathop{\smile\!\!\!\!\!|}^*_{\mathfrak{M}} A_2$, and $\mathrm{tp}(\bar{a}_1/\mathfrak{M}) = \mathrm{tp}(\bar{a}_2/\mathfrak{M})$. We have to show that in some elementary extension of $\mathfrak{N}$ there is $\bar{a} \mathop{\smile\!\!\!\!\!|}^*_{\mathfrak{M}} A_1A_2$ satisfying $\mathrm{tp}(\bar{a}_1/A_1) \cup \mathrm{tp}(\bar{a}_2/A_2)$. By the extension axiom, we may assume that A_1, A_2 and $\bar{a}_1$ (whence also $\bar{a}_2$) are models of T^* containing $\mathfrak{M}$.

Denote T-independence by $\mathop{\smile\!\!\!\!\!|}$. By stability (or merely simplicity) of T, in some $\mathfrak{L}$-elementary $|\mathfrak{N}|$-homogeneous extension $\mathfrak{N}'$ of the $\mathfrak{L}$-reduct of $\mathfrak{N}$ we can find $\bar{a} \mathop{\smile\!\!\!\!\!|}_{\mathfrak{M}} \mathfrak{N}$ realizing $\mathrm{tp}_\mathfrak{L}(\bar{a}_1/A_1) \cup \mathrm{tp}_\mathfrak{L}(\bar{a}_2/A_2)$; moreover for $i = 1, 2$ there are models $\mathfrak{N}_i \supseteq A_i\bar{a}$ which are $\mathfrak{L}$-elementarily equivalent to $\mathfrak{N} \supseteq A_i\bar{a}_i$, such that $\mathfrak{N}_1 \mathop{\smile\!\!\!\!\!|}_{A_1\bar{a}} \mathfrak{N}$ and $\mathfrak{N}_2 \mathop{\smile\!\!\!\!\!|}_{A_2\bar{a}} \mathfrak{N}\mathfrak{N}_1$. Then $\mathfrak{N} \mathop{\smile\!\!\!\!\!|}_{A_1} \mathfrak{N}_1$ and $\mathfrak{N} \mathop{\smile\!\!\!\!\!|}_{A_2} \mathfrak{N}_2$ by transitivity; furthermore $A_1 \mathop{\smile\!\!\!\!\!|}_{\bar{a}} A_2$ implies $A_2 \mathop{\smile\!\!\!\!\!|}_{\bar{a}} \mathfrak{N}_1$, whence $\mathfrak{N}_1 \mathop{\smile\!\!\!\!\!|}_{\bar{a}} \mathfrak{N}_2$. Finally, $\mathfrak{N} \mathop{\smile\!\!\!\!\!|}_{A_1A_2} \mathfrak{N}_1$, whence $\mathfrak{N} \mathop{\smile\!\!\!\!\!|}_{A_1A_2} \mathfrak{N}_1\mathfrak{N}_2$.

Define automorphisms σ_1 on $\mathfrak{N}_1$ and σ_2 on $\mathfrak{N}_2$ such that $(\mathfrak{N}_i, A_i, \bar{a}, \sigma_i)$ is isomorphic to $(\mathfrak{N}, A_i, \bar{a}_i, \sigma)$ for $i = 1, 2$. In particular $(\mathfrak{N}_i, \sigma_i)$ is a model of T^*, for $i = 1, 2$. So σ agrees with σ_1 on $\mathfrak{N} \cap \mathfrak{N}_1 = A_1$ and with σ_2 on $\mathfrak{N} \cap \mathfrak{N}_2 = A_2$, and σ_1 agrees with σ_2 on $\mathfrak{N}_1 \cap \mathfrak{N}_2 = \bar{a}$. Since $\bar{a} \models T$, Lemma 6.3.18 implies that the map $\sigma_1 \cup \sigma_2$ is $\mathfrak{L}$-elementary in $\mathfrak{N}'$, and can be extended to an automorphism σ' of $\mathfrak{N}'$ by homogeneity.

CLAIM. $\mathrm{dcl}_\mathfrak{L}(\mathfrak{N}_1\mathfrak{N}_2) \cap \mathrm{acl}_\mathfrak{L}(A_1, A_2) = \mathrm{dcl}_\mathfrak{L}(A_1, A_2)$.

PROOF OF CLAIM: Consider $e \in \mathrm{dcl}_\mathfrak{L}(\mathfrak{N}_1\mathfrak{N}_2) \cap \mathrm{acl}_\mathfrak{L}(A_1, A_2)$. Then $\mathfrak{N}_1 \mathop{\smile\!\!\!\!\!|}_{A_1\bar{a}} \mathfrak{N}_2$ implies $\mathrm{Cb}_\mathfrak{L}(\mathfrak{N}_1/A_1\mathfrak{N}_2e) \subseteq \mathrm{acl}_\mathfrak{L}(A_1\bar{a})$, and therefore $e \in \mathrm{dcl}_\mathfrak{L}(\mathrm{acl}_\mathfrak{L}(A_1\bar{a}), \mathfrak{N}_2)$ by Lemma 6.3.20. Similarly, $\mathfrak{N}_2 \mathop{\smile\!\!\!\!\!|}_{A_2\bar{a}} A_1$ yields $e \in \mathrm{dcl}_\mathfrak{L}(\mathrm{acl}_\mathfrak{L}(A_1\bar{a}), \mathrm{acl}_\mathfrak{L}(A_2\bar{a}))$. But if $\varphi(\bar{a}, A_1, A_2, e)$ is a formula implying that there is $\alpha \in \mathrm{acl}_\mathfrak{L}(A_1\bar{a})$ and $\beta \in \mathrm{acl}_\mathfrak{L}(A_2\bar{a})$ with $e \in \mathrm{dcl}_\mathfrak{L}(\alpha, \beta)$, then $\bar{a} \mathop{\smile\!\!\!\!\!|}_{\mathfrak{M}} A_1A_2e$ implies that there is $\bar{a}' \in \mathfrak{M}$ with $\models \varphi(\bar{a}', A_1, A_2, e)$ by φ-definability of $\mathrm{tp}(A_1A_2e/\mathfrak{M})$. Hence $e \in \mathrm{dcl}(\mathrm{acl}_\mathfrak{L}(A_1\mathfrak{M}), \mathrm{acl}_\mathfrak{L}(A_2\mathfrak{M})) = \mathrm{dcl}_\mathfrak{L}(A_1A_2)$. ∎

So $\text{tp}_{\mathfrak{L}}(\text{acl}_{\mathfrak{L}}(A_1A_2)/A_1A_2) \vdash \text{tp}_{\mathfrak{L}}(\text{acl}_{\mathfrak{L}}(A_1A_2)/\mathfrak{N}_1\mathfrak{N}_2)$. As σ and σ' agree on A_1A_2,

$$\begin{aligned}\text{tp}_{\mathfrak{L}}(\sigma(\text{acl}_{\mathfrak{L}}(A_1A_2))/\sigma(A_1A_2)) &= \sigma(\text{tp}_{\mathfrak{L}}(\text{acl}_{\mathfrak{L}}(A_1A_2)/A_1A_2)) \\ &= \sigma'(\text{tp}_{\mathfrak{L}}(\text{acl}_{\mathfrak{L}}(A_1A_2)/A_1A_2)) \\ &\vdash \sigma'(\text{tp}_{\mathfrak{L}}(\text{acl}_{\mathfrak{L}}(A_1A_2)/\mathfrak{N}_1\mathfrak{N}_2)) \\ &= \text{tp}_{\mathfrak{L}}(\sigma'(\text{acl}_{\mathfrak{L}}(A_1A_2))/\sigma'(\mathfrak{N}_1\mathfrak{N}_2)).\end{aligned}$$

If follows that $\sigma'\restriction_{\mathfrak{N}_1\mathfrak{N}_2} \cup \sigma\restriction_{\text{acl}_{\mathfrak{L}}(A_1A_2)}$ is $\mathfrak{L}$-elementary in $\mathfrak{N}'$. Moreover, as A_1 and A_2 are closed under σ, so is $\text{acl}_{\mathfrak{L}}(A_1A_2)$. Since $\mathfrak{N}_1\mathfrak{N}_2 \perp_{\text{acl}_{\mathfrak{L}}(A_1A_2)} \mathfrak{M}$, Remark 6.3.19 implies that $\sigma \cup \sigma'\restriction_{\mathfrak{N}_1\mathfrak{N}_2}$ is $\mathfrak{L}$-elementary. Modifying σ' outside $\mathfrak{N}_1\mathfrak{N}_2$, we may therefore assume that σ' extends σ.

As T^* is the model-companion of T_σ, we may embed $(\mathfrak{N}', \sigma')$ in a model $(\mathfrak{N}'', \sigma'')$ of T^*. By model-completeness, the embeddings of $\mathfrak{M}$, A_1, A_2, $\bar{a}$, $\mathfrak{N}$, $\mathfrak{N}_1$ and $\mathfrak{N}_2$ are elementary. It follows that $\bar{a} \perp^*_{\mathfrak{M}} A_1A_2$ in $\mathfrak{N}''$, and $\bar{a} \models \text{tp}(\bar{a}_1/A_1) \cup \text{tp}(\bar{a}_2/A_2)$. Hence $\perp^*$ satisfies the Independence Theorem, and T^* is simple by Theorem 2.6.1.

For the last assertion, consider a finite tuple $\bar{a}$ and a set A with $A = \text{acl}_\sigma(A)$. By supersimplicity of T, there are some $m < \omega$ such that $\bar{a}$ is T-independent of $(\sigma^i(\bar{a}) : i > 0)$ over $A \cup (\sigma^i(\bar{a}) : 0 < i < m)$, and some finite $A_0 \subseteq A$ such that $(\sigma^i(\bar{a}) : i < m)$ is T-independent of A over A_0.

CLAIM. $\bar{a} \perp^*_{A_0} A$.

PROOF OF CLAIM: Put $\text{acl}_\sigma(A_0) = A_\sigma$. Since σ is an $\mathfrak{L}$-automorphism, it is sufficient to show $(\sigma^i(\bar{a}) : i < n) \perp_{A_\sigma} A$ for all $n < \omega$. This is clear for $n \leq m$, so suppose it holds for $n_0 \geq m$. Now

$$\bar{a} \quad \underset{A,(\sigma^i(\bar{a}):0<i<m)}{\perp} \quad (\sigma^i(\bar{a}) : 0 < i \leq n_0)$$

by choice of m; as $n_0 \geq m$, by inductive hypothesis and transitivity

$$\bar{a} \quad \underset{A_\sigma,(\sigma^i(\bar{a}):0<i<m)}{\perp} \quad A, (\sigma^i(\bar{a}) : 0 < i \leq n_0).$$

Hence $\bar{a} \perp_{A_\sigma,(\sigma^i(\bar{a}):0<i\leq n_0)} A$. Since σ is an automorphism, $(\sigma^i(\bar{a}) : 0 < i \leq n_0) \perp_{A_\sigma} A$ by inductive hypothesis, whence $A \perp_{A_\sigma} (\sigma^i(\bar{a}) : 0 \leq i \leq n_0)$ by transitivity again. ■

So T^* is supersimple. ■

Let us note that the restriction on T to be stable (rather than simple) is not very strong, if one wants T_σ to have a model-companion.

PROPOSITION 6.3.23 *Let T be a model-complete unstable theory. Suppose T_σ has the amalgamation property. Then T_σ has no model-companion.*

PROOF: Suppose T^* is a model-companion for T, and let T_S be a Skolemization for (some completion of) T in a language $\mathfrak{L}_S$. Since T is unstable, there is an asymmetric $\mathfrak{L}$-definable relation $\bar{x} < \bar{y}$ ordering arbitrarily long chains. By Ramsey's Theorem there is an $\mathfrak{L}_S$-indiscernible sequence $(\bar{a}_i : i \in \mathbb{Z})^\frown(\bar{b}_i : i \in \mathbb{Z})$ ordered by $<$ in some model $\mathfrak{M}$ of T_S. Let $\mathfrak{M}'$ be the Skolem hull of this sequence, so the map σ mapping $\bar{a}_i$ to $\bar{a}_{i+1}$ and $\bar{b}_i$ to $\bar{b}_{i+1}$ for all $i \in \mathbb{Z}$ induces an automorphism σ' of $\mathfrak{M}'$. Since $(\mathfrak{M}', \sigma') \models T_\sigma$, we can embed it into a model $(\mathfrak{M}^*, \sigma^*)$ of T^*.

Let P be a new unary predicate, $\bar{a}$ and $\bar{b}$ new constant symbols, and T_1 the following expansion of T^*:

1. P is a model of T_S, and σ is an $\mathfrak{L}_S$-automorphism of P.
2. $(\sigma^i(\bar{a}) : i \in \mathbb{Z})^\frown(\sigma^i(\bar{b}) : i \in \mathbb{Z})$ is $\mathfrak{L}_S$-indiscernible and satisfies P, with $\bar{a} < \sigma(\bar{a})$.

If we interpret P by $\mathfrak{M}'$ in $\mathfrak{M}^*$, we get a model of T_1, so this is consistent.

CLAIM. $T_1 \vdash \exists \bar{x}\, [\sigma(\bar{x}) = \bar{x} \wedge \bar{a} < \bar{x} < \bar{b}]$.

PROOF OF CLAIM: Let $(\mathfrak{M}_1, \sigma_1, \bar{a}, \bar{b})$ be any model of T_1, and $\mathfrak{M}_0$ the Skolem hull of $I = (\sigma_1^i(\bar{a}), \sigma_1^i(\bar{b}) : i \in \mathbb{Z})$. Then $\mathfrak{M}_0$ is an $\mathfrak{L}_S$-elementary substructure of $P(\mathfrak{M}_1)$, and a model of T. Moreover $\mathfrak{M}_0$ is closed under σ_1, so if $\sigma_0 = \sigma_1 \restriction_{\mathfrak{M}_0}$, then $(\mathfrak{M}_0, \sigma_0) \models T_\sigma$.

Let $p(\bar{x})$ be the average $\mathfrak{L}_S$-type of $(\sigma_1^i(\bar{a}) : i < \omega)$ over $\mathfrak{M}_0$, i.e. the collection of all formulas satisfied by almost all elements of $(\sigma_1^i(\bar{a}) : i < \omega)$. Since $\mathfrak{M}_0$ is the Skolem hull of I, every $\mathfrak{L}_S(\mathfrak{M}_0)$-formula $\varphi(\bar{x}, \bar{m})$ can be written with parameters $\bar{m} \in (\sigma_1^i(\bar{a}), \sigma_1^i(\bar{b}) : |i| < k)$ for some $k < \omega$; it follows from indiscernibility that either φ or $\neg\varphi$ is satisfied by almost all elements of $(\sigma_1^i(\bar{a}) : i < \omega)$. Hence p is a complete type, and $p = \sigma_0(p)$. Note that $\bar{a} < \bar{x} < \bar{b}$ is a formula in $p(\bar{x})$.

There is an $\mathfrak{L}$-elementary extension $\mathfrak{M}_2$ of $\mathfrak{M}_0$ and an $\mathfrak{L}$-automorphism σ_2 extending σ_0, such that $\sigma_2(\bar{c}) = \bar{c}$ for some realization $\bar{c}$ of $p\restriction_{\mathfrak{L}}$ in $\mathfrak{M}_2$. So $(\mathfrak{M}_2, \sigma_2) \models T_\sigma$; by the amalgamation property for T_σ, we can jointly embed $(\mathfrak{M}_1, \sigma_1)$ and $(\mathfrak{M}_2, \sigma_2)$ over $(\mathfrak{M}_0, \sigma_0)$ in a model $(\mathfrak{N}, \sigma)$. Since T^* is the model-companion of T_σ, we may assume that $(\mathfrak{N}, \sigma) \models T^*$. Then $(\mathfrak{N}, \sigma) \models \exists \bar{x}\, [\sigma(\bar{x}) = \bar{x} \wedge \bar{a} < \bar{x} < \bar{b}]$; as $(\mathfrak{M}_1, \sigma_1)$ is an $\mathfrak{L} \cup \{\sigma\}$-elementary substructure of $(\mathfrak{N}, \sigma)$ by model-completeness of T^*, this sentence holds in $(\mathfrak{M}_1, \sigma_1)$ as well. ■

By compactness, a finite fragment $\Sigma(\bar{a}, \bar{b})$ is sufficient to prove $\exists \bar{x}\, [\sigma(\bar{x}) = \bar{x} \wedge \bar{a} < \bar{x} < \bar{b}]$. But then we may interpret $\bar{b}$ by $\sigma^i(\bar{a})$ for sufficiently large $i < \omega$. Since σ is an $\mathfrak{L}$-automorphism, it must preserve $<$, and $\bar{a} < \bar{x} < \bar{\sigma}^i(\bar{a})$ is impossible. ■

In a similar way, one can show:

EXERCISE 6.3.24 If T is model-complete, unstable, and without the independence property, then T_σ has no model-companion.

As in the case of the generic amalgam of two theories, the existence of a model-companion implies elimination of infinite quantifiers.

PROPOSITION 6.3.25 *Let T be a model-complete theory. Suppose T does not eliminate infinite quantifiers. Then T_σ has no model-companion.*

PROOF: Suppose T_σ has a model-companion T^*, and let $\varphi(x, \bar{y})$ be a formula such that $\exists^\infty x\, \varphi(x, \bar{y})$ is not equivalent to a first-order formula. Then there is a model $\mathfrak{M} \models T$ and a sequence $(\bar{a}_i : i < \omega)$, such that $\varphi(x, \bar{a}_i)$ defines a finite set of size at least i in $\mathfrak{M}$. Embed $(\mathfrak{M}, id)$ into a model $(\mathfrak{N}, \sigma)$ of T^*. Let $\mathcal{U}$ be a non-principal ultrafilter on ω, and $\mathfrak{N}'$ the ultrapower $\prod_\omega \mathfrak{N}/\mathcal{U}$. Let $\bar{a} = (\bar{a}_i : i < \omega)/\mathcal{U}$. Then $\mathfrak{N}' \models \varphi(x, \bar{a}) \to \sigma(x) = x$. But since $\varphi(x, \bar{a})$ defines an infinite set, $\mathfrak{N}'$ embeds into a model of T^* satisfying $\exists x\, [\varphi(x, \bar{a}) \wedge \sigma(x) \neq x]$. So the embedding is not elementary, contradicting model-completeness of T^*. ■

Moreover, under some weak condition the model-companion of T_σ (if it exists) will have the independence property.

PROPOSITION 6.3.26 *Let T be a model-complete stable theory such that T_σ has a model-companion T^*. Suppose (possibly after adding parameters) there are independent a, b in a model $\mathfrak{M}$ of T with* $\mathrm{acl}(a, b) \neq \mathrm{dcl}(\mathrm{acl}(a), \mathrm{acl}(b))$. *Then T^* has the independence property.*

PROOF: Put $p(\bar{x}, \bar{y}) = \mathrm{tp}(\bar{a}, \bar{b})$. Let $(b_i : i < \omega)$ be a Morley sequence in $\mathrm{lstp}(b/a)$, and $(a_I : I \subseteq \omega)$ a sequence of independent realizations of $\bigcup_{i<\omega} p(\bar{x}, b_i)$, all in some model $\mathfrak{M} \models T$. Then the set $\{a_I, b_i : i < \omega, I \subseteq \omega\}$ is independent. Suppose $\varphi(x, a, b)$ isolates $\mathrm{tp}(e/ab)$ for some $e \in \mathrm{acl}(ab) - \mathrm{dcl}(\mathrm{acl}(a), \mathrm{acl}(b))$. For any $I \subseteq \omega$ and $i < \omega$, suppose $e' \models \varphi(x, a_I, b_i)$ and $e' \in \mathrm{dcl}(\mathrm{acl}(a_I), \mathrm{acl}(a_J, b_j : J \neq I, j < \omega))$. Since $\mathrm{Cb}(\mathrm{acl}(a_J, b_j : J \neq I, j < \omega)/e'b_i) \subseteq \mathrm{acl}(b_i)$, Lemma 6.3.20 implies $e' \in \mathrm{dcl}(\mathrm{acl}(a_I), \mathrm{acl}(b_i))$, a contradiction. Hence there is a partial isomorphism σ of $\mathfrak{M}$ with domain $\mathrm{acl}(a_I, b_i : I \subseteq \omega, i < \omega)$, such that σ fixes $\bigcup_{I \subseteq \omega} \mathrm{acl}(a_I) \cup \mathrm{acl}(b_j : j < \omega)$ and moves some element in $\varphi(x, a_I, b_i)$ if and only if $i \in I$. We can then expand $\mathfrak{M}$ to a model of T_σ in such a way that $\forall x\, [\varphi(x, a_I, b_i) \to \sigma(x) = x]$ holds if and only if $i \in I$. Since $\mathfrak{M}$ embeds into a model $\mathfrak{N}$ of T^* and the truth of this formula is preserved, this witnesses the independence property. ■

The following remark gives a sufficient condition for the existence of a model-companion for T_σ.

REMARK 6.3.27 Suppose T is an ω-stable theory of finite Morley rank, such that

1. $RM(ab/c) = RM(a/bc) + RM(b/c)$ for all finite tuples a, b, c.
2. T has the *definable multiplicity property:* whenever a formula $\varphi(x, a)$ has Morley rank n and Morley degree 1, then there is $\psi(y) \in \mathrm{tp}(a)$ such that $\varphi(x, a')$ has Morley rank n and Morley degree 1 for all $a' \models \psi$.

Then T_σ has a model-companion, which is axiomatized by the following axiom scheme:

> If X is a definable set of Morley rank n and Morley degree 1 and Y is a definable subset of $X \times \sigma(X)$ of Morley rank $n + m$ and Morley degree 1, such that the projections of Y to X and to $\sigma(X)$ have Morley rank n and all fibres have Morley rank m, then there is x with $(x, \sigma(x)) \in Y$.

For instance, the theory of algebraically closed fields satisfies the conditions. Hence the theory of algebraically closed fields with an automorphism has a model-companion, called $ACFA$, which is axiomatized by the above scheme.

Even if T is strongly minimal, one does not know a necessary and sufficient condition for the existence of a model-companion for T_σ.

CONJECTURE 6.3.28 *If T is strongly minimal, then T_σ has a model-companion if and only if T has the definable multiplicity property.*

6.4 LOW THEORIES

Recall that a simple theory is *low* if for every formula φ there is a bound n_φ on the length of a dividing φ-chain (and this is strictly stronger than the absence of a dividing φ-chain of length ω). Low theories form a subclass of the category of simple theories which includes all known natural examples. However, Kim and Buechler/Laskowski have constructed a simple non-low theory; in fact Kim and Casanovas have even constructed a supersimple non-low theory. The main consequence of lowness is that for a formula $\varphi(x, y)$ to fork over a set A of parameters becomes a type-definable property in y over A. As a corollary, we shall see that in a low theory strong type and Lascar strong type agree.

DEFINITION 6.4.1 Let $\varphi(x, y)$ be a formula. The rank $D(., \varphi)$ is the least function from the class of all consistent partial types in x to On^+ satisfying for all ordinals α

> $D(\pi(x), \varphi) \geq \alpha + 1$ if there is b such that $D(\pi(x) \wedge \varphi(x, b), \varphi) \geq \alpha$, and $\varphi(x, b)$ divides over the domain of π.

REMARK 6.4.2 A theory T is simple if and only if $D(x = x, \varphi) < \infty$ for all formulas φ.

PROOF: Suppose $D(x = x, \varphi) = \infty$. As for every parameter set there is some ordinal α such that for any partial type $\pi(x)$ over A we have $D(\pi, \varphi) = \infty$ if and only if $D(\pi, \varphi) \geq \alpha$, we can find an ascending sequence $p_0 \subset p_1 \subset \cdots$ of types such that p_{i+1} is a forking extension of p_i via some formula of the form $\varphi(x, a_i)$, for all $i < |T|^+$. Hence $\bigcup_i p_i$ is a consistent type which forks over every parameter subset of size $|T|$, so T is not simple.

The converse follows from the inequality $D(\pi, \varphi, k) \leq D(\pi, \varphi)$, and $D(\pi, \varphi, k) < \infty$ implies $D(\pi, \varphi, k) < \omega$ by compactness. ∎

REMARK 6.4.3 A complete theory T is low if $D(x = x, \varphi) < \omega$ for all formulas φ.

LEMMA 6.4.4 *Let T be a simple theory. Then the following are equivalent:*

1. *T is low.*
2. *For all formulas $\varphi(x, y)$ there is some $k < \omega$ such that if $p(x) \in S(A)$ is any type, $p(x) \cup \{\varphi(x, a)\}$ forks over A, and $(a_i : i < \omega)$ is an A-indiscernible sequence in $\mathrm{tp}(a/A)$, then $\{\varphi(x, a_i) : i < \omega\}$ is k-inconsistent.*
3. *For any formula φ there is some $n < \omega$ such that $D(\pi, \varphi, m) = D(\pi, \varphi, n)$ for all $m \geq n$.*

PROOF: 1. ⇒ 2. Put $k = D(x = x, \varphi) + 1$. If $\{\varphi(x, a_i) : i < k\}$ were consistent, put $p_j(x) = \{\varphi(x, a_i) : i < j\}$ for all $j < k$. Then $\varphi(x, a_j)$ witnesses $D(p_j, \varphi) > D(p_{j+1}, \varphi)$, so $D(p_0, \varphi) \geq k$, a contradiction.

2. ⇒ 3. This is obvious, with $n = k$, as m-forking can always be witnessed by indiscernible m-contradictory sequences, which are k-contradictory by assumption.

3. ⇒ 1. If $D(x = x, \varphi) \geq k$, then this is witnessed by a dividing φ-chain of length k, and there is some $m \geq n$ with $D(x = x, \varphi, m) \geq k$. Hence $D(x = x, \varphi) \leq D(x = x, \varphi, n) < \omega$. ∎

In fact, this shows that if T is low, then for every formula φ there is some $k < \omega$ such that $D(\pi, \varphi) = D(\pi, \varphi, k)$ for all partial types π.

COROLLARY 6.4.5 *If T is low, then for every formula $\varphi(x, y)$ and any A there is a partial type $q(y) \in S(A)$ such that $\varphi(x, a)$ divides over A if and only if $\models q(a)$.*

PROOF: The partial type q will express that there is an A-indiscernible k-contradictory sequence in $\mathrm{tp}(a/A)$, for $k = D(x = x, \varphi) + 1$. ∎

THEOREM 6.4.6 *Let T be a low theory. Then Lascar strong type is the same as strong type, over any set A.*

PROOF: Let A be any parameter set. We have to show that the type-definable equivalence relation $E(x, y)$ describing equality of the Lascar strong types of x and y over A is in fact the intersection of definable equivalence relations.

Fix a complete type $p(x) \in S(A)$, and a symmetric formula $\varphi(x,y) \in E$. Let $\Sigma(a,b)$ be the following condition:

> for all $a' \models \mathrm{lstp}(a/A)$ and all $b' \models \mathrm{lstp}(b/A)$ with $a' \mathop{\smile\!\!\!\!\!\!\;|}_A b'$, the formula $\varphi(x,a') \wedge \varphi(x,b')$ is consistent and does not fork over A.

Note that $\varphi(x,a)$ does not fork over A, so $\Sigma(a,a)$ holds by the Independence Theorem.

CLAIM. $\Sigma(a,b)$ is type-definable over A on realizations a,b of p.

PROOF OF CLAIM: By Lemma 2.5.7 and Theorems 2.5.9 and 2.5.19, the formula $\varphi(x,a') \wedge \varphi(x,b')$ is consistent and does not fork over A for *all* independent realizations $a' \models \mathrm{lstp}(a/A)$ and $b' \models \mathrm{lstp}(b/A)$ if and only if this holds for *some* such a', b'. Hence $\Sigma(a,b)$ is type-defined by

> there exists an independent indiscernible sequence $(a_i b_i : i < \omega)$ in $\mathrm{lstp}(a/A) \cup \mathrm{lstp}(b/A)$ such that $a_0 \mathop{\smile\!\!\!\!\!\!\;|}_A b_0$ and $\bigwedge_{i<\omega} \varphi(x,a_i) \wedge \varphi(x,b_i)$ is consistent,

where the independence can be expressed by $D(a_i/A \cup (a_j b_j : j < i), \psi, k) \geq D(p,\psi,k)$ and $D(b_i/A \cup (a_i, a_j b_j : j < i), \psi, k) \geq D(p,\psi,k)$ for all $i < \omega$, all $\psi \in \mathfrak{L}$ and all $k < \omega$. ■

By Corollary 6.4.5, the negation of Σ is type-definable, so by compactness there is a formula $\vartheta(x,y)$ such that for $a,b \models p$ we have $\Sigma(a,b)$ if and only if $\models \vartheta(a,b)$.

Clearly, Σ is invariant under Lascar strong type. Hence $p(x) \wedge p(y) \wedge E(x,x') \wedge \vartheta(x,y) \vdash \vartheta(x',y)$; by compactness there is a formula $\psi(x) \in p$ such that $\psi(x) \wedge \psi(y) \wedge E(x,x') \wedge \vartheta(x,y) \vdash \vartheta(x',y)$. Hence

$$E^p_\varphi(x,x') := [\psi(x) \leftrightarrow \psi(x')] \wedge \{\psi(x) \rightarrow \forall y\,[\psi(y) \rightarrow (\vartheta(x,y) \leftrightarrow \vartheta(x',y))]\}$$

defines an equivalence relation which is coarser than E. So it has only finitely many classes.

Now suppose $a,b \models p$ and $\models E^p_\varphi(a,b)$ for all $\varphi \in E$. Choose any $\varphi'(x,y) \in E(x,y)$. As E is an equivalence relation, there is a symmetric formula $\varphi \in E$ such that

$$\exists z, z', z''\, \varphi(x,z) \wedge \varphi(z,z') \wedge \varphi(z',z'') \wedge \varphi(z'',y) \vdash \varphi'(x,y).$$

As $E^p_\varphi(a,b)$ holds and trivially $\models \psi(a) \wedge \vartheta(a,a)$, we get $\models \vartheta(b,a)$. So there are a', b' with $E(a,a')$ and $E(b,b')$, such that $\varphi(x,a') \wedge \varphi(x,b')$ contains an element c. As $E(x,y) \vdash \varphi(x,y)$, the choice of φ yields $\models \varphi'(a,b)$; as $\varphi' \in E$ was arbitrary, we get $E(a,b)$.

It follows that E is the conjunction of the definable equivalence relations E^p_φ, for $p \in S(A)$ and $\varphi \in E$. ■

6.5 BIBLIOGRAPHICAL REMARKS

Subsection 6.1.1 is due to Kim [67], who uses ideas of Lascar [88] and Pillay and Poizat [112] (who prove Lemma 6.1.7). Subsections 6.1.2 and 6.1.3 extend known stable results, in particular by Hrushovski [48] and Herwig, Loveys, Pillay, Tanovič and myself [44]. Lachlan's conjecture was proven by Lachlan for superstable theories in [74] and his argument simplified by Lascar [81]; Pillay [103, 104] shows it for one-based theories using the notion of semi-isolation. The proof in the supersimple case is due to Kim [69].

The amalgamation construction in Subsection 6.2.1 is due to Hrushovski, who first used it to obtain exotic stable structures, such as an ω-categorical stable non-superstable structure, a non-locally modular strongly minimal set without definable group [51], or a strongly minimal amalgam of two strongly minimal theories with the definable multiplicity property [50]. He then noticed that it could also produce simple ones [53]; Exercise 6.2.28 and Problem 6.2.29 come from that preprint. The stable method is discussed (not only for ω-categorical structures) in [72, 37, 159], and applied in various contexts in [4, 43, 7, 10, 9, 5, 127, 128, 129]. The simple ω-categorical construction is treated by Evans in [35]; he also adapts it to construct an example without weak elimination of imaginaries. Pantano [101, 102] has further results on the growth rate of algebraic closure in ω-categorical structures. Pourmahdian [130] discusses the relation between the simple and stable construction, proves stable forking for the resulting structures, and provides some conditions under which the construction generalizes to yield non-ω-categorical examples.

Subsections 6.2.2 and 6.2.3 come from [36], except for Theorem 6.2.35, which is due to MacPherson [94], who proved it more generally for ω-categorical groups without the strict order property. CM-triviality was introduced for stable theories by Hrushovski [51], but transfers immediately to our context. The point-line-plane geometry is also considered by Hrushovki [51] and Pillay [105].

Most of the results of Section 6.3 were shown in some form in [27], in particular Corollary 6.3.17 and Theorem 6.3.22; Proposition 6.3.15 appears in Tsuboi [156]. As for conditions under which T_σ has a model-companion, Proposition 6.3.23 and Exercise 6.3.24 are due to Kikyo [62], Proposition 6.3.25 is due to Kudaibergenov. The axiomatization in Remark 6.3.27 is essentially due to Hrushovski [25] (see also [86, 27]); as Morley rank is additive in a strongly minimal theory, this proves one direction of Conjecture 6.3.28. For the converse, Kikyo and Pillay [63] consider a strongly minimal theory T augmented by countably many automorphisms, and show that it allows a model-companion if and only if T has the definable multiplicity property. For a single automorphism, they show that if T is a strongly minimal finite cover of a theory with the definable multiplicity property, then T_σ exists if and only if T has the definable multiplicity property. Note that Hrushovski [50] has conjectured that every strongly minimal set is a finite cover of a set with the definable multiplicity property.

Lowness was defined by Buechler in [17], where he also proves its basic properties and derives the equivalence of strong type and Lascar strong type (Theorem 6.4.6), a result found independently by Shami [139]. The supersimple non-low theory is constructed by Casanovas and Kim in [21].

Bibliography

[1] James Ax. The elementary theory of finite fields. *Annals of Mathematics*, 88:239–271, 1968.

[2] John T. Baldwin. α_T is finite for $\aleph_1$-categorical T. *Transactions of the American Mathematical Society*, 181:37–51, 1973.

[3] John T. Baldwin. *Fundamentals of Stability Theory*. Springer-Verlag, Berlin, Germany, 1988.

[4] John T. Baldwin. An almost strongly minimal non-Desarguesian projective plane. *Transactions of the American Mathematical Society*, 342:695–711, 1994.

[5] John T. Baldwin and Kitty Holland. Constructing ω-stable structures: Fields of rank 2. *Journal of Symbolic Logic*, to appear.

[6] John T. Baldwin and Alistair H. Lachlan. On strongly minimal sets. *Journal of Symbolic Logic*, 36:79–96, 1971.

[7] John T. Baldwin and Niandong Shi. Stable generic structures. *Annals of Pure and Applied Logic*, 79:1–35, 1996.

[8] Andreas Baudisch. Decidability and stability of free nilpotent Lie algebras and free nilpotent p-groups of finite exponent. *Annals of Mathematical Logic*, 23:1–25, 1982.

[9] Andreas Baudisch. Another stable group. *Annals of Pure and Applied Logic*, 80:109–138, 1996.

[10] Andreas Baudisch. A new uncountably categorical group. *Transactions of the American Mathematical Society*, 348:3889–3940, 1996.

[11] Walter Baur. Elimination of quantifiers for modules. *Israel Journal of Mathematics*, 25:64–70, 1976.

[12] Walter Baur, Gregory Cherlin, and Angus Macintyre. Totally categorical groups and rings. *Journal of Algebra*, 57:407–440, 1979.

[13] George M. Bergman and Hendrik W. Lenstra, Jr. Subgroups close to normal subgroups. *Journal of Algebra*, 127:80–97, 1989.

[14] Chantal Berline and Daniel Lascar. Superstable groups. *Annals of Pure and Applied Logic*, 30:1–43, 1986.

[15] Elisabeth Bouscaren, editor. *Model Theory and Algebraic Geometry (LNM 1696)*. Springer Verlag, Berlin, Germany, 1998.

[16] Steven Buechler. *Essential Stability Theory*. Springer-Verlag, Berlin, Germany, 1996.

[17] Steven Buechler. Lascar strong type in some simple theories. *Journal of Symbolic Logic*, to appear.

[18] Steven Buechler. Vaught's conjecture for superstable theories of finite rank. *Annals of Pure and Applied Logic*, to appear.

[19] Steven Buechler, Anand Pillay, and Frank O. Wagner. Elimination of hyperimaginaries for supersimple theories. *Preprint*, 1998.

[20] Enrique Casanovas. The number of types in simple theories. *Annals of Pure and Applied Logic*, 98:69–86, 1999.

[21] Enrique Casanovas and Byunghan Kim. A supersimple non-low theory. *Notre Dame Journal of Formal Logic*, to appear.

[22] Chen C. Chang and Jerome H. Keisler. *Model Theory*. North-Holland, Amsterdam, The Netherlands, 1973.

[23] Olivier Chapuis. Universal theory of certain solvable groups and bounded Ore group-rings. *Journal of Algebra*, 176:368–391, 1995.

[24] Zoe Chatzidakis. Groups definable in ACFA. In Hart et al. [41], pages 25–52.

[25] Zoe Chatzidakis and Ehud Hrushovski. The model theory of difference fields. *Transactions of the American Mathematical Society*, 351(8):2997–3071, 1999.

[26] Zoe Chatzidakis, Ehud Hrushovski, and Ya'akov Peterzil. The model theory of difference fields II. *Preprint*, 1999.

[27] Zoe Chatzidakis and Anand Pillay. Generic structures and simple theories. *Annals of Pure and Applied Logic*, 95:71–92, 1998.

[28] Gregory Cherlin. Superstable division rings. In Angus Macintyre, Leszek Pacholski, and Jeff Paris, editors, *Logic Colloquium '77*, pages 99–111. North-Holland, Amsterdam, The Netherlands, 1978.

[29] Gregory Cherlin, Leo Harrington, and Alistair Lachlan. $\aleph_0$-categorical $\aleph_0$-stable structures. *Annals of Pure and Applied Logic*, 28:103–135, 1985.

[30] Gregory Cherlin and Ehud Hrushovski. Permutation groups with few orbits on 5-tuples, and their infinite limits. *Preprint*, 1995.

[31] Gregory Cherlin and Saharon Shelah. Superstable fields and groups. *Annals of Mathematical Logic*, 18:227–270, 1980.

[32] E. Engeler. A characterization of theories with isomorphic denumerable models. *Notices of the American Mathematical Society*, 6:161, 1959.

[33] Paul Erdös and R. Rado. A partition calculus in set theory. *Bulletin of the American Mathematical Society*, 62:427–489, 1956.

[34] Juri Ershov. Fields with solvable theory (in Russian). *Doklady Akademii Nauk SSSR*, 174:19–20, 1967.

[35] David Evans. α_0-categorical structures with a predimension. *Preprint*, 1998.

[36] David Evans and Frank O. Wagner. Supersimple ω-categorical groups. *Journal of Symbolic Logic*, to appear.

[37] J. Goode. Hrushovski's geometries. In Bernd Dahn and Helmut Wolter, editors, *Proceedings of the 7th Easter Conference on Model Theory*, pages 106–118, 1989.

[38] Rami Grossberg, José Iovino, and Olivier Lessmann. A primer on simple theories. *Preprint*, 1998.

[39] Grzegorczyk, A. Mostowski, and C. Ryll-Nardzewski. Definability of sets in models of axiomatic theories. *Bulletin de l'Académie Polonaise des Sciences*, 9:163–167, 1961.

[40] Bradd Hart, Byunghan Kim, and Anand Pillay. Coordinatization and canonical bases in simple theories. *Journal of Symbolic Logic*, to appear.

[41] Bradd T. Hart, Alistair H. Lachlan, and Matthew A. Valeriote, editors. *Algebraic Model Theory*. Kluwer Academic Publishers, Dordrecht, The Netherlands, 1997.

[42] Ward Henson. Countable homogeneous relational structures and $\aleph_0$-categorical theories. *Journal of Symbolic Logic*, 37:494–500, 1972.

[43] Bernhard Herwig. Weight ω in stable theories with few types. *Journal of Symbolic Logic*, 60:353–373, 1995.

[44] Bernhard Herwig, James G. Loveys, Anand Pillay, Predrag Tanović, and Frank O. Wagner. Stable theories without dense forking chains. *Archive for Mathematical Logic*, 31:297–303, 1992.

[45] Wilfrid Hodges. *Model Theory*. Cambridge University Press, Cambridge, UK, 1993.

[46] Ehud Hrushovski. *Contributions to Stable Model Theory*. PhD thesis, University of California at Berkeley, Berkeley, USA, 1986.

[47] Ehud Hrushovski. Locally modular regular types. In John T. Baldwin, editor, *Classification Theory: Proceedings of the US-Israel Binational Workshop on Model Theory in Mathematical Logic, Chicago, '85, (LNM 1292)*, pages 132–164. Springer-Verlag, Berlin, Germany, 1987.

[48] Ehud Hrushovski. Finitely based theories. *Journal of Symbolic Logic*, 54:221–225, 1989.

[49] Ehud Hrushovski. Pseudo-finite fields and related structures. *Preprint*, 1991.

[50] Ehud Hrushovski. Strongly minimal expansions of algebraically closed fields. *Israel Journal of Mathematics*, 79:129–151, 1992.

[51] Ehud Hrushovski. A new strongly minimal set. *Annals of Pure and Applied Logic*, 62:147–166, 1993.

[52] Ehud Hrushovski. The Mordell-Lang conjecture for function fields. *Journal of the American Mathematical Society*, 9, 1996.

[53] Ehud Hrushovski. Simplicity and the Lascar group. *Preprint*, 1997.

[54] Ehud Hrushovski. The Manin-Mumford conjecture and the model theory of difference fields. *Annals of Pure and Applied Logic*, to appear.

[55] Ehud Hrushovski and Masanori Itai. On model-complete differential fields. *Preprint*, 1998.

[56] Ehud Hrushovski and Anand Pillay. Weakly normal groups. In Ch. Berline, E. Bouscaren, M Dickmann, J.-L. Krivine, D. Lascar, M. Parigot, E. Pelz, and G. Sabbagh, editors, *Logic Colloquium '85*, pages 233–244. North-Holland, Amsterdam, The Netherlands, 1987.

[57] Ehud Hrushovski and Anand Pillay. Groups definable in local fields and pseudo-finite fields. *Israel Journal of Mathematics*, 85:203–262, 1994.

[58] Ehud Hrushovski and Anand Pillay. Definable subgroups of algebraic groups over finite fields. *Journal für die Reine und Angewandte Mathematik*, 462:69–91, 1995.

[59] Ehud Hrushovski and Željko Sokolović. Minimal subsets of differentially closed fields. *Transactions of the American Mathematical Society*, to appear.

[60] Hyttinen. Remarks on structure theorems for ω_1-saturated models. *Notre Dame Journal of Formal Logic*, 36:269–278, 1995.

[61] Nathan Jacobson. *Basic Algebra II (second edition)*. Freeman, New York, USA, 1989.

[62] Hirotaka Kikyo. Model companions of theories with an automorphism. *Journal of Symbolic Logic*, to appear.

[63] Hirotaka Kikyo and Anand Pillay. The definable multiplicity property and generic automorphisms. *Preprint*, 1999.

[64] Byunghan Kim. *Simple First Order Theories*. PhD thesis, University of Notre Dame, Notre Dame, USA, 1996.

[65] Byunghan Kim. Recent results on simple first-order theories. In David Evans, editor, *Model Theory of Groups and Automorphism Groups (LMS LN 244)*, pages 202–212. Cambridge University Press, Cambridge, UK, 1997.

[66] Byunghan Kim. Forking in simple unstable theories. *Journal of the London Mathematical Society*, 57:257–267, 1998.

[67] Byunghan Kim. A note on Lascar strong types in simple theories. *Journal of Symbolic Logic*, 63:926–936, 1998.

[68] Byunghan Kim. Simplicity, and stability in there. *Preprint*, 1999.

[69] Byunghan Kim. On the number of countable models of a countable supersimple theory. *Journal of the London Mathematical Society*, to appear.

[70] Byunghan Kim and Anand Pillay. Simple theories. *Annals of Pure and Applied Logic*, 88:149–164, 1997.

[71] Byunghan Kim and Anand Pillay. From stability to simplicity. *Bulletin of Symbolic Logic*, 4:17–26, 1998.

[72] D.W. Kueker and C. Laskowski. On generic structures. *Notre Dame Journal of Formal Logic*, 33:175–183, 1992.

[73] Alistair H. Lachlan. A property of stable theories. *Fundamenta Mathematicae*, 77:9–20, 1972.

[74] Alistair H. Lachlan. On the number of countable models of a countable superstable theory. In Suppes, editor, *Proceedings of the Fourth International Congress of Logic, Methodology and Philosophy of Science*, pages 45–56. North-Holland, Amsterdam, The Netherlands, 1973.

[75] Alistair H. Lachlan. Two conjectures regarding the stability of ω-categorical theories. *Fundamenta Mathematicae*, 81:133–145, 1974.

[76] Alistair H. Lachlan. Theories with a finite number of models in an uncountable power are categorical. *Pacific Journal of Mathematics*, 61:465–481, 1975.

[77] T. Y. Lam. *A first course in noncommutative rings*. Springer-Verlag, Berlin, Germany, 1991.

[78] Daniel Lascar. Types définissables dans les théories stables. *Comptes Rendus de l'Académie des Sciences à Paris*, 276:1253–1256, 1973.

[79] Daniel Lascar. Définissabilité dans les théories stables. *Logique et Analyse*, pages 489–507, 1975.

[80] Daniel Lascar. Généralisation de l'ordre de Rudin-Keisler aux types d'une théorie. *Colloque C.N.R.S., Clermont*, 249:73–81, 1975.

[81] Daniel Lascar. Ranks and definability in superstable theories. *Israel Journal of Mathematics*, 23:53–87, 1976.

[82] Daniel Lascar. On the category of models of a complete theory. *Journal of Symbolic Logic*, 82:249–266, 1982.

[83] Daniel Lascar. Ordre de Rudin-Keisler et poids dans les théories superstables. *Zeitschrift für mathematische Logik und Grundlagen der Mathematik*, 28:411–430, 1982.

[84] Daniel Lascar. Relations entre le rang U et le poids. *Fundamenta Mathematicae*, 121:117–123, 1984.

[85] Daniel Lascar. *Stability in Model Theory*. Longman, New York, USA, 1987.

[86] Daniel Lascar. Les beaux automorphismes. *Archive for Mathematical Logic*, 31:55–68, 1991.

[87] Daniel Lascar. The group of automorphisms of a relational saturated structure. In *Finite and Infinite Combinatorics in Sets and Logic (Proceedings, Banff 1991)*, NATO ASI Series C: Math. Phys. Sci. 411, pages 225–236. Kluwer Academic Publishers, Dordrecht, The Netherlands, 1993.

[88] Daniel Lascar. Recovering the action of an automorphism group. In Wilfrid Hodges, Martin Hyland, Charles Steinhorn, and John Truss, editors, *Logic: from Foundations to Applications (European Logic Colloquium 1993)*, pages 313–326. Oxford University Press, Oxford, UK, 1996.

[89] Daniel Lascar and Anand Pillay. Forking and fundamental order in simple theories. *Preprint*, 1997.

[90] Daniel Lascar and Anand Pillay. Hyperimaginaries and automorphism groups. *Preprint*, 1998.

[91] Daniel Lascar and Bruno P. Poizat. An introduction to forking. *Journal of Symbolic Logic*, 44:330–350, 1979.

[92] Angus Macintyre. On ω_1-categorical theories of abelian groups. *Fundamenta Mathematicae*, 70:253–270, 1971.

[93] Angus Macintyre. On ω_1-categorical theories of fields. *Fundamenta Mathematicae*, 71:1–25, 1971.

[94] H. Dugald Macpherson. Absolutely ubiquitous structures and $\aleph_0$-categorical groups. *Oxford Quarterly Journal of Mathematics (2)*, pages 483–500, 1988.

[95] Dave Marker, Margit Messmer, and Anand Pillay. *Model Theory of Fields (LN in Logic 5)*. Springer Verlag, Berlin, Germany, 1996.

[96] Alan Mekler. Stability of nilpotent groups of class 2 and prime exponent. *Journal of Symbolic Logic*, 46:781–788, 1981.

[97] Margit Messmer. *Groups and Fields Interpretable in Separably Closed Fields*. PhD thesis, University of Illinois at Chicago, Chicago, USA, 1992.

[98] L. Monk. *Elementary-Recursive Decision Procedures*. PhD thesis, University of California at Berkeley, Berkeley, USA, 1975.

[99] Rahim Moosa. Definable group actions in simple theories. *Preprint*, 1998.

[100] Michael Morley. Categoricity in power. *Transactions of the American Mathematical Society*, 114:514–538, 1965.

[101] M. E. Pantano. *Algebraic Closure in $\aleph_0$-Categorical Structures*. PhD thesis, University of East Anglia, Norwich, UK, 1995.

[102] M. E. Pantano. $\aleph_0$-categorical structures with arbitrarily fast growth of algebraic closure. *Preprint*, 1996.

[103] Anand Pillay. *An Introduction to Stability Theory*. Clarendon Press, Oxford, UK, 1983.

[104] Anand Pillay. Stable theories, pseudoplanes an the number of countable models. *Annals of Pure and Applied Logic*, 43:147–160, 1989.

[105] Anand Pillay. The geometry of forking and groups of finite Morley rank. *Journal of Symbolic Logic*, 60:1251–1259, 1995.

[106] Anand Pillay. *Geometric Stability Theory*. Oxford University Press, Oxford, UK, 1996.

[107] Anand Pillay. Some foundational questions concerning differential algebraic groups. *Pacific Journal of Mathematics*, 179:179–200, 1997.

[108] Anand Pillay. Definability and definable groups in simple theories. *Journal of Symbolic Logic*, 63:788–796, 1998.

[109] Anand Pillay. Forking in the category of existentially closed structures. *Preprint*, 1998.

[110] Anand Pillay. Lecture notes on strongly minimal sets (and fields) with a generic automorphism. *Preprint*, 1999.

[111] Anand Pillay. Normal formulas in supersimple theories. *Preprint*, 1999.

[112] Anand Pillay and Bruno P. Poizat. Pas d'imaginaires dans l'infini! *Journal of Symbolic Logic*, 52:400–403, 1987.

[113] Anand Pillay and Bruno P. Poizat. Corps et chirurgie. *Journal of Symbolic Logic*, 60:528–533, 1995.

[114] Anand Pillay, Thomas Scanlon, and Frank O. Wagner. Supersimple division rings. *Mathematical Research Letters*, pages 473–483, 1998.

[115] Anand Pillay and Željko Sokolović. Superstable differential fields. *Journal of Symbolic Logic*, 57:97–108, 1992.

[116] Bruno P. Poizat. *Déviation des Types*. PhD thesis, Université Pierre et Marie Curie, Paris, France, 1977.

[117] Bruno P. Poizat. Une preuve par la théorie de la déviation d'un théorème de John Baldwin. *Comptes Rendus de l'Académie des Sciences à Paris*, 287:589–591, 1978.

[118] Bruno P. Poizat. Modèles premiers d'une théorie totalement transcendante. In *Théories Stables II*. Institut Henri Poincaré, Paris, France, 1981.

[119] Bruno P. Poizat. Sous-groupes définissables d'un groupe stable. *Journal of Symbolic Logic*, 46:137–146, 1981.

[120] Bruno P. Poizat. Théories instables. *Journal of Symbolic Logic*, 46:513–522, 1981.

[121] Bruno P. Poizat. Groupes stables, avec types génériques réguliers. *Journal of Symbolic Logic*, 48:339–355, 1983.

[122] Bruno P. Poizat. Paires de structures stables. *Journal of Symbolic Logic*, 48:239–249, 1983.

[123] Bruno P. Poizat. Post-scriptum à "Théories instables". *Journal of Symbolic Logic*, 48:60–62, 1983.

[124] Bruno P. Poizat. Une théorie de Galois imaginaire. *Journal of Symbolic Logic*, 48, 1983.

[125] Bruno P. Poizat. *Cours de théorie des modèles*. Nur Al-Mantiq Wal-Ma'rifah, Villeurbanne, France, 1985.

[126] Bruno P. Poizat. *Groupes Stables*. Nur Al-Mantiq Wal-Ma'rifah, Villeurbanne, France, 1987.

[127] Bruno P. Poizat. Corps de rang deux. In *The 3rd Barcelona Logic Meeting*, volume 9, pages 31–44. Centre de Recerca Matematica, Barcelona, Spain, 1997.

[128] Bruno P. Poizat. Le carré de l'égalité. *Journal of Symbolic Logic*, to appear.

[129] Bruno P. Poizat. L'égalité au cube. *Journal of Symbolic Logic*, to appear.

[130] Massoud Pourmahdian. *Simple Generic Structures*. PhD thesis, University of Oxford, Oxford, UK, 1999.

[131] Mike Prest. *Model Theory of Modules (LMS LN 130)*. Cambridge University Press, Cambridge, UK, 1988.

[132] F. P. Ramsey. On a problem of formal logic. *Proceedings of the London Mathematical Society*, 30:427–489, 1930.

[133] Joachim Reineke. Minimale Gruppen. *Zeitschrift für Mathematische Logik*, 21:357–359, 1975.

[134] F. Rowbottom. The Łos conjecture for uncountable theories. *Notices of the American Mathematical Society*, 11:248, 1964.

[135] C. Ryll-Nardzewski. On the categoricity in power $\aleph_0$. *Bulletin de l'Académie Polonaise des Sciences*, 7:545–548, 1959.

[136] G. Schlichting. Operationen mit periodischen Stabilisatoren. *Archiv der Mathematik (Basel)*, 34:97–99, 1980.

[137] A. Seidenberg. An elimination theory for differential algebra. *University of California Mathematical Publications*, 3:31–65, 1956.

[138] Jean-Pierre Serre. *Local Fields*. Springer-Verlag, Berlin, Germany, 1979.

[139] Ziv Shami. A natural finite equivalence relation definable in low theories and a criterion for Lstp=stp. *Preprint*, 1997.

[140] Saharon Shelah. *Categoricity of Classes of Models*. PhD thesis, The Hebrew University, Jerusalem, Israel, 1969.

[141] Saharon Shelah. Stable theories. *Israel Journal of Mathematics*, 7:187–202, 1969.

[142] Saharon Shelah. Every two elementarily equivalent models have isomorphic ultrapowers. *Israel Journal of Mathematics*, 10:224–233, 1971.

[143] Saharon Shelah. Stability, the f.c.p., and superstability; model-theoretic properties of formulas in first-order theories. *Annals of Mathematical Logic*, 3:271–362, 1971.

[144] Saharon Shelah. Uniqueness and characterization of prime models over sets for totally transcendental first-order theories. *Journal of Symbolic Logic*, 37:107–113, 1972.

[145] Saharon Shelah. Differentially closed fields. *Israel Journal of Mathematics*, 16:314–328, 1973.

[146] Saharon Shelah. The lazy model-theoretician's guide to stability. *Logique et Analyse*, pages 241–308, 1975.

[147] Saharon Shelah. *Classification Theory*. North-Holland, Amsterdam, The Netherlands, 1978.

[148] Saharon Shelah. On uniqueness of prime models. *Journal of Symbolic Logic*, 44:215–220, 1979.

[149] Saharon Shelah. Simple unstable theories. *Annals of Pure and Applied Logic*, 19:177–203, 1980.

[150] Saharon Shelah. The spectrum problem I; $\aleph_1$-saturated models, the main gap. *Israel Journal of Mathematics*, 43:355–363, 1982.

[151] Saharon Shelah. The spectrum problem II; totally transcendental theories and infinite depth. *Israel Journal of Mathematics*, 43:355–363, 1982.

[152] Saharon Shelah. Towards classifying unstable theories. *Annals of Pure and Applied Logic*, 80:229–255, 1996.

[153] Lars Svenonius. $\aleph_0$-categoricity in first-order predicate calculus. *Theoria (Lund)*, 25:82–94, 1959.

[154] Wanda Szmielev. Elementary properties of abelian groups. *Fundamenta Mathematicae*, 41:203–271, 1955.

[155] Alfred Tarski. *A Decision Method for Elementary Algebra and Geometry*. University of California Press, Berkeley, USA, 1951.

[156] Akito Tsuboi. Random amalgamation of simple theories. *Preprint*, 1999.

[157] Lou van den Dries. Definable groups in characteristic 0 are algebraic groups. *Abstracts of the American Mathematical Society*, 3:142, 1982.

[158] Robert L. Vaught. A Löwenheim-Skolem theorem for cardinals far apart. In Bar-Hillel, editor, *Logic, Methodology and Philosophy of Sciences*. North-Holland, Amsterdam, The Netherlands, 1961.

[159] Frank O. Wagner. Relational structures and dimensions. In Richard Kaye and Dugald Macpherson, editors, *Automorphisms of First-Order Structures*, pages 153–180. Oxford University Press, Oxford, UK, 1994.

[160] Frank O. Wagner. Groups in simple theories. *Preprint*, 1997.

[161] Frank O. Wagner. *Stable Groups (LMS LN 240)*. Cambridge University Press, Cambridge, UK, 1997.

[162] Frank O. Wagner. Small fields. *Journal of Symbolic Logic*, 63:995–1002, 1998.

[163] Frank O. Wagner. Hyperdefinable groups in simple theories. *Preprint*, 1999.

[164] André Weil. *L'Intégration dans les Groupes Topologiques et ses Applications*. Hermann, Paris, France, 1940.

[165] André Weil. On algebraic groups of transformations. *American Journal of Mathematics*, 77:302–271, 1955.

[166] Carol Wood. Notes on the stability of separably closed fields. *Journal of Symbolic Logic*, 44:412–416, 1979.

[167] Boris Zil'ber. Groups and rings whose theory is categorical (in Russian). *Fundamenta Mathematicae*, 95:173–188, 1977.

[168] Boris Zil'ber. Strongly minimal countably categorical theories (in Russian). *Sibirskiĭ Matematicheskiĭ Zhurnal*, 21:98–112, 1980.

[169] Boris Zil'ber. Totally categorical theories; structural properties and the non-finite axiomatizability. In L. Pacholski, J. Wierzejewski, and A. Wilkie, editors, *Model Theory of Algebra and Arithmetic, (LNM 834)*, pages 381–410. Springer-Verlag, Berlin, Germany, 1980.

[170] Boris Zil'ber. Strongly minimal countably categorical theories II (in Russian). *Sibirskiĭ Matematicheskiĭ Zhurnal*, 25:71–88, 1984.

[171] Boris Zil'ber. Strongly minimal countably categorical theories III (in Russian). *Sibirskiĭ Matematicheskiĭ Zhurnal*, 25:63–77, 1984.

[172] Boris Zil'ber. The structure of models of uncountably categorical theories. In Zbigniew Ciesielski and Czeslaw Olech, editors, *Proceedings of the International Congress of Mathematicians, August 16–24, 1983, Warszawa*, pages 359–368. North-Holland, Amsterdam, The Netherlands, 1984.

Index